TROUBLESHOOTERS' HANDBOOK FOR MECHANICAL SYSTEMS

OTHER McGRAW-HILL HANDBOOKS OF INTEREST

AMERICAN SOCIETY OF MECHANICAL ENGINEERS · ASME Handbooks:
- Engineering Tables
- Metals Engineering—Design
- Metals Engineering—Processes
- Metals Properties

AMERICAN SOCIETY OF TOOL AND MANUFACTURING ENGINEERS:
- Die Design Handbook
- Handbook of Fixture Design
- Manufacturing Planning and Estimating Handbook
- Tool Engineers Handbook

BAUMEISTER AND MARKS · Standard Handbook for Mechanical Engineers
BEEMAN · Industrial Power Systems Handbook
BRADY · Materials Handbook
BURINGTON AND MAY · Handbook of Probability and Statistics with Tables
CALLENDER · Time-Saver Standards
CARRIER AIR CONDITIONING COMPANY · Handbook of Air Conditioning System Design
CARROLL · Industrial Instrument Servicing Handbook
CONSIDINE · Process Instruments and Controls Handbook
CONSIDINE AND ROSS · Handbook of Applied Instrumentation
CROCKER AND KING · Piping Handbook
DUDLEY · Gear Handbook
EMERICK · Handbook of Mechanical Specifications for Buildings and Plants
EMERICK · Heating Handbook
FACTORY MUTUAL ENGINEERING DIVISION · Handbook of Industrial Loss Prevention
FINK AND CARROLL · Standard Handbook for Electrical Engineers
FLÜGGE · Handbook of Engineering Mechanics
HARRIS · Handbook of Noise Control
HARRIS AND CREDE · Shock and Vibration Handbook
HEYEL · The Foreman's Handbook
KALLEN · Handbook of Instrumentation and Controls
KING AND BRATER · Handbook of Hydraulics
KLERER AND KORN · Digital Computer User's Handbook
KOELLE · Handbook of Astronautical Engineering
KORN AND KORN · Mathematical Handbook for Scientists and Engineers
LEGRAND · The New American Machinists' Handbook
MACHOL · System Engineering Handbook
MAGILL, HOLDEN, AND ACKLEY · Air Pollution Handbook
MANAS · National Plumbing Code Handbook
MANTELL · Engineering Materials Handbook
MAYNARD · Industrial Engineering Handbook
MERRITT · Building Construction Handbook
MORROW · Maintenance Engineering Handbook
PERRY · Chemical Engineers' Handbook
PERRY · Engineering Manual
ROSSNAGEL · Handbook of Rigging
ROTHBART · Mechanical Design and Systems Handbook
SHAND · Glass Engineering Handbook
STANIAR · Plant Engineering Handbook
STREETER · Handbook of Fluid Dynamics
TOULOUKIAN · Retrieval Guide to Thermophysical Properties Research Literature
TRUXAL · Control Engineers' Handbook

TROUBLESHOOTERS' HANDBOOK FOR MECHANICAL SYSTEMS

Robert Henderson Emerick, P.E.

Member, ASME
Member, NSPE

McGRAW-HILL BOOK COMPANY

New York St. Louis San Francisco London
Sydney Toronto Mexico Panama

TROUBLESHOOTERS' HANDBOOK FOR MECHANICAL SYSTEMS

 Library of Congress Catalog Card Number 68-28413

19314

34567890 VBVB 754

TO ALL TROUBLESHOOTERS—
WHOEVER YOU ARE, WHEREVER YOU ARE

Preface

Troubleshooting is an exercise in symptom diagnosis. Its purpose is to identify and then eliminate undesirable and sometimes dangerous "bugs" in the performance of a machine or a system.

To accomplish this end, the troubleshooter will carry to the job with him certain appropriate instruments and tools, also, if they are available, applicable machinery catalogs and system drawings. These will make clear to him internal arrangements and normal operating characteristics of the troubled installation.

Finally, and probably most important of all, the troubleshooter should possess an innate ability to reason backward from the trouble he sees and hears to a true determination of its cause. This is seldom easy, since symptoms often offer considerable choice of causes and sometimes more than one cause must be discovered and corrected.

This handbook presents typical troubleshooting problems as the troubleshooter might be expected to meet them. First comes the complaint. This is followed by a listing of possible causes, each of which must be considered and evaluated in the light of data observed on the job. Methods for checking out these causes to establish an individual validity, together with suggestions for their cure, are given under the heading of recommended action.

Each major section of this handbook is introduced by a discussion of the relevant equipment, with emphasis on its characteristics. To illustrate, Sec. 2-7(a) points out significant differences among oil burners operating with forced-draft, induced-draft, or natural-draft furnaces. Understanding of these differences is essential to arriving at a sound diagnosis and in deciding on a proper cure.

Since a troubleshooter's time is often needed on other work, probably profit-making work which troubleshooting is not, fast solving of the problems encountered is of high importance. Toward this end, each major section includes a Troubleshooters' Field Guide, designed to summarize much of what is detailed in the text and to point out at a glance the possible causes for a described symptom. The field guides also indicate how to check each possible cause for validity and suggest means for correcting what is found to be wrong.

The author acknowledges with gratitude the priceless cooperation of equipment manufacturers who furnished photographs, drawings, and detailed data concerning their products. Without these helps, the troubleshooter would find it difficult to establish a base of normal performance from which to direct his investigations, and against which to measure the success of his efforts.

Robert Henderson Emerick

Contents

Preface *vii*

DIVISION 1. Troubleshooters' Lexicon of Instruments and Tools . . . 1

1-1 The Lexicon 1
1-2 Troubleshooters' Instruments, Listed and Described 22
1-3 Troubleshooters' Tools, Listed and Described 31
List of Suppliers 36

DIVISION 2. Boilers, Heaters, and Firing Equipment, Troubles and Corrections . . . 37

2-1 Cast-iron Boilers, Domestic 37
2-2 Steel Boilers, Low-pressure, Domestic 41
2-3 Warm Air Central Furnaces, Domestic and Commercial 42
2-4 Space Heaters, Direct-fired 45
2-5 High-pressure Boilers, Fire-tube 46
2-6 High-pressure Boilers, Water-tube 49
2-7 Firing Equipment, Oil Burners 55
2-8 Firing Equipment, Gas Burners 63
2-9 Firing Equipment, Coal, Hand 74
2-10 Firing Equipment, Coal, Stokers 76
2-11 Firing Equipment, Coal, Pulverized 78
2-12 Questions and Answers 80

DIVISION 3. Hydronic Heating Systems, Finding and Correcting Trouble Causes . . . 83

3-1 Gravity Hot Water Heating Systems 83
3-2 Converting Gravity Systems to Forced Circulation 92

3-3 Forced Circulation Hot Water Heating Systems 94
3-4 Questions and Answers 108

DIVISION 4. Steam Heating Systems, Handling Dissatisfactions in Operation . . . 115

4-1 Steam Generated on the Premises 115
4-2 Steam from High-temperature–High-pressure Water 124
4-3 Steam from a District Steam System 128
4-4 Questions and Answers 135

DIVISION 5. Electric Space Heating Systems, Sources of Trouble . . 137

5-1 Resistance Systems 137
5-2 Heating by Light 139
5-3 Infrared Systems 141
5-4 Questions and Answers 141

DIVISION 6. Domestic Hot Water Systems, Troubles and Relieving Them . . . 147

6-1 Direct-fired Heaters 147
6-2 Unfired Water Heaters 161
6-3 Electric Water Heaters 166
6-4 Solar Water Heaters 168
6-5 Hot Water Distribution Systems 170
6-6 Questions and Answers 172

DIVISION 7. Warm Air Heating Systems, Helping Dissatisfied Users . 175

7-1 Gravity Systems 175
7-2 Forced Circulation Systems 176
7-3 Questions and Answers 189

DIVISION 8. Fuel and Ash Handling Systems, Making the Equipment Work . . . 193

8-1 Coal Delivery, Storage, and Furnace Feeding 193
8-2 Fuel-oil Problems and Solutions 195
8-3 Fuel Gas, Natural 198
8-4 Safety in Handling Gases 199
8-5 Gas Fuel, Liquefied Petroleum Gases 200
8-6 Gas Fuel, City Artificial Gas 201
8-7 Ash Handling 202
8-8 Questions and Answers 210

DIVISION 9. Comfort Air-conditioning Systems, What to Do with Bad Performance 211

9-1 Domestic Installations 211
9-2 Commercial Installations, Mechanical 216
9-3 Commercial Installations, Absorption 229
9-4 Commercial Installations, Ammonia Absorption 231
9-5 Steam Jet Systems 235
9-6 Evaporative Cooling 236
9-7 Water Problems in Refrigeration Systems 237
9-8 Special Complaints in Air-conditioning Systems 239
9-9 Heat Pump Installations 241
9-10 Ice Storage Systems for Air Conditioning 244

DIVISION 10. Ice Making and Cold Storage, Correcting Machinery Troubles 255

10-1 Refrigerants and Their Properties 255
10-2 Safety in Handling Refrigerants 258
10-3 Ice Making and Handling 260
10-4 Cold Storage Plants 267
10-5 Locker Plants 269
10-6 Servicing Hermetically Sealed Motors 273
10-7 Questions and Answers 275

DIVISION 11. Ventilation and Ventilating Systems, Troubles and Treatments 277

11-1 Ventilation Purposes and Methods 277
11-2 Troubles with Natural Ventilation Systems 290
11-3 Troubles with Mechanical Ventilation Systems 293

DIVISION 12. Humidity Control, How to Increase It or Decrease It . 305

12-1 Relative Humidity, Its Meaning and Measurement 305
12-2 Symptoms of Unsatisfactory Humidity 309
12-3 Correcting Low Humidity 310
12-4 Correcting Excessive Humidity 317

DIVISION 13. Compressed-air Equipment and Systems 325

13-1 Basic Facts for the Troubleshooter 325
13-2 Compressed-air Complaints and Their Treatment 331

DIVISION 14. Engines and Turbines, Types and Characteristics . . 341

14-1 Steam Engines, Characteristics 341
14-2 Steam Engine Troubles 344
14-3 Diesel Engines, Characteristics 348
14-4 Diesel Engine Troubles 352
14-5 Steam Turbines, Characteristics 357
14-6 Steam Turbine Troubles 360
14-7 Gas Turbines, Characteristics 365
14-8 Gas Turbine Troubles 369
14-9 Gasoline Engines, General 379

DIVISION 15. Pumps and How to Improve Performance 381

15-1 General Factors Affecting Pump Performance 381
15-2 Types and Troubles of Reciprocating Pumps 383
15-3 Types and Troubles of Rotary Pumps 391
15-4 Troubles of Centrifugal Pumps 398
15-5 Troubles with Shallow-well Pumps 412
15-6 Troubles with Deep-well Pumps 413
15-7 Common Troubles with Wells and Water Systems 418

DIVISION 16. Electric Driving Motors 423

16-1 Types and Troubles of Electric Motors Driving Mechanical Equipment 423

DIVISION 17. Incinerators and Fireplaces, Correcting Nuisances . . 433

17-1 Incinerator Terminology 433
17-2 Waste Classifications 434
17-3 Incinerator Classifications 436
17-4 Incinerator Testing 438
17-5 Particular Incinerator Troubles and Their Correction 439
17-6 Fireplaces and Their Troubles 445

DIVISION 18. Piping and Miscellaneous. 455

18-1 Correcting Troubles with Indoor Piping 455
18-2 Correcting Troubles with Outdoor Piping 457
18-3 Correcting Troubles with Underground Piping 459
18-4 Handling Water Hammer in Pipe Lines 465

18-5 Septic Tank Troubles and Cures 469
18-6 Complaints from Outdoor Swimming Pools 478
18-7 Complaints from Indoor Swimming Pools 485
18-8 General Safety Precautions for Troubleshooting 488
General References 497

Indexes 501

Index of Troubleshooters' Field Guides 503

Subject Index 505

TROUBLESHOOTERS' HANDBOOK FOR MECHANICAL SYSTEMS

DIVISION 1

Troubleshooters' Lexicon of Instruments and Tools

1-1. THE LEXICON

Absolute Humidity: Defined as the weight of water vapor in a unit volume, commonly expressed as grains per cubic foot (7,000 grains = 1 lb).

Absolute Pressure: This is the sum of the gage reading and the atmospheric pressure. At sea level, the atmospheric pressure commonly used is 14.7 psi. This atmospheric value decreases with elevation; for example, at 2,000 ft the pressure is 13.664 psi; at 10,000 ft, only 10.1 psi; and at 16,000 ft, 7.7 psi.

Absolute Temperature: The observed temperature plus 459.67 F or, on the centigrade scale, the observed temperature plus 273.15 C.

Absolute Zero: A temperature of −459.67 F or −273.15 C. This is the base point from which all absolute temperatures are calculated.

Activated Alumina: An aluminum oxide that is used as a drying agent and absorbs moisture readily. The moisture is removed by heat.

Activated Carbon: A form of carbon so treated as to make it capable of adsorbing certain odors or specific vapors.

Aerosol: The term for heterogeneous assemblages of minute particles suspended in the air. These particles may be solid or liquid and as small as 0.01 microns or less. Fog, dust, and smoke may be termed aerosols.

Air Binding or Air Bound: A condition in which a bubble or other pocket of air is present in a pipeline or item of equipment and, by its presence, prevents or reduces the desired flow or movement of the liquid or gas in the pipeline or equipment. For example, air in the suction side of a pump may cause the pump to lose its fluid and stop pumping.

Air-Fuel Ratio: The amount of air, in cubic feet or pounds, being furnished for combustion per cubic foot or per pound of the fuel. For example, 12 lb of air per pound of coal is a 12 to 1 ratio and gives approximately perfect combustion.

Algae: These are plant growths that develop to form a scum on the surface of water-recirculating equipment, such as cooling towers and condenser water lines. Their presence interferes with the exchange of heat from the water.

Alkalinity: A condition of water expressing the amount of carbonates, bicarbonates, hydroxides, silicates, or phosphates in solution, and measured in parts per million (ppm) or grains per gallon as calcium carbonate.

Ambient Temperature: The temperature of the air in the area of study or consideration. For example, the ambient air temperature in the boiler room is 90 F.

Analysis, Proximate: This term applies to a solid fuel, and the analysis is directed at finding the percentages of moisture, volatile matter, fixed carbon, and ash per unit weight of the sample. Example, the proximate analysis of a coal might show the sample to contain moisture, 3 percent; volatile matter, 10 percent; fixed carbon, 78 percent; ash, 9 percent.

Analysis, Ultimate: A chemical analysis of the fuel, determining the percentage of the chemical constituents in percent of the weight of the unit sample. An ultimate analysis determines the percentages of sulfur, hydrogen, carbon, nitrogen, oxygen, and ash. The ultimate analysis is required to determine a heat balance and the efficiency of the combustion.

Angle of Repose: When coal, ashes, and other loose materials are assembled in a pile, the maximum angle made by the sides of the pile with the ground or horizontal surface on which the pile is placed, and maintained without sliding downward of the material, is termed the angle of repose. Example, a snowslide indicates the angle of repose of the snow on a mountainside has been exceeded.

Ash-free Basis: When fuels are delivered on an ash-free basis, it means that the percentage of the ash has been deducted and the other constituents have their percentages recalculated on a 100 percent total without the ash.

Bagasse: A fuel produced as a by-product of the abstraction of juice from sugar cane. The dried cane is usually fed into a specially designed furnace by means of overfeed stokers.

Barometric Pressure: The atmospheric pressure as indicated by a barometer. Barometers are calibrated in inches of mercury. At sea level, the normal barometric pressure is accepted as 30 in., although the exact pressure under standard conditions of air temperature of 70 F and 60 percent relative humidity is 29.921 in. of mercury.

Beaded Tube End: When a tube is fitted into a tube sheet and rolled tight by an expander, the end of the tube projecting slightly from the face of the tube sheet is folded back against the sheet as a circular bead of metal.

Belled Tube End: This is the same as a flared tube end. The open end of the tube is expanded to a diameter greater than the diameter of the tube as manufactured.

Blast-furnace Gas: A by-product from a blast furnace which is the result of burning coke in the furnace with limited air. The gas is combustible but of relatively low calorific measurement.

Blow: In warm air distribution systems, this term refers to the distance the air travels from its outlet to the point where its velocity has fallen to 50 fpm. Also, this aspect is frequently called "throw."

Blowback: The difference in pressure between a safety valve's opening pressure and its automatic closing pressure after the excess pressure has been released. This difference is about 3 percent of the opening pressure as a general average, although as much as 5 or 6 percent is sometimes observed.

Blowtorch Effect: In gas- or oil-burning furnaces, the impingement of the flame on any surface, such as a tube or refractory wall, has the ultimate effect of that surface being burned as by a blowtorch. This is a combustion condition to be avoided as destructive to the surface.

Boiler Horsepower: The amount of heat required to evaporate 34.5 lb of water an hour, from water at 212 F into steam at 212 F. This amount of heat is equivalent to 33,475 Btu per hour.

Breeching: The metal duct that carries the smoke and gases of combustion from a furnace to the stack or chimney for ultimate discharge to the atmosphere.

Brine: In refrigeration work, brine is *any* liquid cooled by the refrigerant and used to pick up heat without undergoing any change in state, other than temperature. The brine must have no flash point or, alternatively, a minimum flash point of 150 F, as determined by the American Society for Testing and Materials (ASTM).

Caking Coals: Coals that become soft under the usual furnace temperatures and merge into undesirable masses of coke. These coals can cause considerable trouble in an underfeed or chain grate stoker by clogging air-supply apertures, or tuyeres.

Calorie: The calorie represents one-hundredth of the heat required to lift temperature of 1 g of water from 0 to 100 C. Thus it is approximately equal to the amount of heat required to heat 1 g of water 1 deg C at a constant atmospheric pressure. One calorie is the equivalent of 0.003968 Btu.

Calorific Value: The heat value of a fuel, expressed in either Btu per pound or calories. *Note:* 1 lb = 453.59 g, approximately.

Carry-over: The condition of water boiling over into the steam lines from a steaming boiler. It can cause damage to some equipment, such as turbine blades and engine cylinders.

Caustic Embrittlement: A term used in some instances to describe a form of metal failure, apparently from fatigue, at joints and tube ends in steam boilers. The elimination of riveting from boiler-drum construction has done much to eliminate this embrittlement. The cause has been thought to lie in the chemical composition of the boiler water being such as to react with the steel under stress and produce fatigue cracks.

Centigrade: A temperature scale on which the melting and freezing point of ice is rated as zero and the boiling point of water is 100, all at atmospheric pressure. The absolute temperature on the centigrade scale is taken as −273.15; consequently, on this scale water boils at 373.15 C absolute. See *Kelvin* also.

To convert centigrade to the more common Fahrenheit temperature, multiply the centigrade reading by $9/5$ and add 32. Thus the boiling point of water, Fahrenheit, is $100\text{ C} \times 9/5 + 32 = 212\text{ F}$. By reversing the procedure, Fahrenheit may be converted to centigrade: $(212 - 32) \times 5/9 = 180 \times 5/9 = 100\text{ C}$.

Change of State: The change of a material from solid to gaseous, from gaseous to solid, from solid to liquid, etc. The opposite of steady state.

Chimney Effect: The upward movement of warm air or gas, compared with the ambient air or gas, due to the lesser density of the warmed air or gas. Chimney effect may be a cause of uneven heating in buildings two or more stories high.

Coal Gas: A fuel formed by the distillation of coal, usually in a retort or in a coke oven. It is the basic gas in some city artificial-gas systems and shows an average calorific value of 500 to 550 Btu per cu ft.

Coefficient of Performance: In reference to heat pumps, this term represents the ratio of energy fed into the pump compared with the output, all in Btu. To illustrate, if 5 kw are being taken by the compressor and another 10 kw are going into the booster heaters, the total input is 15 kw. Now if the heat delivered actually totals 60,000 Btu, the coefficient of performance is $60{,}000/(15 \times 3414) = 1.17$, approximately. In short, the heat pump is putting out more energy than is being put into it, the difference being made up from the machine's use of the latent heat of vaporization in the refrigerant.

Colloidal Fuel: A mixture of fuel oil and powdered coal.

Conductivity: The rate of heat flow through a homogeneous material, expressed in Btu per hour per square foot, per degree Fahrenheit of temperature difference.

Convection: The movement of a fluid set up by a combination of differences in density and the force of gravity. For example, in a vertical tank, water made warm at the bottom will rise and displace the cooler water at the top, the latter sinking to the bottom as the result of its greater density. Gravity hot water heating systems circulate their hot water in accordance with the convective laws. Gravity warm air heating systems operate similarly.

Deaeration: Term used to identify the process of removing air and gases from boiler feedwater before the water reaches the boilers.

Degasification: The removal from water of a gas such as carbon dioxide. A feature of the ion-exchange method of water softening.

Degree Day: For each calendar day, the number of heating degree days is the difference in degrees between the mean temperature for that day and 65 F. For example, a mean for the day of 50 F produces 65 − 50 or 15 degree days. The figure of 65 F is the dividing temperature between heat needed and not needed.

Degrees of Superheat: Generally applied to steam, the term may be appropriate to any gas. Each of these fluids normally has a pressure that is characteristic of its temperature, and when the temperature is higher than is normal for the existing pressure, the increase of temperature is expressed as degrees of superheat.

Desuperheater: A device designed to provide saturated steam from a superheated supply. Usually functions by means of water sprays injected into the superheated steam.

Dew Point: The temperature of the air at which any reduction of the ambient temperature would produce some condensation of the air's water-vapor content. Another definition: the dew point is the temperature at which the air's relative humidity becomes 100 percent.

Downfeed Steam Heating System: The steam supply main is higher than the radiation that the steam serves, and the steam must flow downward to reach the radiators. For one-pipe systems, the steam going to the radiator and the condensate leaving the radiator descend in the same pipe. For two-pipe systems, the condensate leaves the radiator by its own pipe, the supply steam entering the radiator from the opposite end.

Dry Steam: Steam that has a temperature consonant with its pressure and contains no more than ½ of 1 percent moisture.

Dulong's Formula: An equation by which the heating value of a fuel may be approximated, if the ultimate analysis is available. The formula is:

$$\text{Btu per pound} = 14{,}540\ C + 62{,}000\ (H - 1/8\ O) + 4{,}050\ S$$

In this formula, C is carbon, H is hydrogen, O is oxygen, and S is sulfur, each expressed in its percentage of the 1-lb weight of the fuel.

For accuracy, the calorific determination is preferred.

Dutch Oven: A furnace that is mounted ahead of, and projects from, the front wall of the boiler. This type of furnace is commonly seen in bagasse-burning installations.

Eductor (May Be Termed an Injector in Some Instances): This device uses a flow of water, steam, or air to induce a secondary flow of water or some other liquid sucked from a pool or vessel that is in the process of being emptied. When the prime fluid is steam and the secondary fluid is water and the mixture is discharged into a boiler, the device is termed an *injector.*

Efficiency: In boiler operation, efficiency is the ratio of the energy fed to the boiler to the energy obtained from the boiler. Example, a pound of coal with a calorific value of 14,000 Btu per lb is burned under a steam boiler to produce 11,000 Btu per hour in steam output. Then the efficiency of output to input is 11,000/14,000=80 percent efficiency, approximately.

Other forms of efficiency apply to dust collection, transmission of power, generation of power, etc.

Ejector: Similar to an eductor in principle.

Enthalpy: Term used instead of the longer "total heat" or "heat content." It is expressed in Btu per pound.

Evasé Stack: An expanding vertical duct on the outlet of a fan or blower, used to convert discharge velocity into static pressure. When used on induced-draft fans, the stack is much shorter than would be the case with another form of stack. In some designs, the fan is built into the base of the stack, and outside air from the fan discharge induces the flue gas into the stack and carries it upward to the atmosphere.

Fahrenheit: A temperature scale, most common in the United States, on which the boiling point of water is considered to be 180 deg above the freezing point of water and the zero is 32 deg lower than the freezing point. Thus the original range of the Fahrenheit scale is 212 deg between what Fahrenheit thought was the coldest day possible and the boiling point of water.

The absolute temperature for the Fahrenheit scale is −459.688 deg below the Fahrenheit zero. This figure is usually approximated for practical computations as −460 F. Thus water boils at 672 F, absolute (460 deg below zero + 212 deg above zero).

For converting Fahrenheit to centigrade, see *Centigrade.*

Fan Performance Curves: Very useful in checking the operation of a fan or blower, these curves are provided by the fan manufacturer and show the fan discharge pressure, power input, discharge volumes in cubic feet, speeds, etc. Where fan malfunction is suspected, the troubleshooter can find help in identifying the trouble by comparing the present operating data with these curves.

Feedwater Treatment: In order to eliminate scale formations in a boiler, the feedwater usually requires treatment. Prime treatment is that of softening, but it may include deaerating for the removal of oxygen and filtering for the removal of oxides and other materials in suspension. The troubleshooter should be able to recognize the symptoms of scale, odor, foaming, and oxygen corrosion and, when these are present, recommend that a specialist in feedwater treatment be brought in to prescribe elimination procedures.

Fire Point: The lowest temperature at which fuel oil will emit enough vapor to continue burning when subjected to ignition.

Flame Detector: A device that monitors the flame in a furnace that is burning oil, gas, or pulverized coal fuel. Failure of the flame results in a signal and the actuation of various protective controls on the fuel feed to prevent explosion from unpurged re-ignition.

Flash Point: The temperature at which sufficient vapor is emitted from fuel oil to produce a momentary flame when subjected to ignition, this flame immediately expiring.

Flooded Refrigeration System: An arrangement by which only part of the circulated refrigerant is evaporated, the remainder being separated from the vapor and then recirculated.

Fly Ash: Combustion ash so fine that it is carried up and into the atmosphere by the movement of the flue gases. It can become a neighborhood nuisance by settling on surfaces in the area after it loses its velocity.

Foaming: A condition of the boiler water in the steam-releasing areas. Surface tension of the water, resulting from undesirable materials in solution, causes the formation of bubbles that persist and, going out with the steam, carry boiler water into the steam lines and steam-using equipment.

Formula for Gravity, API: The American Petroleum Institute (API) has established the formula for calculating the gravity of a fuel oil as:

$$\text{Degrees API} = \frac{141.5}{\text{specific gravity at } {}^{60}\!/_{60}\text{ F}} - 131.5$$

The symbol $^{60}/_{60}$ F is interpreted as the ratio of the weight of a given volume of oil at 60 F to the weight of the same volume of water at 60 F. For example, assume an oil at 60 F weighing 7 lb per gal and water at 60 F weighing 8.33 lb per gal; then $^{7}/_{8.33} = 0.84$. Substituting 0.84 in the equation gives an answer of 37, approximately, putting this oil in the range of Commercial Standard No. 2.

Fouling: A condition of the flue-gas passages in a boiler or furnace that adversely affects the transfer of heat. Usually in the form of soot or scale.

Freeze-up: A condition of ice formation on a refrigeration system at the expansion device. As a result of this ice, operation of the system is unsatisfactory. The expansion valve may freeze either closed or open, the former stopping circulation, the latter eliminating the control.

Fuel-oil Burners, Types:

a. *Pressure-atomizing or gun-type burners* are designed to atomize the oil for combustion under an oil-supply pressure of 100 psig. Equipment typically includes a horizontal air tube with the oil pipe centered in this tube.

b. *Rotary burners* employ a centrifugal thrower ring that mixes the oil and the air. The flame of the rotary burner generally is vertical.

c. *Vaporizing burners,* also called *pot burners,* use the heat of combustion to vaporize the oil in the pool or pot beneath the vaporizer ring, and this vapor rising through the ring ignites and maintains combustion in the burner. Fuel enters the pot by gravity through a float valve, and the rate of firing is controlled by the adjustable positions of the float valve.

Fuel Oil, Commercial Grades:

Grade	Gravity, API	Btu/gal
1	38–45	137,000–132,900
2	30–40	141,800–135,800
4	12–32	153,300–140,600
5	8–20	155,900–148,100
6	6–18	157,300–149,400

Fume: Any kind of noxious vapor arising from a process of combustion or chemical reactions. Includes smoke, odorous materials, metallic dusts.

Fume Afterburners: Units designed to consume combustible fumes by means of a direct-fired combustion chamber through which the fumes must pass on their way to the stack and the atmosphere. Afterburners usually are fired with gas, although oil may be used in certain instances. Operating temperature is conventionally from 1200 to 1400 F and is sometimes reached with the help of a platinum catalyst.

Fusible Plug: A protective device, commonly installed in fire-tube boilers and other temperature-pressure equipment, that melts under excessive temperatures and thus prevents the buildup of dangerous pressures. The plug is hollow, threaded, and filled with a meltable material.

Fusion: A term used to describe the melting of ash or the melting of ice. The temperature of fusion is the temperature at which the change from solid to liquid takes place.

Gas, Carbureted Water: An artificial gas formed by passing steam through a bed of glowing coke and thereafter enriching the gas so formed with petroleum vapor. For additional data, see Table 2-5.

This gas is a common one in city gas systems where coal is an economical fuel.

Gas, Coke-oven: A coal gas formed as a by-product in the manufacture of coke. See Table 2-5.

Gas, Liquefied Petroleum: A gas fuel that is stored in liquid form and is converted into gas as it leaves the storage tank by a pressure regulator that steps down the storage pressure on the liquid at the tank outlet and thereby permits the liquid to assume its normal gaseous state at the existing temperature and reduced pressure. Of the two kinds of liquefied petroleum in general use, at atmospheric pressure propane goes from liquid to gas at −51 F; butane, at 15 F. See Table 2-4 for the characteristics of these gases.

Gas, Natural: The constituents of natural gases vary considerably, depending on the gas field from which the gas is obtained. Natural gas is the major fuel in the gas category. Refer to Table 2-5.

Gravity, Specific: The ratio of the gravity of a given substance and volume to the gravity of a standard substance at the same volume, at a standard temperature. For liquids, the standard substance and temperature is water at 39.2 F, this being the point of water's greatest density. For gases, the standard substance is air, at the same temperature as the gas under consideration. The specific gravity of solids, particularly woods, is referred to water as 1.

Grindability: A descriptive term of a characteristic of coal that is important to pulverized-coal systems. The standard from which grindability is measured is the number 100, representing a very soft, easily ground, bituminous coal. As the coal samples become harder, the grindability number decreases.

Grindability is usually stated on the Hardgrove scale, and the formula applicable is:

$$\text{Grindability} = 6.93\,W + 13$$

The *W* in this equation is the weight in grams of material in a 50-g sample that will pass through a 200-mesh sieve. The other figures are constants.

Pulverizer mills generally may be expected to show 100 percent capacity with coals having a grindability index of 55. As the index falls to 25 or 30, the mill capacity likewise falls to about 65 percent.

Hardgrove Scale: A measuring standard for the grindability of coal. See *Grindability*.

Hardness: The degree of hardness, applied to water, is based on the amount of calcium and magnesium salts in the water, expressed as grains per gallon or parts per million of calcium carbonate. Note that 1 grain of hardness per U.S. gallon = 17.1 ppm.

In a publication by the Water Conditioning Foundation of Glenview, Illinois, entitled the "Recommended Standards for Household Water Supplies," maximum allowable hardness is 3.5 grains per gal.

Head, Dynamic: Total pressure, static and velocity, expressed as height in inches or feet, of a column of liquid, usually water or mercury.

Head, Net Positive Suction (npsh): Npsh is a most important factor in the performance of a pump. It may be defined as the *absolute* pressure available at the inlet to a pump. A formula for computing the npsh for any liquid is npsh equals pressure on surface of liquid to be pumped, as atmospheric, minus vapor pressure of the liquid, minus the friction and turbulence losses in the full length of the suction piping, plus or minus the difference in elevation between the liquid being pumped and the center line of the pump. If the liquid is higher than the center line, the difference is added; if below, subtract.

Heat Balance: The name given to a procedure by which the efficiency of a combustion process is determined, all heat losses being added together and the total being subtracted from 100 percent. The remaining figure is the efficiency.

The losses are heat loss in dry gas, percent; loss in moisture in the fuel, percent; moisture in the air, percent; loss by radiation from hot surfaces, percent; other losses not accounted for, 1.5 percent. Also, loss due to the burning of hydrogen in the fuel, percent; loss in unburned carbon in the ash (coal fuel), percent; and finally, the loss from the incomplete combustion of the carbon, percent.

Heat, Latent: The heat, in Btu, required for a material to change its state, as ice to water and water to water vapor or steam.

Heat, Sensible: Heat that changes the temperature of a material but does not change its form.

Heat, Specific: The heat absorbed or relinquished by a unit of a substance in changing its temperature 1 deg F.

The common unit is the pound, and the unit expressing heat is the

British thermal unit for liquids and solids. The specific heat of water is 1; the specific heat of air is 0.24, also taken on the basis of the pound. To convert cubic feet of air to pounds, refer to the standard tables of the American Society of Heating, Refrigerating and Air-Conditioning Engineers (ASHRAE) appearing in its Guide.

Heat-transmission Coefficient: A figure that may be used in calculating the total heat transfer through a material and that represents all phases affecting the transfer, such as conduction, radiation, convection, and the surface values. It is usually expressed as the number of Btu (or the percentage of 1 Btu) transmitted per degree Fahrenheit difference in temperature, per square foot of area, for the given material and its thickness, over a time of 1 hour. The symbol for this coefficient is U.

High Side: Refers to high pressure on the discharge side of a refrigerant compressor, as compared with the low side, or suction side. The condenser and all piping up to the expansion valve on the evaporator are on the high side.

Higher Heat Value: A standard recommended by the American Society of Mechanical Engineers (ASME), the higher heat value of a fuel includes the heat value of the hydrogen in the fuel. A lower heat value, recognized in certain other countries, modifies the total heat in the fuel to reflect losses resulting from the hydrogen burning to form water, this water then absorbing sufficient heat to be converted into superheated steam and so lost.

Hogged Fuel: Wood that has been chipped and shredded, usually by a machine called a "hog."

Hot-gas Line: The gas from a refrigerant compressor going directly to the evaporator for the purpose of defrosting the latter.

Hydronics: A term for the techniques of heating or cooling by a liquid.

Ignition Arch: A refractory arch arranged in a furnace above the point of coal entry into the furnace for the purpose of radiating heat immediately against the green coal and thus promoting early ignition. This arch is seen most frequently over chain-grate or traveling-grate stokers.

Intercooling: Cooling a gas or air between stages of the compressor. Usually involves a shell and tube heat exchanger, with water or another cooling fluid circulating through the exchanger.

Internal Treatment: Feedwater treatment in which the chemicals are introduced directly into the boiler, rather than into the water before it enters the boiler.

Ion: An atom which carries an electric charge, either positive or negative.

Isobaric: The word applies to a change carried through a constant pressure. Isobars are lines on a map that connect points of equal barometric pressure. Superheating steam is an isobaric procedure.

Isothermic: This word describes a condition of change accomplished at constant temperature.

Joint, Expansion, Bellows: An item of equipment used to compensate for the expansion and contraction of a run of piping. The device is built with a flexible metallic bellows that stretches or is compressed as necessary to accept the movements of the piping.

Joint, Expansion, Slip: Performs the same service as the bellows joint but accomplishes its purpose by means of a cylinder that moves in and out of the main body of the device. This form of joint incorporates a packed gland around the slip which may develop some leakage after a time and could require either repacking or a simple taking up on the gland nuts.

Jumper Tube: A bypass of pipe or hose used by a serviceman to reroute or otherwise change the flow direction temporarily of a liquid or gas.

Kelvin: A temperature scale on which the lowest possible temperature, designated "zero," is 273.15 deg below the zero on the centigrade scale. To convert centigrade to Kelvin (C to K), add 273.15 to the centigrade reading. Thus the boiling point of water in the atmosphere which is 100 on the centigrade scale becomes 373.15 K.

Kerosene: This petroleum product is a liquid fuel having an average latent heat of vaporization of 105 to 110 Btu per pound and a specific heat of 0.50. Sometimes called coal oil.

Lagging: A term sometimes used to designate the insulating covering on piping and ducts. More appropriately, it applies to the insulation on machinery casings.

Leak Detector: See Sec. 1-3. *Troubleshooters' Tools.*

Ligament: The minimum cross section between two adjacent holes in a header, shell, or tube sheet. Frequently referenced in the design of heat-exchange equipment.

Lignite: A coal of high moisture content and low calorific value, generally less than 8300 Btu per pound. May require predrying before being used as a fuel.

Liquid-ash Removal System: An arrangement of piping by which molten ash is removed continuously or intermittently, as desired, from the bottom of a furnace. The operating medium is generally compressed air with pneumatic controls.

Liquid Line: The refrigerant piping that conveys liquid refrigerant from the condenser to the expansion valves.

Live Steam: Steam that has not as yet given up any of its energy. Example, steam as it emerges from the boiler.

Load Factor: The ratio of the average load during a given period and the normal maximum, or full capacity, load rating of the unit under consideration. Applicable also to the complete system.

Log Sheet: A daily or hourly record of operating data commonly maintained in utility and industrial plants. Very useful to the troubleshooter in locating areas of trouble.

Long-flame Burner: An oil or gas burner in which the mixture of fuel and air is delayed long enough to produce a long flame from the burner nozzle. Can be a source of trouble if the flame impinges on either refractory or tubes surfaces.

Low Heat Value of a Fuel: See explanations under *Higher Heat Value.*

Low Side: In a refrigeration circuit, the low side includes all parts of the circuit between the evaporator outlet and the suction inlet of the compressor.

Mean Indicated Pressure: In engine technology, the average net pressure per square inch acting on the piston during the power stroke. It is determined by the use of an indicator while the engine is operating.

Mechanical Equivalent of Heat: One Btu = 778.2 ft-lb of mechanical energy.

Medium, Cooling: A fluid that picks up heat in various sources and carries it to heat-exchange equipment where the heat is removed and disposed of. Liquids in this kind of circuit include chilled water, brine, ice (which removes heat by melting), and solid CO_2 (which removes heat by vaporizing).

Medium, Heating: A substance used to convey heat from the heat source to the heat application. A heating medium may be water, steam, air, gas, or a proprietary liquid such as Dowtherm or Humble-Therm 500.

Mercury Column: A measure of pressure. One inch of mercury at 62 F equals 0.4897 lb per sq in. and is the equivalent of a water column 13.57 in. high. At 32 F, the inch of mercury (chemical symbol Hg) exerts a pressure of 0.49112 psi.

Meta-anthracite: A coal classification of the highest rank with fixed carbon of 98 percent or higher and volatile matter 2 percent or less, when computed on a moisture-free and ash-free basis.

Micron: A unit of length equal to 1/25,400 in. and to 1/1,000,000 m. This unit is highly useful in expressing the diameter of dust particles and in stating the capacity of a filter.

Miniature Boiler: A descriptive term for a fired pressure vessel that does not exceed the following values: inside diameter of shell, 16 in.; overall length or height, 3½ ft; heating surface, 20 sq ft; maximum allowable working pressure, 100 psig.

Mollier Diagram: A chart which presents graphically heat and temperature values that appear elsewhere in the steam tables. It is useful in calculating heat-energy problems.

Natural Circulation: Another term for the movement of fluids in a system resulting from differences of density rather than from the actions of a pump or blower. Refer to *Convection.*

Nitrogen: This is an inert gas making up approximately 79 percent by volume of the atmosphere and used extensively as a purging medium for refrigeration systems and for combustible-gas systems. It is stored commercially in tanks and cylinders.

Normal Barometer: The atmospheric pressure reading at sea level. This is 29.9 in. of mercury, or 14.7 psi, approximately. In practical computations, a barometer value of 30 in. is generally satisfactory, since the pressure may vary almost from minute to minute as the result of weather fluctuations.

Overfire Air: Secondary air introduced into a furnace above the grate for the purpose of completing combustion and producing turbulence in the furnace. This turbulence increases the efficiency of the combustion process.

Oxygen: An elemental gas that constitutes approximately 21 percent of the atmosphere by volume, the remainder being nitrogen as previously noted. Actually, other elements are present in the atmosphere, such as argon and helium, but they appear to exert no effect on reactions between the air and other materials and therefore are generally ignored.

Ozone: An unstable compound of oxygen produced by the electric field immediately surrounding an electrode operating at high voltage.

The presence of ozone to a troubleshooter suggests that a high-voltage discharge has taken place, such as a lightning stroke, and thus may have significance in the correction of the trouble.

Packaged Air Conditioner: An air-conditioning cabinet that contains all parts needed to discharge conditioned air, such as the compressor, condenser, evaporator, receiver, filters, and fans or blowers. This arrangement is distinguished from the separation of compressor and condenser, usually placed outdoors, and the air-handling and evaporator units mounted indoors. The separate arrangement is generally called a "split system."

Packaged Steam Generator: A steam generator that is shipped complete from the factory of the builder, with all auxiliary equipment in place, such as fans, pumps, controls, and firing equipment. The buyer connects the electric, water, and fuel services and the installation is then ready for operation.

Particle Size: In evaluating the efficiency of a filter, electrostatic or otherwise, it is customary to indicate the percentage of particles of a specified micron diameter and larger that are caught by the filter and removed from the air stream.

Peak Load: The maximum load carried by a system or a unit of equipment over a designated period of time.

pH, or pH Value: A term based on the hydrogen ion concentration in water, which denotes whether the water is acid, alkaline, or neutral. A pH value of 8 or more indicates a condition of alkalinity; of 6 or less, acidity. A pH of 7 means the water is neutral, neither alkaline nor acid. The hydrogen ions are counted in their concentration on the basis of grams per liter.

Plenum: An enclosed space that functions as a reservoir for air or gas and from which the air or gas is bled off through one or more ducts.

Porcupine Boiler: A vertical boiler in the form of a shell or cylinder from which a number of short, dead-end tubes project for extra heating surface.

ppm: The abbreviation for "parts per million." Frequently used in indicating the hardness of water in terms of calcium carbonate.

Preheated Air: Air that is warmed above the atmospheric temperature for the purpose, in combustion, of promoting faster ignition of the fuel and more complete burning.

Pressure Vessel: A closed vessel of any shape that confines some kind of fluid, either liquid or gas, at some pressure above atmospheric. When the vessel is not subjected to direct heating, such as an oxygen flask, it is termed an "unfired pressure vessel." This distinguishes it from a fired boiler or heater.

Primary Air: Combustion air introduced into a furnace in such a way as to make initial contact with the fuel, usually being mixed with the fuel either immediately before it ignites or simultaneously with the ignition. In pulverized coal systems, the primary air enters the pulverizer mill where it is mixed with the coal and proceeds to the burners as a mixture.

Psychrometric Chart: A presentation of the characteristics of atmospheric air for various temperatures, humidities, and pressures by means of curves and lines. Essential for calculating air-conditioning design.

Pulsation: A panting of the flames in a furnace, indicating cyclic and rapid changes in the pressure in the furnace.

Purge: The evacuation of air, or any other designated gas, from a duct line, pipeline, container, or furnace. Purging may be accomplished in some instances simply by use of a fan or blower, in others by driving out the air or gas by means of an inert gas, such as nitrogen, under higher pressure.

Quality of Steam: Expresses the dryness of saturated steam as a percentage of perfect dryness. For example, steam that contains 1 percent moisture has an expressed quality of 99 percent; 3 percent moisture would indicate 97 percent dry saturated. The quality is an indication of how well the separator in the boiler drum is removing moisture before the steam leaves the boiler. Complaints of water hammer at times have been traceable to steam that was excessively wet.

Radiation, Equivalent Direct (EDR): Heat delivery in the amount of 240 Btu, the amount of heat assumed to be delivered by one square foot of direct radiation under standard conditions of 215 F steam temperature and 70 F ambient.

Radiation Loss: See *Heat Balance.*

Radiation, Thermal: The transmission of heat from a hot surface to a cooler one in the form of invisible electromagnetic waves, which on being absorbed by the cooler surface, raise the temperature of that surface.

Raw Water: Water, from any source, that requires treatment before it can be used in steam generation or in a process.

Reaction: The process by which the bringing together of two or more materials produces a new substance or a release of energy. Reactions are encountered most often in chemistry.

Rectifier, Electric: The electric rectifier is a device for converting alternating current to direct current by means of a mercury arc or electronic functions.

Rectifier, Refrigeration: In ammonia-absorption refrigerating systems, the rectifier is a combination of heat exchanger and condenser assigned to the early cooling of the refrigerant gas and to the removal by condensation of water which might be entrained with the refrigerant.

Refractory: A material particularly able to maintain its strength and shape under exposure to high heat. Commonly used in furnaces as walls and floors and, where needed, as baffles. When a refractory failure is observed by the troubleshooter, he should look for overheating, fusion, spalling, or deformation under the load. In some furnaces, the formation of slag on refractories ultimately produces failure. A check of furnace temperatures is indicated.

Refrigerant: Any substance that performs a cooling function by absorbing heat when it expands or vaporizes.

Refrigeration, One Ton of: The standard ton of refrigeration is represented by the absorption of 288,000 Btu per 24 hours, or 12,000 Btu in 1 hour. In another definition, it is the amount of heat removed from 1 ton (2,000 lb) of water in order to convert the water into ice. Since the latent heat of freezing is 144 Btu per lb, the total is obviously 288,000 Btu.

Relative Humidity: The weight of water vapor present in air (or a gas) expressed as a percentage of the maximum weight of water vapor possible in the considered air (or gas) at the given temperature and pressure. Relative humidity appears as curved lines on the psychrometric chart. See *Absolute Humidity.*

Residual Fuel: Crude petroleum after the removal of volatile hydrocarbons and water.

Retarders: Sometimes called "spinners," these are helical strips or ribbons of metal centered in horizontal or vertical fire tubes of a fire-tube boiler for the purpose of increasing the wiping effect on the inner surfaces of the tubes, by the flue gases on their way to the stack. The tubes cannot be cleaned by scraping or brushing until these strips are removed. Retarders also increase the boiler frictional resistance to the flue gases.

Retort: This is a trough or channel built into an underfeed stoker, through which the stoker ram pushes green coal into the fire. The coal enters the fire from below, hence the name "underfeed."

Return, Dry: In a steam heating system, the condensate return piping is "dry" because it is above the water line in the boiler.

Return, Wet: The condensate return is below the water level of the boiler and therefore is full of water at all times.

Reversed Return System: This is a two-pipe arrangement with the return piping placed so that each circuit of hot water from the boiler to the radiator and back to the boiler is of approximately the same length as every other circuit. The intent of this layout is to keep pipe friction equal and thus avoid any short circuiting and its consequent uneven heating. A common arrangement with gravity hot water systems, it has been superseded in new construction by forced circulation, in which protection against short circuiting is achieved by means of balance cocks on the return from each circuit. This plan saves installation costs for the piping eliminated.

Ringlemann Chart: A card bearing four rectangles, each rectangle being crossed by black lines to show four patterns of smoke density. The card itself is white, and the rectangles may or may not be pierced by sighting holes as an aid to matching a smoke plume with the rectangles.

Run-of-mine Coal: Bituminous coal delivered as it comes from the mine without screening.

Safe Working Pressure: The maximum working pressure for a given vessel, boiler, flask, or cylinder, allowable under the ASME Code. The safe working pressure, in pounds per square inch gage, may be stamped on the unit by the ASME or by the National Board of Boiler and Pressure Vessel Inspectors.

Saturation: A term used to describe a condition of air or gas in which the maximum possible amount of water vapor is present in the mixture for the existing temperature and pressure. The saturation line on the psychrometric chart is useful in air-conditioning design for the determination of the dew point. Saturation also describes a condition of 100 percent relative humidity. See *Dew Point.*

Secondary Air: Air introduced into a furnace above or around the flames as may be necessary to promote combustion. This air is in addition to the primary air which enters either as a mixture with fuel or as blast underneath a stoker. See *Primary Air.*

Secondary Combustion: The burning of fuel in areas or spaces beyond the outlet of a furnace and where combustion is not intended by the design.

Slack: A coal of fine size, often screenings. Maximum size is not likely to exceed 2½ in.

Slag Tap Furnace: A furnace for burning pulverized fuel in which the ash puddles in the bottom of the furnace in a molten state and is removed periodically or continuously, depending on the design of the system, while still in the molten condition.

Solute: The material dissolved in a solution.

Solution, Strong: In absorption refrigeration systems, the liquid entering the generator is strongly charged with the refrigerant and consequently is known as the strong solution or "strong liquid." In the generator, the refrigerant is boiled out of the solution and proceeds as a vapor to the next unit in the system, usually a rectifier. See *Rectifier.*

Solution, Weak: What is left in the generator after the refrigerant has been boiled off is termed the weak solution or "weak liquid."

Specific Gravity: This term applies to the density, or weight in pounds per cubic foot, of a given substance, referred to a standard substance having a specific gravity of 1.

For liquids the standard is water; for gases, air. To illustrate, the specific gravity of propane in the liquid form is 0.508, indicating that a cubic foot of liquid propane at 60 F weighs a little more than one-half a cubic foot of water. In the vapor phase, also compared with air at a common temperature of 60 F, the density of the propane vapor becomes 1.552, or more than half again as heavy as air.

Specific Heat: The amount of heat, expressed in Btu, required to elevate one pound of a material one degree Fahrenheit. The specific heat of water is 1 for all practical purposes, although the exact description is $1/180$ of the heat needed to raise the temperature of 1 lb of water from 32 F to 212 F. The specific heat of air is 0.24; steam varies but may be assumed at 0.49 at lower temperatures, 0.50 at 300 F. Refer to the steam tables for exact values.

Specific Volume: The reciprocal of density. Whereas density refers to the weight per cubic foot, specific volume is based on the number of cubic feet to 1 lb. This number of cubic feet, compared with the number of cubic feet for 1 lb of water where liquids are involved and with air for gases and vapors, indicates the specific volume of the subject liquid or gas. See *Specific Gravity* for an illustration.

Spectroscopic Analysis: The identification of materials by a common microscope is not always satisfactory, and where possible, the use of a spectroscope will eliminate doubts. In the spectroscope, the color of each element may be determined by light rays passing through, or by reflection from, the material's surface. Troubleshooters may be obliged from time to time to call for a spectroscopic analysis of material samples.

Standard Air: This is air at 70 F, at sea level, and at 29.92 in. barometric pressure. Under these conditions, it weighs 0.078 lb per cu ft, and looking at it from the volume viewpoint, 12.82 cu ft are needed to weigh 1 lb.

Superheat: To superheat steam or a gas is to add heat beyond that normal to its existing pressure, thereby elevating its temperature without increasing the pressure. The "degrees of superheat" represent the difference in temperature between that normal for the pressure and the final temperature. Commonly encountered with steam installations, the term is appropriate also to some conditions of the refrigerant in a refrigeration system.

Temperature, Critical: The temperature at which the properties of a liquid and its vapor phase are identical. For example, the critical temperature of water and steam is 705.4 F. At this temperature, the pressure is 3,206.2 psia, and the volume of 1 lb of water is the same volume as that of 1 lb of steam. The enthalpy, or total heat content, of the water is 902.7 Btu per lb, and that of the steam which it becomes is the same 902.7 Btu. There is no latent heat needed to turn the water into steam; the change of state results wholly from the combination of temperature and pressure.

Temperature, Dry-bulb: The temperature read on the conventional thermometer.

Temperature, Wet-bulb: The wet-bulb thermometer is kept moist by a wick and water reservoir, and the temperature read is generally several degrees lower than the dry-bulb reading as the result of water evaporation around the bulb. The difference in readings points to the relative humidity in the air. These two temperatures are essential in the design of air-conditioning and refrigeration systems.

Tertiary Air: The injection of air into a furnace in addition to the primary and secondary air. The purpose of the tertiary injection is generally the establishment of greater furnace turbulence and the reduction of smoke. This is not a universal practice, but will be encountered from time to time.

Therm: A unit of heat having a value of 100,000 Btu. Common in gas combustion studies.

Thimble: A short piece of pipe or duct that acts as a sleeve through which a smaller pipe or duct transits a wall. A boiler breeching commonly enters a masonry stack through a thimble, and similarly, steel stacks passing through roofs are kept out of contact with the roof by a thimble. Roof thimbles perform a fire-prevention function in separating the roof and the hot stack.

Tie Rod: The sections of cast-iron sectional boilers are held in tight contact by means of tie rods that pass entirely through the sections. Pulling up too tightly by wrench on the tie rods is a cause of cracked sections, a fact of interest to the troubleshooter.

Torch: Combustible material on a metal rod, such as oil-soaked rags, used to light off oil and gas burners. The torch is extinguished by being plunged into a prepared receptacle.

Turbidity: A term that refers to the presence of fine suspended matter in water, in such a degree as to reduce or eliminate the clarity of the water. Raw water from a river or lake may show a high turbidity from the presence of mud.

Tuyeres: Castings appearing as components of underfeed stokers and designed to admit air to the green coal moving through the retorts.

Underwriters' Loop: Also called the Hartford return, this is an arrangement of the return piping connections to a steam boiler. It is used to balance pressures between the supply and return sides of the boiler and thus prevent boiler water from backing out of the boiler and into the returns.

Vacuum Heating System: A two-pipe steam system in which a vacuum pump is placed on the condensate return. By means of the pump, a relatively wide differential in pressures and temperatures across the heating units is possible, thereby increasing the heat output from the individual radiation. Also, on startup, the pump rapidly removes air from the system and thus speeds up delivery of heat.

Vapor: The gaseous form of a substance that, under other conditions of pressure, temperature, or both, is a solid or a liquid.

Vapor Barrier: A material that is impervious to the passage of water vapor through it. Vapor barrier service is of top importance in protecting other materials, especially insulation, against the deteriorating entry of the vapor. Since vapor pressure increases when temperature increases, the flow is from the warmer to the cool zones; consequently the barrier should be placed against the warm zone. Without the barriers, condensation of the vapor occurs, sometimes resulting in extensive damage to buildings.

Vapor Heating System: A heating system in which steam is circulated at atmospheric pressure or less. Returns to a receiver or direct to the boiler are by gravity. Sometimes called "atmospheric systems," vapor systems are less frequently offered by designers than formerly. Their effectiveness is greatest during mild or tempering weather.

Vertical Firing: Oil, gas, or pulverized-coal burners so arranged that the fuel is discharged from the burner in a vertical direction, either upward from low-set burners or downward from burners in the top of the furnace.

Viscosity: The characteristic of fluids, both liquid and semiliquid, that causes them to resist changes in their shape. It is generally expressed in terms of Seconds Saybolt Universal or Seconds Saybolt Furol. In either case, the seconds represent the time required by a given liquid to pass through a standard orifice at a temperature of 100 F for the Universal and 122 F for the Furol. The orifice of the latter is somewhat larger, since this instrument is intended for the heavier liquids.

The tendency of most liquids, particularly fuel and lubricating oils, to increase in viscosity with a falling thermometer may require the provision of an auxiliary heating system if the material is to be successfully pumped.

Waste Fuel: Any combustible substance that is a by-product of a process or product. For example, bagasse is an otherwise waste by-product of a sugar mill.

Waste Heat: Heat created or released as a by-product of some process or of some manufacturing or generating activity and usually is wasted if no useful application is found. As an example, the heat released by the operation of a diesel engine in its exhaust is wasted to the atmosphere unless some application is found. Common applications are the heating of water and the passing of the exhaust gases through the heat exchange surfaces of an unfired steam boiler, thereby generating steam for heating. Less common is the use of the waste heat in an absorption refrigeration or air-conditioning system.

Water Gas: A basic gas in commercial gas works formed by the reaction of an incandescent bed of carbon when injected with live steam. The gas consists mainly of carbon monoxide and hydrogen, and its calorific value is comparatively low, being about 300 Btu per cu ft. General practice is to enrich the gas by mixing with another, frequently petroleum gas if it is available.

Water Hammer: In water lines, a loud thumping noise that results from a sudden stoppage of the flow, thereby converting flow momentum into an instantaneous increase in pressure.

In steam lines, water of condensation may be picked up and carried through the steam main at high velocity, and when this straight-line velocity impacts on change-of-direction fittings, a loud metallic knocking results.

Windbox: A plenum from which air for combustion, such as primary air, is supplied to a stoker or to gas or oil burners.

1-2. TROUBLESHOOTERS' INSTRUMENTS

(a) ***Field Instruments.*** These are the checking and measuring instruments that the troubleshooter will need from time to time to determine conditions on the job. Basic instruments will have to answer mainly questions of temperature and pressure. Less often needed will be instruments to determine such values as CO_2, speed of rotating machinery in revolutions per minute, the amount of positive or negative draft, and elapsed time. Each category will be discussed from the viewpoint of form, calibrations, and applications.

(b) ***Thermometers, Insertion Type.*** In form, straight stem is most useful, the bulb being inserted in an existing thermometer well.

Calibrations depend on the normal operating temperatures of the fluid being checked. For industrial steam, without superheat, 400 F top will do for lines operating at pressures of 225 psig or less. For steam heating from a low-pressure boiler with a 15 psig limit, 250 F is enough.

Oil temperatures, both lubricating and fuel, will generally range from

FIG. 1-1. Clip-on type of thermometer for pipelines. (*Abrax Instrument Corporation.*)

100 to 150 F. Gas fuel comes in at the ambient temperature usually, but flue gases enter the stack at anywhere from 300 to 550 F, depending on the installation and the presence or absence of economizers, air preheaters, or waste-heat recovery equipment.

(c) ***Thermometers, Clip-on, for Pipes.*** Figure 1-1 illustrates the clip-on thermometer. The standard calibration covering a range of from 0 to 300 F is likely to be most useful, but other calibrations are available if ordered.

(d) ***Thermometers, Conventional Wall Type.*** At least one of these is essential for noting ambient air conditions, particularly in boiler rooms where the temperature of the room may affect the natural draft through a boiler furnace or heater. It is also needed for checking the temperatures in rooms or spaces and the discharge of air from registers in a warm air heating system. This is a dry-bulb thermometer.

(e) ***Thermometers, Wet-bulb Type.*** Resembling the dry-bulb thermometer in its calibrations, the wet-bulb instrument incorporates a small water well and wick for the purpose of determining relative humidity in the space. The temperature is lower on the wet-bulb instrument as the result of chilling by evaporation, and the amount of temperature depression is translatable into relative humidity. See *Psychrometric Chart.*

(f) ***Thermometers, Pyrometer Type.*** Figure 1-2(a) shows a probe type instrument suitable for easy handling in making contact with a hot surface or in reaching the heart of an air stream. Calibrations are available up to 2000 F and down to sub-zero temperatures, as necessary to meet the service conditions, as standard practice. The market also offers optical and platinum thermocouple instruments for higher temperatures. For determining flame temperatures, a range extending to about 3000 F will be necessary in some high-pressure steam generators.

Figure 1-2(b) illustrates a serviceman's pocket thermometer ranging from −40 to 210 F. This range is most helpful in troubleshooting when both heating and air conditioning are involved. This thermometer is inexpensive.

(g) ***Pressure Gages.*** Pressure gages carry particular identities, depending on the service to which they are attached. Common designations are as follows:

The standard pressure gage is calibrated in pounds per square inch, beginning at zero pounds and extending to whatever maximum pressure is stipulated by the buyer to the gage maker.

Pressure gages operate on one of three principles. The Bourdon gage is arranged with the pressure being introduced into a C-shaped tube (this is the Bourdon tube), and the pressure tends to straighten out the tube. Movement of the tube is indicated by a pointer, the position of the pointer showing the pressure, in pounds, that holds it in that position.

The bellows gage actuates its pointer from a bellows movement. It generally is limited in use to relatively low pressures, up to about 50 psi.

The third arrangement is termed a diaphragm gage. The pressure leans against a diaphragm that in turn is checked in its movement by a spring. The deflection of the spring indicates the pressure being exerted on the diaphragm. Diaphragm gages find their most common application in low-pressure and vacuum service. Calibration is generally in inches of water rather than in pounds, because of the small range of travel over which they function.

For the troubleshooter, the type of pressure gage taken to the job is dictated by the pressure or vacuum values that he is obliged to measure and verify.

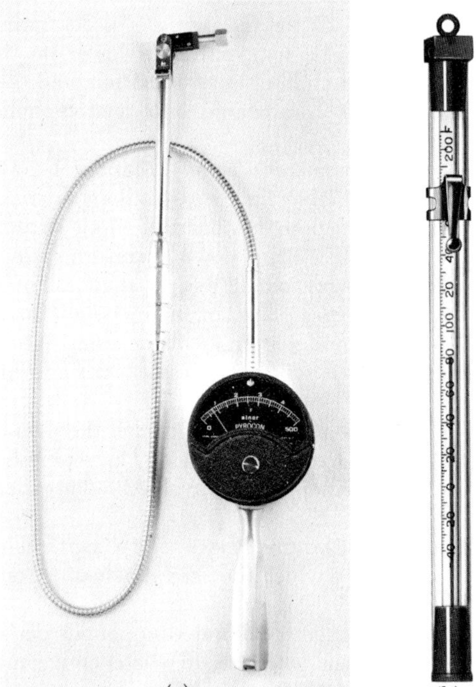

(a) (b)

FIG. 1-2. (a) Probe-type thermometer. Trade name Pyrocon. (*Alnor Instrument Company.*) (b) Serviceman's pocket thermometer. (*Marsh Instrument Company.*)

(h) ***Vacuum Gages.*** Vacuum gages may resemble the standard pressure gage in appearance, but the calibration begins at atmospheric pressure, which is the zero point on the gage, and extends to 30 in. water gage of negative pressure. Thirty inches is usually considered to be the commercial maximum vacuum, but the true maximum for any given place and time coincides with existing barometric pressure. Thus for a time and area showing a 29.90 barometer, the maximum vacuum is also 29.90 in. In short, eliminate all the air pressure, whatever it happens to be, and the result is the maximum vacuum.

One form of vacuum gage is the manometer or U tube, used for determining the furnace draft in a boiler. The manometer is termed a draft gage by most users and is a unit in any combustion testing kit.

(i) ***Compound Gages.*** These instruments are calibrated to indicate pressure and vacuum. The same pointer swings over the dial from one to the other as appropriate. For the troubleshooter working on low-pressure steam heating systems or process applications, the compound gage is a most handy device to take to the job. The calibrations for this work generally range from 30 in. on the vacuum side of the dial to 15 psi on the pressure side.

(j) ***Altitude Gages.*** Designed for hot water heating systems, the altitude gage is available with the dial calibrated to show the head of water in feet, *above the gage*, or the pressure in pounds, or both. Some gages include not only the head or pressure but also the temperature of the water.

A prime use of the altitude gage in the troubleshooter field kit is to check the accuracy of the gage normally found mounted on the boiler. Alternatively, a pressure gage may be so used, and the pressure-gage reading converted into feet by the troubleshooter. An altitude gage therefore may be regarded as optional for field checking.

(k) ***Gages for Corrosive Fluids.*** Any of the described gages may be used for corrosive fluids, ammonia being the most common, *provided* the manufacturer has been informed of the intended service and has seen to it that corrosion-resistant materials have been incorporated in the construction of the gage.

(l) ***Draft Gages.*** Available in the form of U tubes, inclined tubes which are also called manometers and liquid gages, the instrument is used to measure draft, in inches of water, at boiler outlets, in the base of a chimney, etc. Figure 1-3 shows an easily carried draft gage for the troubleshooter's kit.

(m) **CO_2-indicating Gage.** The CO_2 gage that indicates the percentage of carbon dioxide in the flue gas is a *must* instrument. Each fuel has an optimum percentage of CO_2 representative of perfect combustion. The measurement by Orsat is a little too difficult to

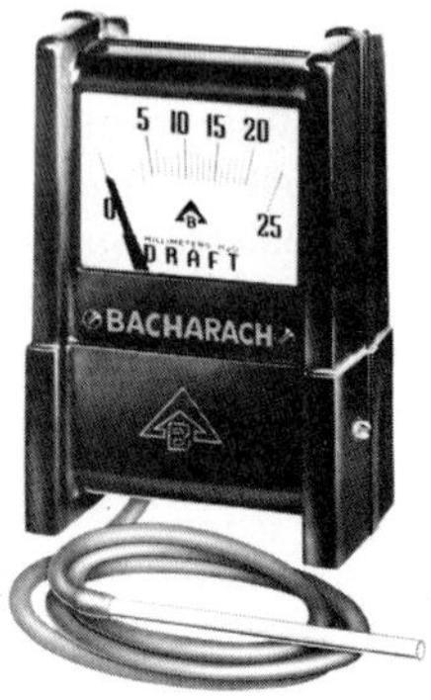

FIG. 1-3. Draft gage. (*Bacharach Industrial Instrument Co.*)

handle in the field, and a unit such as that shown in Fig. 1-4(a) serves very well. Figure 1-4(b) pictures a complete combustion testing kit.

(n) ***Explosimeter.*** This instrument is not a fixture in the kit, but the troubleshooter should have one in the shop or at least know where one is available at short notice. This instrument indicates the flammable or explosive character of the air in which the work must be carried on. It is particularly important in determining the state of the air before entering manholes and tunnels. Gases accumulate in these spaces from the dissociation of steam, the seepage of gasoline and oils from the surrounding earth, and the degeneration of various materials.

The explosimeter may appear to be fairly expensive, but an equal price put on a life would indicate that life is regarded as very cheap. This writer has used the explosimeter and worked in atmospheres that were shown to be 100 percent explosive. This is hard on the nerves and dictates the use of nonsparking tools and the wearing of rubber-soled shoes. Figure 1-5 illustrates a form of explosimeter easily portable.

(o) ***Manifold for Testing Refrigeration Systems.*** The arrangement of valves and gages illustrated in Fig. 1-6 is handy to have when checking conditions as follows:

1. Testing for a low charge of refrigerant in the system.
2. Evaluating the condition of the compressor valves, both high side and low side.
3. Adding refrigerant to determine correct amount.
4. Testing the thermostatic expansion valves.
5. Correctly adding oil to the compressor.
6. Bench testing of pressure controls.

How to use the instrument in these ways is described in instructions issued by the manufacturer.

(p) ***The Velometer.*** Illustrated in Fig. 1-7, the velometer is used to measure air flow. The instrument is particularly useful in the warm air heating field, being used to measure the warm air volume being emitted from registers. Unbalanced discharge across the face of a register is readily discovered with the velometer.

(q) ***Humidistat.*** An instrument such as that illustrated in Fig. 1-8

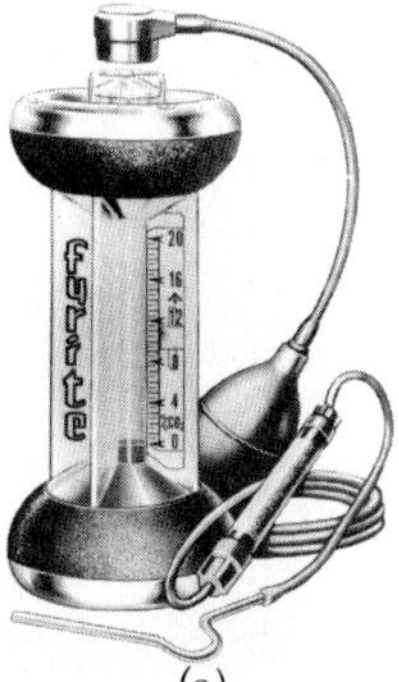

(a)

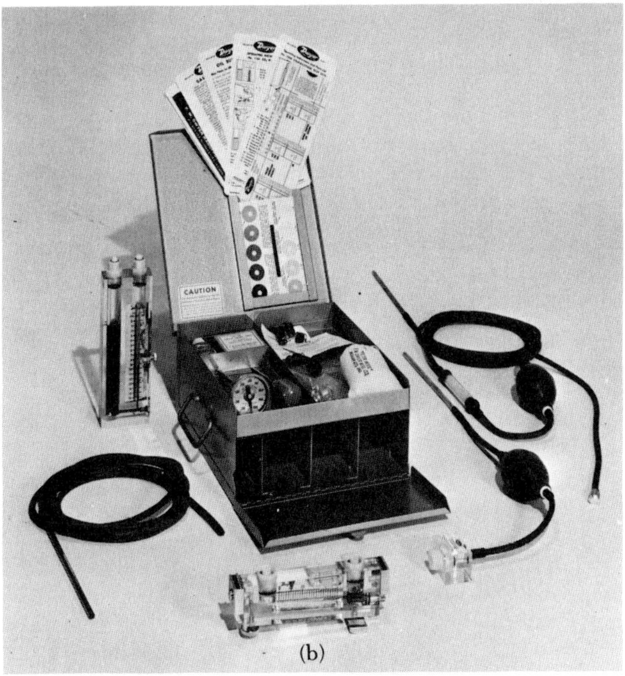

(b)

FIG. 1-4. (a) CO_2 indicating gage. (*Bacharach Industrial Instrument Co.*) (b) Complete combustion testing kit. (*F. W. Dwyer Manufacturing Company.*)

FIG. 1-5. Explosimeter. (*The Mine Safety Appliances Company.*)

is a lot easier to handle out on the job than is a sling pyrometer. All that is necessary is to hang up the humidity indicator where the air is representative of the area, and the relative-humidity percentage appears in a few seconds. Other shapes and sizes are commercially available. Use of the humidistat is essential whem making a check on

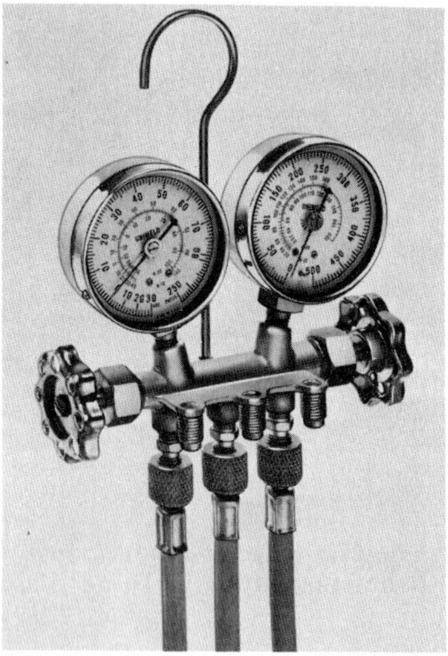

FIG. 1-6. Refrigerant system testing and charging manifold. (*Uniweld Products, Incorporated.*)

the performance of an air-conditioning coil. The sensitivity of these instruments is readily shown by simply breathing upon the element.

(r) ***The Ultrasonic Detector.*** This item, shown in Fig. 1-9, is designed to locate leaks and friction areas by ultrasonic methods, particularly in potentially hazardous areas which may be encountered in chemical plants, petroleum refineries, and certain industrial activities. Ultrasonic sound is measured by a meter and operates on small batteries.

Since the total weight is little more than 6 lb, the instrument is readily portable. Cheaper instruments are available for those troubleshooters who have limited need for leak and friction detection.

(s) ***Stethoscope.*** An inexpensive doctor's stethoscope is an excellent helper in checking on the internal performance of gears, pumps, and pipelines. It is a great improvement over such makeshifts as screwdrivers, oil can nozzles, and other sound-conducting rods and bars.

(t) ***Dial Indicator.*** This is a must for checking rotating alignments and clearances.

(u) ***Refrigeration-testing Pocket Thermometer.*** Figure 1-2(b) shows a form of pocket instrument calibrated for the refrigeration range of −40 to 210 F, which is very useful to the air-conditioning serviceman.

(v) ***Voltmeter—Ammeter.*** When selecting these instruments for the field kit, consider the voltage range expected to be encountered. Services from 110 to 220 volts are frequent, as is a low-voltage control

FIG. 1-7. Velometer. (*Alnor Instrument Company, Division of Illinois Testing Laboratories, Inc.*)

FIG. 1-8. Humidistat. (*Abbeon Supply Company.*)

FIG. 1-9. Electronic, ultrasonic leak detector in portable form. (*Hewlett-Packard Company.*)

circuit, anywhere from 6 to 24 volts. More than one unit may be required. See Fig. 1-10.

(w) ***Ringlemann Smoke Comparison Chart.*** Not strictly an instrument, the Ringlemann smoke density indicator is available in the form of a card 5 × 3½ in. Smoke clouds are viewed through a horizontal opening in the middle of the card and can be directly compared with the four steps of the Ringlemann scale. The full-size chart is issued by the U.S. Bureau of Mines.

(x) ***Flame Mirror.*** This is a mirror on the end of a long rod, useful to the troubleshooter in looking around corners. It is comparable in design and purpose to the dentist's mirror.

(y) ***Tachometer.*** This instrument is extremely useful for checking the rpm of any rotating shaft end available for contact. Figure 1-11 illustrates a multiple-speed instrument in its carrying case.

(z) ***Wattmeter.*** For measuring power demands of the equipment. Good to have, but not often necessary.

(aa) ***Ohmmeter.*** With this instrument, it is possible to determine if a circuit is open or shorted by measuring the resistance and comparing this with what it should be. See Fig. 1-10.

(bb) ***Megger or Megohmmeter.*** Useful in testing the refrigeration compressor. Follow directions in connecting to the compressor and ground.

1-3. TROUBLESHOOTERS' TOOLS

(a) *About Field Tools.* The troubleshooter's tool box, the kind he can carry under his arm, is, or certainly should be, stocked with the conventional items of wrenches, hammers, screwdrivers, micrometers, feelers, tape, a spring-loaded steel scale, files, a knife, a small magnifying glass, pliers, and such other small items as may be dictated by the nature of the particular job at hand. If the area atmosphere in which the work is to be done happens to be potentially explosive, all tools must be of nonsparking materials, and the troubleshooter himself is advised to replace his leather shoes with ones having rubber soles.

Larger tools that go out on trouble jobs only for particular kinds of trouble will be discussed in the following paragraphs.

(b) *Soldering Sets.* For use with both soft and hard, or silver, solder. Should be on hand for replacing tubing and stopping leaks in copper piping systems.

(c) *Welding Sets.* Gas-fired or electric welders as suitable. The former is probably best suited for domestic applications; the electric, for industrial work. "Standard Procedure Specifications for Pipe Weld-

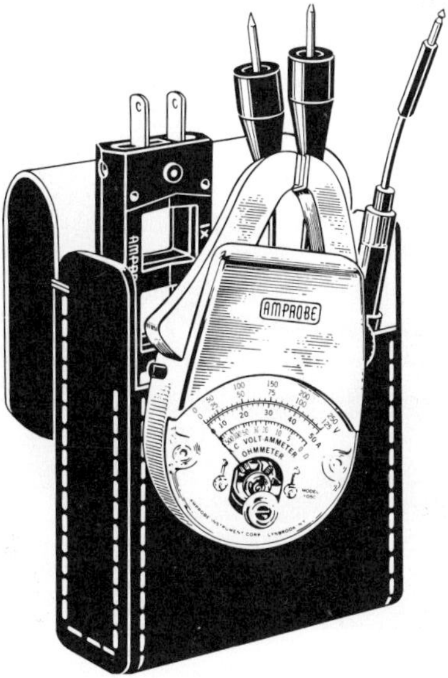

FIG. 1-10. Voltmeter, ammeter, and ohmmeter in carrying case. (*Amprobe Instrument.*)

FIG. 1-11. Tachometer in carrying case. (*James G. Biddle Company.*)

ing" have been developed and proven by the National Certified Pipe Welding Bureau, and the troubleshooter should become familiar with these specifications. On some jobs, welding work will not be accepted unless it is performed by a certified welder. Certification is obtainable from the various insurance companies, such as Hartford Steam Boiler Inspection and Insurance Company, from the U.S. Coast Guard, through the National Certified Pipe Welding Bureau, and from other authorities.

(d) ***Battery Light.*** Most jobs seem to occur in areas of poor illumination. A good, strong-battery portable electric lamp is a most handy tool to have around.

(e) ***Extension Cord and Lamp.*** A 100-watt lamp, or larger, on the end of a 25-ft extension cord has the advantage of not running down as

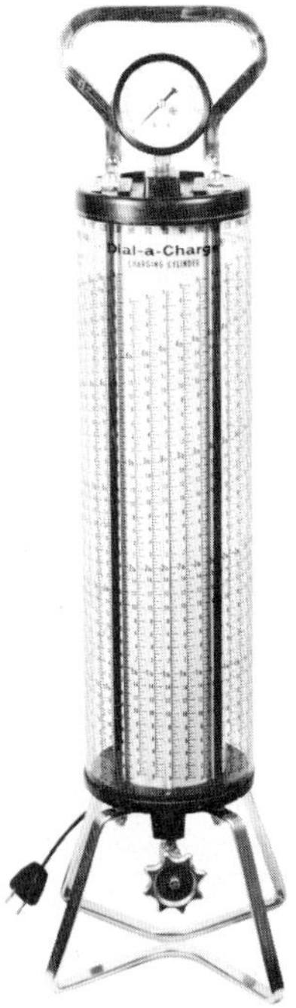

FIG. 1-12. Temperature-adjustable charging cylinder. (*Robinair Manufacturing Corporation.*)

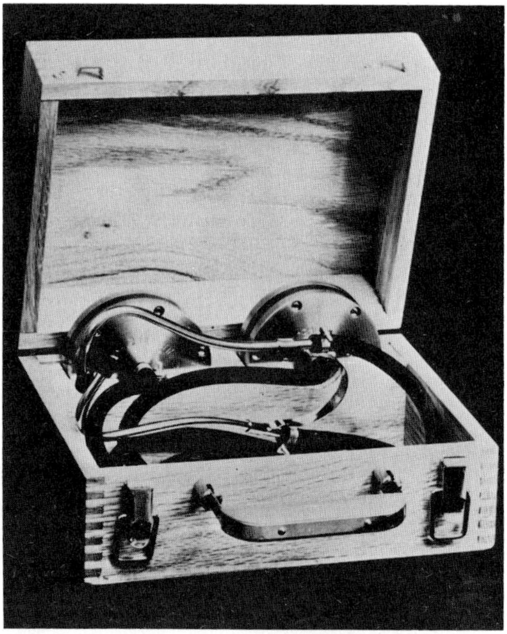

FIG. 1-13. Stethoscopic type of underground leak detector. (*Schneider Manufacturing Company.*)

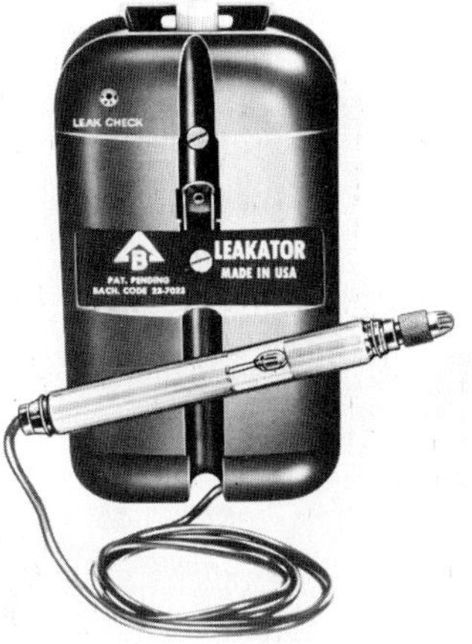

FIG. 1-14. Refrigerant leak detector; weight: 1½ lb. (*Bacharach Industrial Instrument Co.*)

the battery lamp does. Use it in preference to the battery if practical.

(f) ***Refrigerant-charging Cylinder.*** A fairly recent development is the temperature-adjustable charging cylinder shown in Fig. 1-12. By means of an electric heating element, the temperature of the refrigerant in the cylinder may be increased as necessary to match the existing pressure in the system being charged. Standard sizes of these cylinders are 2½, 5, and 10 lb.

(g) ***Portable Electric Drill.*** With changeable bits, the drill carried in the truck can be a very useful tool.

(h) ***Vise.*** Usually carried in the truck.

(i) ***Snake.*** This is a plumber's tool. Not needed by other mechanical servicemen.

(j) ***Leak Detectors.***

1. Figure 1-13 shows a stethoscopic form of leak detector in its carrying case. Because of differences in hearing sensitivity among individuals, the manufacturer recommends checking each suspected area of an underground leak by at least two persons. This unit is designed to locate underground leaks of water, steam, and some gases.

2. Halogen-sensitive detectors are illustrated by Figs. 1-14 and 1-15. This type of instrument employs a probe that, when it encounters a leak, extinguishes a normally lighted neon lamp or otherwise produces a signal on the dial. These are battery-operated instruments very light of weight and small of size, thus being well suited to portable service.

FIG. 1-15. Halogen-type leak detector for refrigeration systems. (*General Electric Company.*)

3. Figure 1-9 shows an electronic, ultrasonic translator which detects leaks by translating ultrasonic frequency into audible sounds. This unit is approved by the Underwriters Laboratories for service in hazardous locations.

(k) ***Fuses.*** An assortment of cartridge fuses of various capacities should go along on every service call. If the existing fuses are not already blown, there is a good chance one or more will be in the course of checking the electrical system for shorts and overloads.

(l) ***Refrigerant Charges.*** The various forms of the Freon family may be taken along in 50-lb drums. The type of refrigerant used should be asked before leaving on a call, and a drum taken along just in case makeup or even a complete charge is needed.

(m) ***Capacitor Tester.*** This item will be particularly useful on air-conditioning systems that are suspected of having been knocked out by lightning. A bad capacitor does not always show up in the visual inspection, and without the tester, the true condition of the unit is subject to doubt.

(n) ***Capacitor Replacement Kits.*** Available as start-and-run kits, these replacement packages handle a range of voltages from 110 to 440. The kit includes resistors rated 2 watts for start and 1 watt for run, terminal connectors, and wire jumpers. Very handy to have in preference to a trip back to the shop or to a supply house.

List of Suppliers

The following manufacturers of instruments and tools issue catalogs that can be of value to the troubleshooter. Even if no immediate use for a given item is needed, having the catalog and pertinent data on hand against an urgency of tomorrow or next week or even next year is recommended.

Abrax Instrument Corporation, 179–180 Jamaica Ave., Jamaica, New York 11432

Alnor Instrument Company, Room 516, 420 N. LaSalle St., Chicago, Illinois

American Instrument Company, 8030 Georgia Ave., Silver Spring, Maryland

Amprobe Instruments Division of Soss Manufacturing Company, Lynbrook, New York

Bacharach Industrial Instrument Co., 200 N. Braddock Ave., Pittsburgh, Pennsylvania 15208

Delcon Division of Hewlett-Packard Company, 333 Logue Ave., Mountain View, California

F. W. Dwyer Manufacturing Company, P.O. Box 373, Michigan City, Indiana

Gelman Instrument Company, P.O. Box 1448, Ann Arbor, Michigan

General Electric Company, Troup Highway, Tyler, Texas

Leak-Tec Division, American Gas and Chemicals, Inc., 511 E. 72 St., New York, New York 10021

Marsh Instrument Company, Skokie, Illinois 60076

Mine Safety Appliances Company, 201 N. Braddock Ave., Pittsburgh, Pennsylvania 15208

Schneider Manufacturing Company, Reading, Massachusetts

Uniweld Products, Inc., 2850 Ravenswood Rd., Fort Lauderdale, Florida 33312

DIVISION 2

Boilers, Heaters, and Firing Equipment

2-1. CAST-IRON BOILERS, DOMESTIC[*]

(a) *Sectional Type for Steam or Hot Water, Discussion.* These units are designed for low pressure, meaning a maximum of 15 psig for steam and 30 psig for water. They are available from the smallest of sizes to commercial and industrial units that count their Btu input in the millions. They are fired variously with coal, gas, or oil and operate as designed on either natural or power draft. Their sectional design enables them to be installed as replacement units in existing buildings, the sections usually being so dimensioned as to let them pass readily through standard doorways. Since the expected life of a domestic heating boiler is 20 years, at least one replacement is likely in the life of any building; consequently, a piecemeal erection facility is an asset in these circumstances.

(b) *Complaint.* The cast-iron sections in several areas of the boiler develop cracks after from 1 to 2 years operation. Replacement of these sections is expensive and keeps the boiler out of service during the period of the work.

Cause

Boiler sections are held together in airtight contacts by means of single or multiple draw rods. Since cast iron is not elastic, drawing up too tightly by the erector can result in one or more cracked sections under the added strains of expansion and contraction as the boiler alternately heats and cools.

Recommended Actions

1. Check the strain on the draw rods by strain wrench and compare this with the recommendation of the boiler manufacturer. Slack off on the nuts if strain is found excessive.

[*] See Table 2-3, p. 59.

2. Under no circumstances use an extension on the wrench handle. If gas tightness is not achieved with normal setting up of the nuts on the draw rods, renew the asbestos seals in the leaky joints.

(c) ***Complaint.*** Boiler is rusting over-rapidly and indicates that an unnaturally short life is in prospect.

Cause

Rusting is expedited by letting the boiler stand idle in a humid environment. This is a frequent condition in domestic cellars or boiler rooms during the summer nonoperating season.

Recommended Actions

1. Ventilate the boiler room or cellar. If air movement by natural draft through doors and windows is not practical, provide a small exhaust fan. The amount of current it uses is insignificant compared with the cost of a new boiler.

2. Set a small tray of slaked lime in the boiler furnace and close the stack damper and all air access openings to the furnace.

3. If boiler contains a submerged domestic hot water coil, enough heat is generated to depress the humidity to satisfactory levels.

(d) ***Complaint.*** When firing boiler with bituminous coal, considerable smoke and occasional puffbacks soil the premises.

Cause

Bituminous coal is a high volatile coal, and the sudden release of gas in large volumes is often too great to be handled efficiently by the boiler firebox and the chimney.

Recommended Actions

1. Compare the actual area of the flue in square inches with the area recommended by the boiler manufacturer. If area is smaller than recommended, extra friction is depressing draft. In this circumstance, consider installing a draft-inducer fan.

2. Switch fuel from bituminous to anthracite, coke, or gas. If gas is adopted, make sure that a suitable draft hood is provided between the chimney and the boiler.

(e) ***Complaint.*** When firing with oil, smoke and odors emerge from chimney and become a nuisance in the neighborhood.

Causes

1. Inadequate combustion air supply, poor draft.

2. Burner nozzle capacity is too large for the boiler furnace and is impacting flame on the refractories.

3. Nozzle is giving a flame of the wrong shape for the furnace that contains it.

Recommended Actions

1. Check draft with draft gage. Note if draft at whatever level indi-

cated is steady or pulsing. Pulsing draft is from a partial obstruction or a pulsing flame.

2. Examine chimney for clearness of obstructions, such as bird nests, dislocated flue tile, or heavy coating of soot. Probe by lowering a brick on a line and swinging this brick back and forth to shake loose soot or obstructions. Vacuum clean if necessary.

3. Compare burner nozzle size and type with nozzle recommended by boiler manufacturer. Replace with new nozzle if necessary.

4. Clean or replace fuel-oil filter in oil line to burner.

5. Check oil pressure to burner. If too low, atomization of oil will be incomplete and smoke will follow. In absence of information from manufacturer concerning the correct pressure, assume a minimum of 100 psig.

(f) ***Complaint.*** Oil-fired furnace puffs back and is otherwise characterized by rough ignition.

Causes

1. Puffbacks will occur when the chimney draft is too low or the ignition spark is weak.

2. Ignition electrodes are set with an improper gap.

3. Oil contains water.

4. Oil vapors collecting in furnace are not vented and, when ignited by spark, do so explosively.

Recommended Actions

1. Check presence of an oil pool in the bottom of the furnace. If found to be present, light with a piece of burning paper and make sure the oil is consumed before attempting burner ignition. Alternatively, oil may be mopped out of furnace. If pool is allowed to burn along with the regular burner, dangerous overheating will result plus considerable probability of an explosion.

2. Check oil storage tank for water in bottom and pump out if found.

3. Inspect chimney for obstructions and undersize flue. See that chimney cap, if any, does not obstruct the draft.

4. Check electrode gap for its proper setting.

5. Check voltage of ignition transformer.

(g) ***Complaint.*** Oil burner either will not start or, after ignition, fails to hold the flame and cuts out.

Causes

1. Check fuel oil for water.

2. Check stack switch for correct operation.

3. Voltage fluctuations may be affecting burner motor, causing it to cut out from either overvoltage or low voltage.

4. Porcelain insulators have developed cracks or have shorted.

5. Ignition transformer is defective.

(h) ***Complaint.*** Oil-burner installation operates but with excessive noise, such as a grinding, hum, or rattling.

Causes

1. Burner motor is loose on its mounting with some knocking.
2. The pump and motor are misaligned and squealing.
3. Pump gears are worn, producing grinding sounds.
4. Burner fan is rattling on its shaft.

Recommended Action

Check each of the possible causes as given above. The nature of the trouble indicates the steps to be taken to correct it. For example, realign the pump and motor.

(i) ***Complaint.*** Gas-fired boiler fails to meet its rating in steam or hot water.

Causes

1. Gas pressure is lower than burners are designed to handle.
2. Flue-gas passages through boiler are sooted.
3. Calorific value of gas fuel is lower than fuel on which the boiler is rated.
4. Calibration of steam or water thermostat is faulty, causing burners to cut out prematurely.

Recommended Actions

1. By means of pressure gage, verify gas-supply pressure.
2. Compare gas Btu value with value on which boiler capacity is rated.
3. Clean flue-gas passages through boiler to improve heat exchange.
4. Prove out performance of thermostats and all other automatic control equipment.
5. Verify free venting of flue gas to atmosphere. Check for draft interferences, such as trees overhanging the vent outlet, the presence of a decorative outlet, or an excessive number of vent turns and horizontal runs. If downdrafts are observed from any of the noted conditions, correct as completely as possible.

General Precautionary Notes

1. When boiler has been totally shut down with the pilot extinguished, make sure that the manual gas shutoff valve and the pilot shutoff valve have been closed for at least 5 minutes before attempting to relight the system.

2. After the pilot has been relighted, allow 2 minutes to elapse before pressing the reset button and putting the burners on automatic control. After these 2 minutes, open the manual gas shutoff valve. Then open the automatic gas shutoff valve and verify its proper performance. Make sure this suggested sequence is in accordance with the manu-

facturer's directions for this particular burner, and where there is a difference, follow the manufacturer.

3. Never leave a gas-fired unit unattended for a single minute while it is on manual control. Failure to observe this caution may result in fire or explosion.

(j) ***Complaint.*** On a warm air, forced circulation heating system, smoke and odors issue from the warm air supply registers and foul the furniture and drapes in a room.

Causes

1. Stop the burner, either oil or gas. Probabilities are that the heat exchanger has developed a crack or other leak.

2. Dirt in the filter has become ignited and is emitting smoke and smells.

Recommended Actions

1. Inspect filter. If trouble source is here, replace the filter with a new one.

2. Remove heater casing and inspect the heat exchanger. Cracked or ruptured exchanger will require welding or complete replacement, depending on the extent of deterioration found.

2-2. STEEL BOILERS, LOW-PRESSURE, DOMESTIC

(a) ***Discussion.*** Steel boilers for low pressure, 15 psig steam and 30 psig water, are available in horizontal, return-tubular forms, also Scotch boilers, vertical, tubeless, and tubed, and in firebox or locomotive types with water legs enclosing the furnace.

(b) ***Complaints.*** Complaints with the various fuels noted under Sec. 2-1 may be duplicated in the steel boilers, plus two other conditions peculiar to the steel:

The steel shows cracks in some areas after from 1 to 4 or 5 years of operation. These may be called fatigue cracks or cracks indicative of caustic embrittlement.

Cause

A reaction with the boiler water of the steel shell, usually occurring in the areas of rivets or welded joints.

Recommended Action

Analyze the boiler water and change it with additives as prescribed by boiler-water chemical-treatment specialists.

(c) ***Complaint.*** In fire-tube steel boilers, the spinners become fouled with soot and create excessive draft loss through the boiler. This has the effect of reducing boiler capacity.

Cause

These spiral ribbons, or spinners, produce a whirling motion of the flue gases, thereby improving the efficiency of heat exchange to the boiler. In time, accumulations of soot on these ribbons partially obstruct the gas flow and, in addition, form insulating layers on the tubes, decreasing heat exchange.

Recommended Action

Inspect the spinners for cleanliness and, if too foul to be cleaned in place by air jets or vacuum nozzles, pull the spinners and clean both them and the fire tubes thoroughly before returning the boiler to service.

(d) ***Operational Complaints.*** Refer to the appropriate sections for cast-iron boilers and to Table 2-3.

2-3. WARM AIR CENTRAL FURNACES, DOMESTIC AND COMMERCIAL

(a) ***Discussion.*** Originally designed strictly for heating with ducts attached for distributing the warm air throughout a building, a large percentage of these units are now arranged for accommodating summer cooling coils as well. Because of the wide difference of discharge air volumes on the heating and cooling cycles (the cooling air volume being three to four times that of heating), use of a common fan calls for a driving motor of two, and sometimes three, speeds.

The fan and air filter become sources of trouble, in addition to the firing equipment, and will be considered in that order. The cooling coils, when provided and mounted in the air stream, present their own particular complaints which will be examined in the sections dealing with summer air conditioning.

(b) ***Complaint.*** Fan is extremely noisy with a clattering sound.

Causes

1. Fan is usually set-screwed to its shaft, and the screw has come loose.

2. Filter has vibrated out of its usual position and is being rubbed intermittently by falling against the fan.

Recommended Action

Shut down heater and examine fan and filter. Reestablish fan on its shaft and filter within its brackets as may be found necessary.

(c) ***Complaint.*** Heater is not providing sufficient heat.

Causes

1. Discharge temperature of air is too low.

2. Volume of discharge is too small.

3. Supply ducts are not insulated, and the loss of heat is affecting the temperature of the warm air issuing from the registers and diffusers.

4. Air filter is dirty and preventing effective passage of the air through it.

Recommended Actions

1. Take the temperature of the warm air discharge from the heater. If too low, adjust oil or gas burners to a higher rate of combustion.

2. With an anemometer or velometer, measure output from each register or diffuser and compare with the cubic feet per minute called for by the design. Add up the total of the discharges and check this against the rated output of the fan. If fan is delivering less air than its nameplate calls for, check speed in rpm with a shaft tachometer. Discuss any underspeed condition with the fan or heater manufacturer; a replacement may be required.

3. Clean the filter or replace it if it is of the throwaway type.

4. If duct exposure is the cause of the trouble, insulate the supply ducts. Insulate returns also if heater is barely able to reach its rated Btu capacity and heat conservation is necessary.

(d) ***Complaint.*** Heater short-cycles. That is, it cuts in and out more frequently than it should.

Causes

1. Control thermostat is so located as to be influenced by drafts, windows, or reflected heat.

2. Thermostat or stack limit switch is defective.

3. If burning oil, water in oil will interrupt flame, and safety cutout then takes charge.

Recommended Actions

1. Check oil for water and, if water is found, pump storage tank dry.

2. Evaluate position of environmental control thermostat. Relocate if influenced by nearby factors.

3. Verify correct functioning of stack switch.

4. Test out thermostat and limit switch for correct operation. Replace if necessary.

(e) ***Complaint.*** Oil-fired unit is producing smoke and odors.

For causes and recommended actions, see Sec. 2-1(e).

(f) ***Complaint.*** Ignition is accompanied by furnace puffs or flarebacks.

For causes and recommended actions, see Sec. 2-1(f).

(g) ***Complaint.*** Burner does not start or quickly cuts out.

For causes, see Sec. 2-1 (g).

(h) ***Complaint.*** Burner is noisy.

For causes and recommended action see Sec. 2-1 (h).

(i) ***Complaint.*** Instead of being direct-fired, central unit obtains its heat from steam coils, as shown in Fig. 2-1. Heat is insufficient.

Causes

1. The obvious cause in this instance is the waterlogged condition of

the lower half of the pipe coils. The condensate from the steam is expected to return to the boiler by gravity, but inspection of the sketch in Fig. 2-1 shows the water level of the boiler to be above the lower steam coils. This condition eliminates approximately half of the heating capacity of the coils.

2. Other possibilities, *in addition* to the flooded coil, are an undersized coil, a fan that is not delivering its rated cfm, incorrectly sized air ducts, and a clogged filter.

Recommended Actions

1. Disconnect the condensate return to the boiler and lead it instead to a receiver, from which it is relayed to the boiler by pump. The pump may be controlled by a float switch or by the boiler water level controller.

2. Check the steam pressure and temperature against the heating requirements of the building. Verify the adequacy of the steam coil on the basis of the normal steam conditions. This may be done by weighing the condensate or by checking the coil manufacturer's capacity charts.

3. Measure the cfm output of the fan by anemometer or velometer. If under speed, take proper steps to correct. Compare air volume with the cross-sectional area of the ducts. If undersized, excessive velocity will develop noise and frictional resistance that may reduce the effective output of the fan. As a rough guide, air velocity of more than 1,200 fpm is subject to suspicion, suggesting undersized ducts.

4. Remove, clean, or replace the filter.

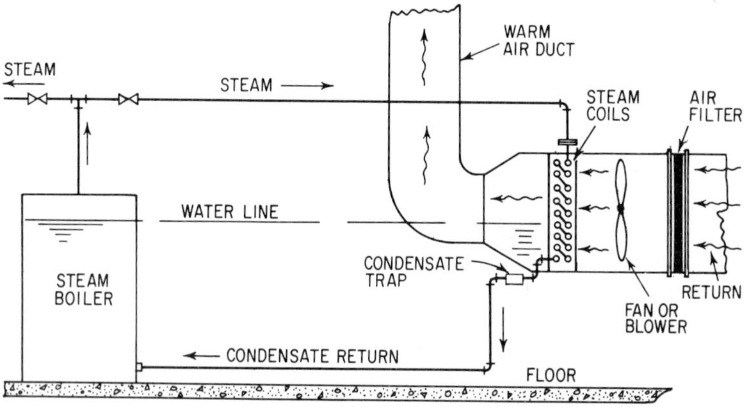

FIG. 2-1. Flooded steam coils (no scale). *Note:* Same condition would exist with Hartford return.

2-4. SPACE HEATERS, DIRECT-FIRED

(a) ***Discussion.*** These units, burning oil or gas, generally are found in cabinet form with a blower enclosed or as floor heaters in which the unit is suspended under the floor and the warm air discharge rises through a grille or grating at floor level.

Usual method of firing is by oil or gas, the former fuel being fed by a float valve to a pot burner. Gravity moves the oil from the storage tank to the float valve and thence into the pot.

All units are fitted with a safety thermostat that cuts off the fuel when furnace temperatures become excessive. Ignition may be either electrical or manual.

(b) ***Complaint.*** The difference in temperature between floor and ceiling is unreasonably wide, with all the heat rising to the ceiling. Temperature difference may be as much as 20 deg, with 60 F at the floor level and 80 F just under the ceiling.

Causes

1. Heated air naturally rises, its place at low levels being taken by inflowing cooler air.

2. The condition is aggravated by a ceiling of above-average height, such as 10 to 14 ft.

Recommended Actions

1. Placement of an ordinary fan of the 10- or 12-in. size at a position about 8 to 9 ft above the floor, trimmed to discharge slightly downward, will improve conditions by between 5 and 10 deg.

2. Reduce along-the-floor drafts caused by loosely fitting doors and windows. Do this with weatherstripping or by the addition of storm windows or storm doors.

3. If there is an open fireplace in the room or in an adjacent room with an open doorway, make sure the chimney damper is closed during periods of fireplace idleness.

(c) ***Complaint.*** The furnace is hard to light. Torches dropped into the pot burner of the oil-fired unit smoke a little and then go out.

Causes

1. Draft is poor, with little or no difference in temperature between the bottom and the top of the chimney.

2. Chimney is cold, and fire pot is cold. Loss of heat from torch is so rapid that the flame expires before the oil is heated enough to vaporize.

Recommended Actions

1. If possible, remove furnace breeching from chimney and build a small fire of paper or excelsior in the bottom of the chimney. This will warm the base of the chimney enough to establish a draft. Then quickly replace the breeching and drop a new torch into the fire pot.

2. Alternatively, set up a small paper or excelsior fire in the fire pot. When this fire is nearly dead, restore the oil feed. The pot should be warm enough to ignite the oil. Note that this expedient tends to foul the bottom of the pot with charred materials and to generate smoke. If repeated very often, the furnace must be taken out of service and cleaned.

(**d**) ***Complaint.*** The heater warms only the room it is in.

Causes

1. Warm air rises but moves very little laterally.

2. As a general rule, warm air from a space heater cannot be expected to fan out through more than one doorway. This tendency, if not modified, pretty well limits space heating units to a single room.

Recommended Actions

1. A cabinet type of space heater, if equipped with a fan that discharges horizontally along the floor, will provide an expanded boundary of comfort.

2. For a floor furnace that discharges only vertically, the stationing of a fan at floor level can spread the rising currents of warm air horizontally, sometimes even through doorways to adjacent rooms.

2-5. HIGH-PRESSURE BOILERS, FIRE-TUBE

(**a**) ***Discussion.*** "High pressure" refers to working pressures in excess of 15 psig for steam and 30 psig for water.

Fire-tube boilers are built of steel and are available in the following forms: steam generators fired with oil or gas; horizontal return tubular boilers fired with oil, gas, or coal; Scotch marine boilers with one or more furnaces fired with oil, gas, or coal. Note that these Scotch boilers are sometimes built with furnaces at each end. Also common in the fire-tube classification is the locomotive type of boiler, burning oil, gas, or coal fuel. Vertical fire-tube boilers, although still being built and operating in many industries, are met relatively infrequently and usually in small sizes.

In public utility plants and large industrials, regular boiler-room attendants maintain a high level of performance, and trouble seldom develops to a serious extent. However for those plants in which boiler maintenance is predominantly casual, the troubleshooter's work will be expedited by the following listing of complaints and their handling.

(**b**) ***Complaint.*** Occasional peak demands for steam are not met adequately by the boiler, and steam pressure drops so rapidly that certain processes must be stopped until the pressure can be restored. The boiler is a fire-tube steam generator.

Cause

This type of boiler has limited reserve capacity, and any steam demand that exceeds the designed capacity by more than 10 percent is likely to go unsatisfied. The drop in pressure is the result of the steam leaving the boiler faster than it is being generated.

Recommended Actions

1. Consider the installation of a steam accumulator. To perform satisfactorily, there must be a need for steam at a pressure below that of the boiler's regular output. In operation, steam at the regular pressure enters the accumulator where it heats the water until the heat value per pound in the accumulator is equal to, or nearly equal to, the boiler water. Then when the boiler pressure falls due to a heavy demand for steam, the reduction in pressure over the surface of the water in the accumulator causes a reverse action to take place, the heat stored in the water enabling it to flash into new steam at the reduced pressure. Thus the accumulator stores heat or releases it in response to the fluctuation of pressure at the boiler outlet.

Accumulators may be vertical or horizontal, Fig. 2-2 illustrating the latter form.

2. An alternate action is to install a peak-load boiler, one that comes on the line for this service exclusively. The growth pattern in some industries is such that the purchase and installation of a new boiler releases an otherwise too-small boiler from regular service to the status of a peak-load standby.

(c) ***Complaint.*** Boiler is producing wet steam.

Causes

1. Boiler is being driven above its rating, thereby setting up a velocity of steam flow, through the steam outlets, that is carrying drops of water with it.

2. Boiler water is contaminated with some form of foreign material that increases surface tension; for example, oil. When this is happening, the boiler is said to be priming. Another promoter of moisture carry-over is foaming, likewise the result of contaminating materials.

3. Steam-release section of boiler is small, and the dry pipe is not separating moisture from the steam as it should. This condition is generally coupled with a rate of steam generation in excess of that for which the boiler is designed.

Recommended Actions

1. Check boiler water for purity and absence of oils and soaps. If present, blow down boiler to clear, using both a low and surface blow if both are available.

2. Reduce steam output to a maximum rating.

3. If steam is still wet at maximum rating, shut down boiler and inspect condition of the dry pipe. Discuss the situation with the boiler manufacturer.

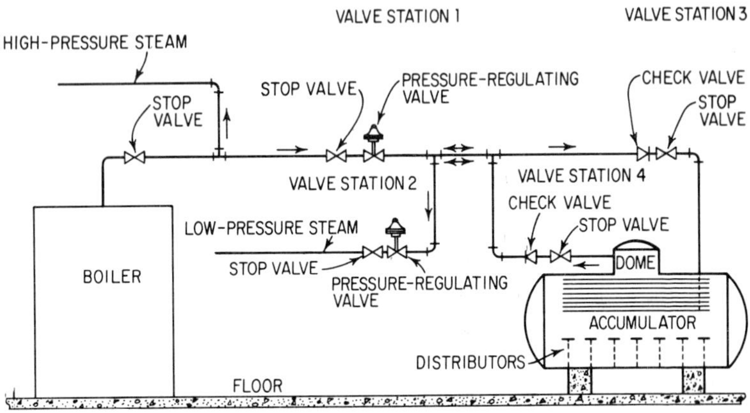

FLOW NOTES:

1. BOILER TO ACCUMULATOR THROUGH STATIONS 1 AND 3
2. BOILER TO LOW-PRESSURE HEADER THROUGH STATIONS 1 AND 2
3. ACCUMULATOR TO LOW-PRESSURE HEADER THROUGH STATIONS 4 AND 2

FIG. 2-2. Steam boiler and accumulator piping arrangement.

4. Provide and install a steam separating drum, or "dome," on the boiler outlet to return excess water to the boiler. This dome is standard design with some makes of boilers.

(**d**) ***Complaint.*** Boiler smokes excessively on oil fuel.

Causes

1. Insufficient mixing of the fuel and air in the boiler furnace. Mixture is too rich.
2. Oil is too heavy for the firing equipment installed.

Recommended Actions

1. Adjust burners.
2. Check viscosity of oil. Increase firing temperature.
3. Switch to a lighter oil.

(**e**) ***Complaint.*** Gas-fired boiler does not meet its rating.

Causes

1. Gas pressure is too low.
2. Calorific value of gas, in Btu per cubic foot, is less than the value of the gas on which the boiler rating is based.
3. Draft fan is under speed or otherwise failing to provide its rated discharge.

Recommended Actions

1. Check gas pressure. Set up regulator, if necessary.
2. Verify calorific value of gas with laboratory test by supplier.
3. Check fan speed with tachometer. Measure discharge pressure

with gage. Correct speed and output as necessary if found lacking. Communicate with boiler manufacturer.

(f) ***Complaint.*** Coal-fired boiler smokes excessively. Fire tubes soot up heavily.

Causes

1. Incomplete combustion in the boiler furnace is producing both smoke and soot.
2. Draft is inadequate.
3. Grade of coal being burned is high in volatile matter.
4. Spinners in fire tubes may be introducing excessive resistance to movement of flue gases.

Recommended Actions

1. Check draft over fire. If less than 0.15 in. water gage, look for draft interference or restrictions.
2. Consider using coal with less volatile matter.
3. Examine spinners for distortion. Measure draft at boiler outlet to breeching.
4. Check breeching for possible horizontal, or even down, pitch. If dependent on natural draft, breeching must have an upward pitch to the stack of at least 1 in. to the foot.

2-6. HIGH-PRESSURE BOILERS, WATER-TUBE

(a) ***Discussion.*** The water-tube boiler presently is being built in sizes ranging from the packaged, shipped-complete, ready-to-hook-up design all the way to the units rated at far beyond a million pounds of steam an hour each. Pressure-wise, operating pressures beyond the critical of 3,206 psi are being furnished in steadily increasing frequency, and 5,000 psi has been reached in several instances.

Steam temperature of 1200 F in these supercritical boilers has been achieved by means of special steels; for example, the main steam piping at Eddystone Station of the Philadelphia Electric Company is formed of stainless steel. For the troubleshooter, it is well to remember that steel begins to glow in the dark at 800 F or a little above. It is also well to remember that a steel pipe that looks gray and cold may in fact have a surface temperature of from 400 to 700 F. In short, a *gray* pipe can be dangerous, and contacts with it should be avoided until its true temperature is established.

Water-tube boilers appear in two basic forms: the straight tube as illustrated in Fig. 2-3 (a), and the bent tube shown in various encountered drum arrangements in Fig. 2-3(b) and (c).

A recent development brought to America from France is the once-through drumless boiler, in which the feedwater enters one end of a continuous tube and emerges as steam at the other. The absence of

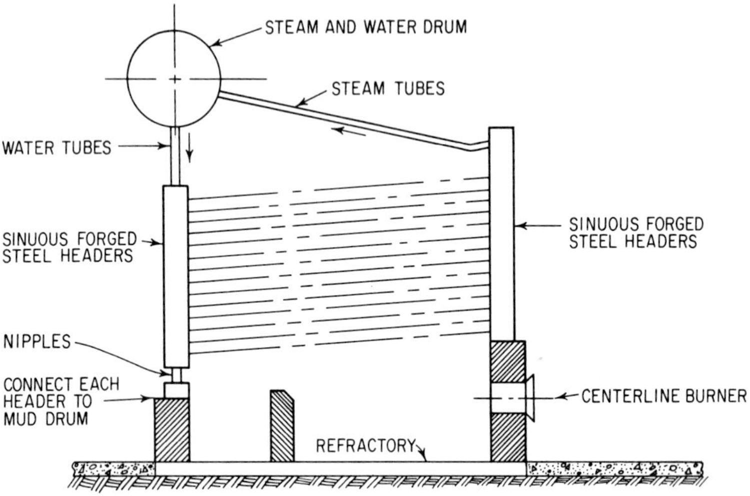

(a)

NOTES:

1. BOILER MAY BE FIRED FROM OTHER END IF DESIRED
2. IF BRICKED UP TOO TIGHT, STEAM TUBES MAY BUCKLE OUT OF DRUM
3. EXPANSION AND CONTRACTION OF TUBES THROUGH BAFFLES MAY CAUSE LEAKY BAFFLES IF VERTICAL. HORIZONTAL PREFERRED.

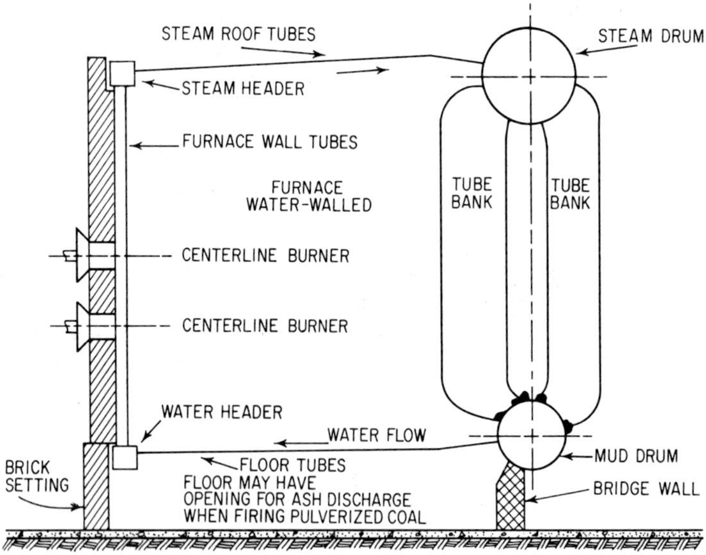

(b)

NOTES:

1. DARK AREAS AROUND TUBES AT MUD DRUM MARK COMMON POINTS OF CORROSION UNDER SOOT DEPOSITS
2. IF FLOOR TUBES BURN OUT, CHECK FOR INADEQUATE WATER CIRCULATION

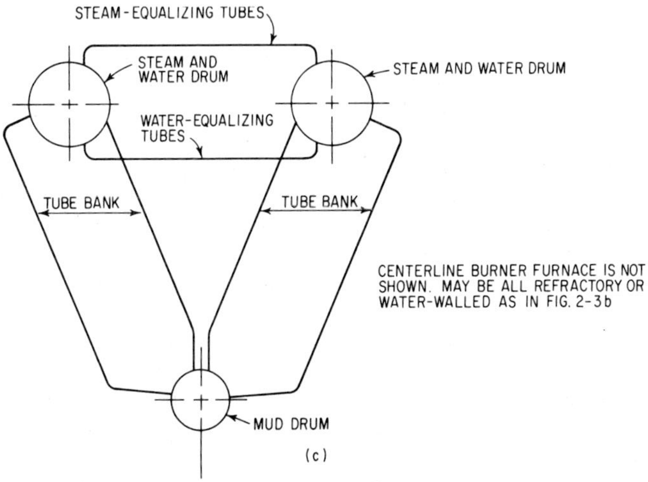

FIG. 2-3. (a) Straight-tube boiler (no scale). (b) One form of two-drum boiler (no scale). (c) Form of three-drum boiler (no scale).

water storage in this design makes it quite susceptible to fast burnouts if the water feed is interrupted for more than a few seconds. Because of this characteristic, the once-through boiler must be accorded almost continuous supervision, and feedwater purity and control are particularly critical.

Since high-pressure boilers are operated in most states by licensed engineers, the trouble symptoms described will be of the unusual variety, the operating people being familiar with the ordinary troubles that occur from time to time in a high-pressure boiler room. A list of the states requiring licensed personnel appears on Table 2-1, and a list of the cities on Table 2-2.

(b) ***Complaint.*** When boiler is operating at rated output, considerable vibration is observed. This is regarded as being self-induced. If not eliminated soon, it may cause fatigue failures in some tubes.

Cause

The flow of combustion gases around tubes of circular cross section produces vortices in the flow, first on one side of the tube, then on the other. The result is a pulsation that sets up a vibration in the gases, and the vibrating gas in turn delivers the vibrating urge to the steel walls of the gas passes. The frequency of the vibrations becomes a function of the width of the enclosure.

Recommended Actions

1. The simplest correction is accomplished by the installation of a "splitter" or baffle in the gas stream in the gas passage following the tubes over which the gas has been passing. In more complicated situations, analysis must be made of the frequency of the vibrations in order to deresonate the gas stream from the formation of the vortices.

2. This writer has eliminated vibrations in a preheated air duct by the use of stiffening angles welded to the duct in appropriate locations. However, this expedient is not likely to succeed with combustion gas, according to field reports brought in by others. Duct stiffening, therefore, should be undertaken only with the knowledge that it is experimental.

(c) ***Complaint.*** In a straight-tube, sectional-header boiler, the development of leaks around the tubes, at the place where these enter the transverse steam drum, has become a chronic trouble. Tubes have been rerolled or replaced several times, but leaks appear soon after.

Cause

Reports from other instances of this type of leakage indicate that the steam drum is being locked in position by the brick setting. Expansion therefore is entirely passed on to the tubes, and as these alternately cool and heat in response to the operating schedule of the boiler, leakage becomes inevitable.

Recommended Action

Break loose the brick setting where this encloses the drum and rebuild it with sufficient clearance allowed around the drum to permit normal expansion. This will enable the drum to move with the tubes, and the joints will no longer be strained.

(d) ***Complaint.*** Corrosive materials accumulate at the junctures of steam-generating tubes and the drums, particularly at the mud drum. This is the result of vanadium and sulfur in the fuel becoming moisture-rich from the water vapors in the combustion gases. At times of low loads, flue gases may cool below the dew point, creating some sulfuric acid.

Causes

As enumerated in the complaint.

Recommended Actions

1. Change to a fuel that contains less sulfur. For various reasons, this change may not be economical. Then other treatments must be considered.

2. Wash down the corrosive deposits periodically. This action does not stop the production of the acids, but it does delay the development of serious destruction to the tubes and drums. Water may be simply hot, about 180 F, and delivered against the deposits under pressure.

Table 2-1. States Requiring Boiler Operators to Be Licensed

U.S. States and Canadian areas	Power boilers	Low-pressure boilers	Unfired pressure vessels
Arkansas	Yes	No	No
District of Columbia	Yes	Yes	Yes
Louisiana	Yes	No	No
Maine	Yes	Yes	No
Massachusetts	Yes	Yes	No
Minnesota	Yes	Yes	No
Montana	Yes	Yes	Yes
New Jersey	Yes	Yes	No
Ohio	Yes	Yes	Yes
Oklahoma	Yes	No	No
Canadian Provinces:			
Alberta	Yes	No	No
British Columbia	Yes	Yes	No
Manitoba	Yes	Yes	Yes
New Brunswick	Yes	Yes	No
Newfoundland	Yes	Yes	No
Northwest Territory	Yes	Yes	Yes
Nova Scotia	Yes	Yes	No
Ontario	Yes	Yes	Yes
Quebec	Yes	Yes	No
Saskatchewan	Yes	Yes	No
Yukon Territory	Yes	Yes	No

Alternatively, solutions of soda ash are used. *Caution:* If the boiler furnace is refractory lined, care must be exercised to limit the impact of the water on the refractory. Shielding the brick with tarpaulin may be feasible in some instances.

3. Certain types of inhibitors have been used with considerable success. Reductions in dew point corrosion up to 80 percent have been reported. Cost of inhibitors for large boilers is not excessive in view of the destruction avoided. One report indicates that $25 a week will do it.

(e) ***Complaint.*** Steam tubes show some distortion, with a few having permanent sets.

Causes

Tubes have been overheated. Boiler has been operated at rates in excess of its rating for long periods.

Recommended Action

If tubes are not in actual contact with each other, the fact of a slight bend is not serious and does not call for peremptory replacement. If

Table 2-2. U.S. Cities Requiring Boiler Operators to Be Licensed

U.S. City	Power boilers	Low-pressure boilers	Unfired pressure vessels
Buffalo, N.Y.	Yes	Yes	Yes
Chicago, Ill.	Yes	No	No
Dearborn, Mich.	Yes	Yes	No
Denver, Colo.	Yes	No	No
Des Moines, Iowa	Yes	Yes	No
Detroit, Mich.	Yes	Yes	No
East St. Louis, Ill.	Yes	No	Yes
Greensboro, N.C.	Yes	Yes	Yes
Kansas City, Mo.	Yes	No	Yes
Los Angeles, Calif.	Yes	No	No
Memphis, Tenn.	Yes	Yes	No
Milwaukee, Wis.	Yes	Yes	No
Mount Vernon, N.Y.	Yes	No	No
New Orleans, La.	Yes	Yes	No
New York, N.Y.	Yes	No	Yes
Niagara Falls, N.Y.	Yes	Yes	No
North Miami Beach, Fla.	Yes	Yes	Yes
Oklahoma City, Okla.	Yes	Yes	No
Omaha, Nebr.	Yes	Yes	No
St. Louis, Mo.	Yes	No	No
Seattle, Wash.	Yes	Yes	No
Spokane, Wash.	Yes	Yes	No
Tacoma, Wash.	Yes	Yes	No
Tampa, Fla.	Yes	Yes	Yes
Tulsa, Okla.	Yes	Yes	No
Yonkers, N.Y.	Yes	No	No

tubes are in contact, or are likely to enter into contact under subsequent loads, replacement is necessary to avoid a later burnout while the boiler is steaming.

Determine cause of overheating. Water circulation may be at fault in this particular area of the boiler, with sluggish flow calling for improvement. In some instances, the best treatment might be the installation of a circulating pump to move the water faster within the boiler. A pump is likely to be needed at high ratings in those boilers operating at pressures approaching the critical.

(f) ***Complaint.*** Several failed boiler tubes have been found to contain corrosion tumors on their inner surfaces. Evidence indicates these tumors, if given time, will perforate the tubes.

Cause

Faulty feedwater treatment.

Recommended Actions

1. Make an analysis of the feedwater and turn this over to a boiler water specialist.

2. Check all possible entries of oxygen into the feedwater and condensate systems. Verify the performance of the deaerator, if one is being used, and the feedwater heaters. Examine all glands with steam sealing. Investigate vent condensers and the steam condensers wherever used.

3. Replace all tubes that on inspection show definite tumors.

2-7. FIRING EQUIPMENT, OIL BURNERS

(a) ***Discussion.*** Basic factors affecting the application of an oil burner to a particular boiler include the nature of the draft available, that is, whether it is natural, induced, or forced.

Both natural- and induced-draft arrangements must produce a negative pressure condition in the firebox or combustion chamber sufficient to draw in adequate secondary air for combustion and to overcome the frictional resistance of the boiler passages, the breeching, and the stack and eject the flue gases into the atmosphere.

Forced draft is produced by a blower or fan that is integral with the burner, furnishes both primary and secondary air for combustion, and has enough pressure to overcome the frictional resistance of the passages between the firebox and the atmosphere.

The forced-draft burner requires particularly designed firebox and boiler passages, so that no leakage of gases is possible. This kind of burner can operate with natural draft, but a natural-draft burner cannot operate with a forced-draft kind of arrangement, since the burner fan has neither the capacity nor the delivery pressure needed by the sealed-up boiler. In troubleshooting, therefore, the first essential is to recognize the burner for what it is—a natural-draft, an induced-draft, or a sealed-in, forced-draft type.

Adequate flame-inspection ports are notable by their absence from the forced-draft boilers; consequently, information on the burner performance must be collected by instruments. The following list is considered minimal for a forced-draft installation:

A small, portable, CO_2 indicator.

A draft gage.

A pressure gage, suitable for measuring oil pressures up to 125 psig for high-pressure burners and from 1 to 15 psig for low-pressure burners.

A vacuum gage. Trouble in the fuel-oil supply line is sometimes indicated by the vacuum conditions.

A combination voltmeter and ammeter.

A flame mirror.

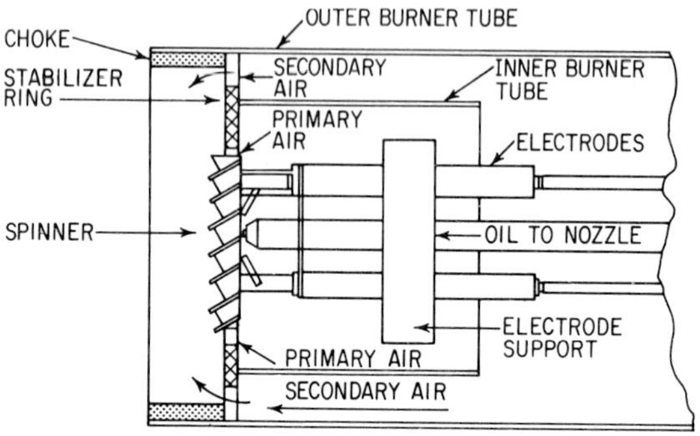

NOTE:

FLAME RETENTION RESULTS FROM THE PRIMARY AIR BEING DELIVERED IN A MULTIPLE PATTERN THAT INCLUDES A PASSAGE THROUGH THE CENTER OF THE BURNER TUBE AND PORTS AROUND THE BURNER NOZZLE. SECONDARY AIR INTRODUCED FROM THE PERIPHERY OF THE BURNER TUBE SHAPES THE FLAME AND COMPLETES COMBUSTION. THIS TYPE OF BURNER NEEDS NO COMBUSTION CHAMBER.

FIG. 2-4. Oil burner with flame retention (no scale).

Most of these instruments are usable with other forms of burners as well.

The "flame-retention" oil burner is a form of forced-draft equipment. The name reflects the ability of the device to continue in operation despite fluctuations in furnace conditions that might extinguish a conventional burner. Construction of this burner provides primary air through the center of the burner tube and thence through a number of small ports around the burner nozzle. The velocity of this air is designed to mix adequately with the oil spray, but so controlled as not to tear apart, or blow off, the burner flame. Secondary air enters the combustion from the periphery of the burner tube and envelopes the flame. The necessary excess air comes from this source.

Swirlers or spinners are arranged to give the mixture a rotating movement off the nozzles. Figure 2-4 illustrates.

In any type of burner, an important element for the troubleshooter to consider goes by the name of "flame speed." When the flame is stable, it appears to be stationary, holding a position of about ½ in. off the nozzle. However if the oil-air mixture is entering the furnace at too fast a rate, combustion is smoky and inefficient and may be lost entirely. Conversely, a very slow speed might result in a flashback, a non-ignited drool, and a furnace puff.

The flame speed for any burner becomes a function of the combined oil pressure, the primary air-fuel ratio, the furnace temperature, and, of course, the configuration of the nozzle. The job of the troubleshooter is to discover which of these, or combination of them, is behind the malperformance of the flame and to make corrections accordingly.

(b) ***Complaint.*** Burner is hard to light. Ignition may be quite rough.

Causes

1. For a burner with both low-fire and high-fire operating characteristics, low fire is starting with too much air.
2. Transformer is delivering a weak spark.
3. Electrodes are set incorrectly or are heavily sooted.

Recommended Actions

1. Adjust the low-fire air shutter until the lighting off becomes smooth. Set fire about ½ in. ahead of spinner for a flame-retention burner. Aim for a CO_2 reading between 10.5 and 12.0 percent. When burner goes to high fire, adjust to clear up smoke. Try for 11.0 to 12.0 percent CO_2.
2. Verify condition of transformer. Replace with a proved unit if necessary.
3. Clean all electrodes.

(c) ***Complaint.*** There is excessive vibration in the burner and blower.

Causes

1. Blower wheel is out of balance, because of manufacturing defect or from damage during transit and erection.
2. Blower is out of balance from accumulations of soot, or some foreign material or object has become stuck in the wheel.

Recommended Action

Remove blower and inspect, subject to both static and dynamic balance. If out of balance and condition is inherent rather than caused by removable foreign material, return blower to factory for replacement by a new unit. Check balance of new unit before reinstalling in the burner.

(d) ***Complaint.*** Fuel-oil feed to burner is erratic, sometimes cuts out momentarily.

Causes

1. Poor vacuum on supply piping.
2. Excessive vacuum observed at pump suction.
3. Fuel feed pipe too small for length.
4. Lift height is critical, or water in the line.
5. In winter, line may be frozen.

Recommended Actions

1. Poor vacuum indicates possibility of air leak into line. Presence of air can sometimes be recognized by a singing in the line.

2. Need for an excessive vacuum indicates possibility of a partially clogged line or a fouled strainer. Check both. Also check suction lift; this could be too much for the pump. A too long supply pipe also can require an excessive vacuum to move the oil, particularly if the long line is combined with a high suction lift.

3. Observe flame interruptions. If puffs in the furnace are accompanied by a white smoke, look for water. White smoke also can be caused by excess air and too high a condition of draft. For water, drain the system; for excess draft and air, adjust the controls.

4. If lift-length combination is too great for dependable operation with the burner pump, install a new booster pump close to the storage tank.

(e) ***Complaint.*** Operation of oil burner is accompanied by excessive smoke and soot.

Causes

1. Burner nozzle or nozzles show wear, an incorrect spray angle, or possibly accumulations of dirt.

2. Rate of combustion is too great for the volume of the combustion chamber.

3. Wrong oil pressure, usually too low.

4. Combustion air is insufficient for good combustion.

5. Poor draft, resulting from a too small chimney, from obstructions in the chimney, or from downdraft effects.

Recommended Actions

1. Check CO_2. If too low, reduce the excess air.

2. Check the oil pressure. Correct as necessary.

3. Try increasing combustion air until chimney clears.

4. Measure the draft and watch the indicator for a fluctuating condition. Fluctuations may be resulting from downdrafts caused by trees or adjacent buildings. If so, correct by stack extension, by installation of an induced-draft fan, or by suitable chimney top arrangements to prevent the entry of downdrafts.

5. Remove, inspect, and check all burner nozzles. Replace as necessary to provide the correct atomization.

6. Calculate the burning rate, in pounds per hour, against the combustion chamber volume. If boiler has previously operated satisfactorily, chances are the combustion chamber is adequate. In cases of doubt, query the boiler-burner manufacturer.

(f) ***Complaint.*** Oil consumption is excessive.

Causes

1. Nozzle is oversize.

2. Boiler passes are heavily sooted, obstructing heat exchange.

3. Inefficient combustion resulting from either excess air or too little

air. The former is shown by high stack temperatures in combination with low CO_2; the latter, by smoke and soot.

4. Furnace condition shows broken refractory, allowing heat leaks from the combustion spaces.

5. Boiler needs some baffling. Original baffles deteriorated, and hot gases are short-circuiting to stack.

6. Poor draft conditions.

Recommended Actions

1. Check for cleanliness of gas passages and improve as needed.

2. Check combustion by means of the CO_2 indicator and adjust combustion air to produce 11 to 12 percent CO_2.

3. Verify the suitability of the nozzles. If existing nozzle is oversize or defective in any manner, replace with the proper unit.

4. Inspect furnace refractory and baffles. Repair or replace as found necessary.

5. If all furnace and boiler checks are negative, look for an installation too small for the actual load. This may call for a complete load recalculation for comparison with the capacity of the boiler and burner.

(g) **Complaint.** Burner short-cycles.

Cause

Limit switch is set too low, or thermostat make-and-break is set too close.

Recommended Action

Reset the limit switch or the thermostat contactors as required for proper operation.

(h) ***Complaint.*** Oil burner produces annoying smoke and odors.

See Sec. 2-1(e).

(i) ***Complaint.*** Burner and fan run noisily.

See Sec. 2-1(h).

Table 2-3. Troubleshooter's Field Guide for Cast-iron Boilers, Steam and Hot Water

Possible cause	How to check for cause	Recommended ways to correct
COMPLAINT: Cracks have developed in the sections of a cast-iron boiler. REFERENCE: Sec. 2-1(b).		
1. Drawrods holding the sections together have been pulled too tight.	*a.* Check the tension on the rods with strain wrench. Compare with recommendation of boiler manufacturer.	*a.* Slack off on the rods. Recaulk between the sections. Replace cracked sections.

Table 2-3. Troubleshooter's Field Guide for Cast-iron Boilers, Steam and Hot Water (*Continued*)

Possible cause	How to check for cause	Recommended ways to correct
COMPLAINT: Boiler shows excessive rusting. REFERENCE: Sec. 2-1(c).		
1. Boiler has been allowed to stand idle for 1 or more months on end in a humid atmosphere.	*b.* Check relative humidity of atmosphere at boiler's location.	*b.* Dehumidify the atmosphere. During seasonal shutdowns, set tray of slaked lime in the firebox. After scraping rust, paint with red lead or bitumastic. Ventilate the room to relieve humidity.
COMPLAINT: Bituminous coal smokes and sometimes puffs back. REFERENCE: Sec. 2-1(d).		
1. Chimney and firebox are too small to handle high-volatile coal.	*c.* Check capacity of chimney, both in cross-sectional area and in draft developed.	*c.* Switch to another fuel, such as anthracite, oil, or gas. Install a draft inducer.
2. There are too many bends in breeching for available draft.	*d.* Compare draft, in. W.G., with bends × 0.05 in. W.G. per bend friction.	*d.* Straighten breeching. Install draft inducer.
COMPLAINT: Oil-burning boiler offends the neighborhood with smoke and odors. REFERENCE: Sec. 2-1(e).		
1. Poor draft and/or inadequate combustion air supply.	*e.* Check draft with draft gage. Should have 0.15 to 0.20 in. water gage in firebox.* See that boiler room is not air-starved.	*e.* Improve draft and combustion air as necessary. Consider draft inducer.

* Overfire draft is not necessary for a flame-retention type of burner. This burner is strictly a forced-draft operating unit, in which the burner fan supplies all combustion air, both primary and secondary, and the furnace is sealed against air exchange with the surrounding atmosphere.

Table 2-3. Troubleshooter's Field Guide for Cast-iron Boilers, Steam and Hot Water (*Continued*)

Possible cause	How to check for cause	Recommended ways to correct
2. Burner nozzle size or angle is incorrect for firebox and boiler capacity.	*f.* Check burner nozzle against the boiler manufacturer's recommendation.	*f.* Replace nozzle with proper size.
COMPLAINT: Oil-fired furnace lights off at times with flarebacks or furnace puffs. REFERENCE: Sec. 2-1(f).		
1. Poor draft.	*g.* Check as above.	*g.* Improve draft.
2. Weak spark from ignition transformer.	*h.* Check transformer. Observe spark for position relative to burner discharge.	*h.* Replace transformer. Adjust spark gap. Clean porcelain.
COMPLAINT: Oil burner is hard to light off and frequently cuts out shortly after ignition. REFERENCE: Sec. 2-1(g).		
1. Water in the fuel oil.	*i.* Run off sample of oil between storage tank and burner. If milky, look for water or air.	*i.* Probe storage tank for water. Pump out if found. For air, look for air leak in oil supply to burner pump.
2. Stack switch is set too low.	*j.* Check switch setting.	*j.* Reset switch to correct temperature.
3. Porcelain insulators have cracks or are sooted.	*k.* Inspect.	*k.* Clean or replace as indicated.
4. Defective ignition transformer.	*l.* Check transformer for voltage.	*l.* Replace if defective.
COMPLAINT: Oil burner installation operates with excessive noise, such as a grinding or rattling. REFERENCE: Sec. 2-1(h).		
1. If knocking is present, burner motor may be loose on its mounting.	*m.* Check tightness of motor holding-down bolts.	*m.* Tighten bolts. Realign motor and blower if necessary.

Table 2-3. Troubleshooter's Field Guide for Cast-iron Boilers, Steam and Hot Water (*Continued*)

Possible cause	How to check for cause	Recommended ways to check
2. Pump gears are worn.	*n*. Listen to pump operation, using a sound rod or stethoscope.	*n*. Replace pump.
3. Burner fan is loose on its shaft.	*o*. Check keyway or setscrew.	*o*. Tighten as necessary.
COMPLAINT: Gas-fired boiler fails to meet its rating. REFERENCE: Sec. 2-1(i).		
1. Gas pressure is lower than burner design calls for.	*p*. Check gas pressure at main valve. Contact gas company to learn if low pressure is temporary or permanent.	*p*. If pressure cannot be increased, replace the burner with one suitable for the low pressure.
2. Flue gas passages of boiler are sooted.	*q*. Inspect.	*q*. Clean as necessary. Adjust burner to provide a clean flame.
3. Boiler thermostat is incorrectly set and cutting out burners too soon.	*r*. Check setting of the thermostat.	*r*. Adjust thermostat.
4. Fuel gas burners are not suitable for the kind of gas being burned.	*s*. Verify the type and model of burner against the kind of gas, its calorific value, and other characteristics.	*s*. Replace burners with correct kind. If gas calorific value is low, choose new burners to handle an increased volume of gas to make the boiler rating.
5. Downdrafts may be affecting burner performance.	*t*. Check draft diverter for amount and frequency of downdrafts.	*t*. Eliminate excessive downdraft by modifying the vent for increased height. Consider a draft inducer.

CAUTION: See Sec. 2-1(j) before undertaking any field work on a gas-fired installation. Neglect of the precautionary safety measures outlined in this section can result in serious explosions.

2-8. FIRING EQUIPMENT, GAS BURNERS

(a) *Discussion.* The type of gas burner installed under any form of boiler depends on the kind of gas it is expected to burn, that is, natural gas, manufactured gas, or the propane and butane forms of liquefied petroleum gas. The characteristics of these gases are given in Tables 2-4 and 2-5.

Table 2-4. Characteristics of Propane and Butane Gases

Property	Propane	Butane
Boiling temp.°F at atmospheric pressure	−51	15
Weight per gal at 60 F, liq form, lb	4.24	4.84
Vapor pressure at 100 F, psig	195	65
Volume of gas, cu ft per liq gal	36	32
Heating value, Btu:		
Per cu ft	2550	3200
Per gal	91,800	102,400
Per lb	21,650	21,500
Specific gravity:		
Liquid	0.509	0.576
Gas	1.52	2.00
API at 60 F	147	111

Table 2-5. Characteristics of Other Fuel Gases

Kind of gas and values	Specific gravity*	Heating value Btu per cu ft
Carbureted water gas	0.63	550
Coke-oven gas:		
Max	0.44	645
Avg	0.40	559
Min		466
Natural gas:		
Max	1.21	1548
Avg	0.72	1090
Min	0.57	977

* Specific gravity is referred to air.

The burners themselves can be classified as aspirating or raw-gas burners. The design of the former provides primary air for combustion prior to ignition, so that a premixed fuel emerges from the burner tip.

The advantages of this design include ease of adjustment, since the air volume aspirated by the velocity and expansion of the gas is readily controlled by the gas pressure.

The major disadvantage is probably the tendency to flashback to the mixing orifice if the gas pressure falls much below the normal pressure rating of the burner. Up to 10 psig is needed.

With the raw-gas burner there is no premixing of gas and primary air. The raw gas burns at a tip formed of multiple ports, and air under pressure from a fan or blower forms what might be called on-the-spot mixing with the gas as it leaves the port. The fan therefore is an essential item of equipment for these burners; whereas the aspirating burner operates on atmospheric air pressure.

Flame-retention characteristics are achieved by surrounding the main burner port by a number of relatively small ports, a common practice with high-heat-release installations. Because of frictional resistance to gas flow through the small ports, velocity is low, and the flame is seldom "blown off" as happens at times with the main port. These small ports act as pilots in circumstances that lose the ignition of the main burner and reestablish the main flame.

Figure 2-5 illustrates the principles of the aspirating burner; Fig. 2-6, a flame-retention type of pressure gas burner.

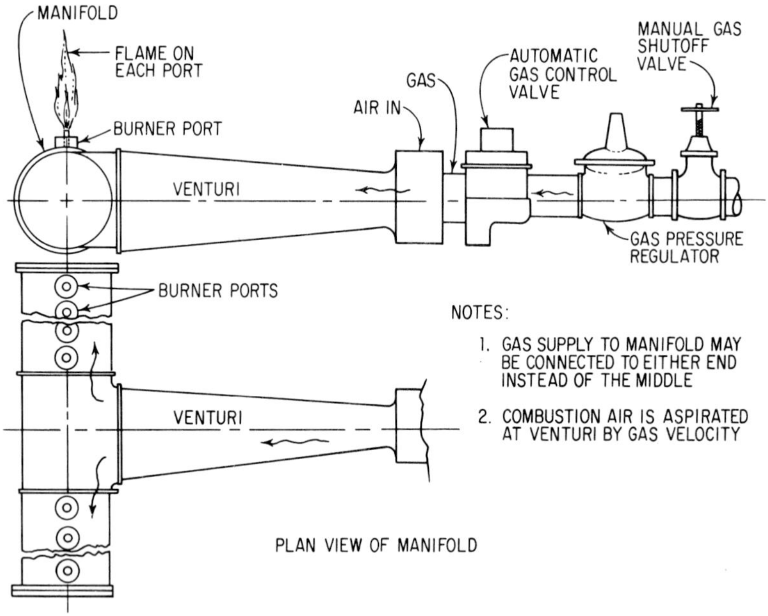

FIG. 2–5. Venturi connection to manifold burner (no scale; schematic only).

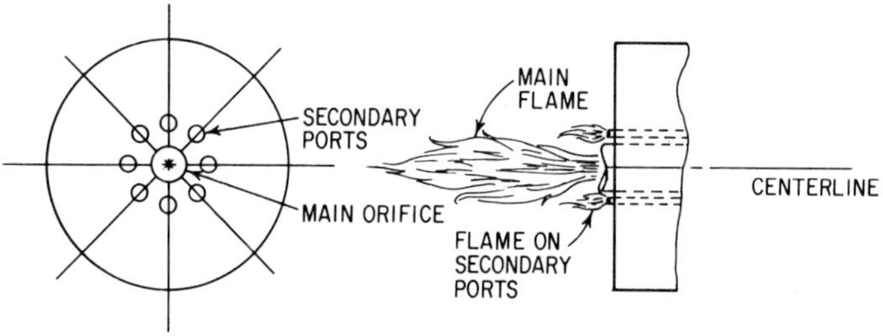

FIG. 2–6. Schematic sketch of flame-retention burner (no scale). *Note:* Number of flame-retention ports varies with burner manufacturer.

(b) ***Burner Applications.*** Aspirating burners cannot be used with forced-draft installations, since the volume and pressure of air necessary to maintain the desired furnace conditions would produce positive furnace pressures. These positive pressures decrease the aspirator capacity and promote flashbacks.

The negative-pressure furnace conditions produced by natural draft, on the contrary, increase the effectiveness of the aspirator and retard the tendency to flashback.

Figure 2-7 shows a venturi type of atmospheric burner in which the aspirated air is drawn into the expanding gas flow from a pressure spud, or orifice. The enlargement of the venturi body permits the mixture of gas and inspirated air to lose velocity and regain static pressure, prior to reaching the burner area of ignition.

This arrangement depends on the availability of gas pressures from 5 to 10 psig, for burning manufactured gas and natural gas respectively. It is not a practical design for service with liquefied petroleum gas, since the combustion ratio of air to gas goes as high as 30 to 1, and that is beyond the capacity of the inspirator. The venturi type of mixer is common on domestic installations.

Under forced-draft firing, the swirlers that surround the nozzle ports

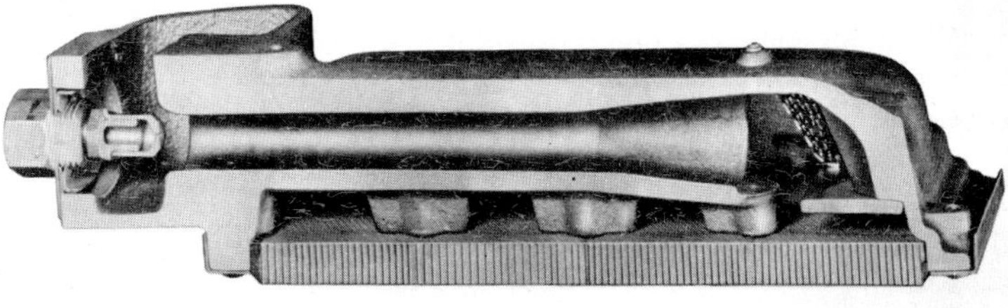

FIG. 2–7. Venturi used with an atmospheric gas burner.

reduce the velocity of the primary air to a degree that makes a flame blowoff unlikely if not impossible. This is because the flame propagation rate is faster than the velocity of the air.

The secondary air, under pressure from the blower, emerges from nozzles arranged around the periphery of the burner. There it surrounds the flame, providing whatever excess is needed for the combustion conditions.

Complaints, causes, and remedies will be examined in subsequent paragraphs under subheads.

(c) ***Complaint.*** With aspirating burners, burners lose ignition.

Causes

1. Burner is being supplied with gas at a pressure above its design pressure. Since the amount of air aspirated is a function of the gas pressure, excess pressure brings too much air, and the flame dies.

2. If liquid is in the fuel line, the continuity of the fuel supply is interrupted.

3. Pilot is out, and automatic gas valve acts to stop flow of fuel to main burner.

Recommended Actions

1. Readjust the gas pressure to the burners.

2. If liquid is found to be present (water of condensation, oil, tar, etc.), install a separator or filter in the fuel line.

3. Clean pilot and relight.

(d) ***Complaint.*** Flashback with aspirating burner.

Causes

1. Gas pressure is too low. Velocity of gas at ports is lower than the velocity of the flame propagation, and the flame travels backward to the mixing orifice.

2. Primary air reaching nozzle is insufficient.

Recommended Actions

1. If pressure of gas cannot be corrected by local adjustments, notify gas company. If a multiple-burner installation, shut down one or more of the burners in order to build up pressure on those remaining in operation.

2. Check for a clogged fuel line, a fouled orifice, or a dirty filter.

3. Clean and check the position of the venturi throat with respect to the gas-supply nozzle. Adjust venturi as necessary.

(e) ***Complaint.*** Burners are not producing enough heat.

Causes

1. Volume of gas reaching burner is below burner rating.

2. Calorific value of gas is less than design ratings.

3. Obstructions in burner port.

4. Obstructions in fuel line to burner.

5. Gas pressure is lower than design of burner calls for.

Recommended Actions

1. Check fuel line for obstructions, low pressure, and volume reaching burner. Check flow on meter in cubic feet per minute.

2. Examine and clean burner ports to remove lint, insects, or other forms of obstruction.

3. Obtain calorific rating of gas from the gas supplier.

(f) ***Complaint.*** Burners pulsate.

Cause

Insufficient draft. When oxygen in furnace is exhausted, flame will pant, sometimes go out and relight from the pilot.

Recommended Actions

1. Reduce rate of firing until flame steadies.

2. Check draft at draft diverter. Verify functioning of barometric damper.

3. Clean chimney of soot and remove other obstructions such as birds' nests, tree branches, and leaves.

4. Examine furnace room for easy supply of combustion air. Starvation of air because of a closed or partially closed furnace room can cause burners to pant.

5. Examine primary air shutter for an incorrect setting and adjust as necessary to steady flame.

(g) ***Complaint.*** Stack temperature above the draft diverter is much too high. Normal temperature does not exceed 400 F plus the room air temperature.

Causes

1. Inadequate combustion air. Air starvation in furnace room.

2. Overfiring. Burner is handling a greater Btu input than its design calls for.

3. On warm air heating systems involving the forced circulation of air through ducts, blower is running too slow, filter is clogged, or heat exchanger is burned out or defective.

Recommended Actions

1. For natural and manufactured gases, refer to the meter and, with a watch, count the number of seconds necessary for one cubic foot of the gas to pass through the meter. Then determine the hourly input of Btu to the burner by this equation:

$$\text{Btu per hour input} = \frac{3{,}600 \text{ seconds} \times \text{Btu per cu ft}}{\text{seconds per cu ft}}$$

Example: Heating value of gas is 1000 Btu per cu ft.

$$\text{Then, } \frac{3{,}600 \times 1000}{45 \text{ seconds per cu ft}} = 80{,}000 \text{ Btu per hour.}$$

If the burner is rated at any value less than 80,000 Btu per hour input, overfiring is taking place.

2. Verify the full availability of combustion air to the burner. On power burners, make sure the blower is operating at its design speed.

3. Examine and clean air filters.

4. If all other checks prove negative, calculate the load on the system and compare with burner output. Then compare the circulating fan output, in cfm, with the required cfm to meet design conditions.

(h) ***Complaint.*** Soot and smells are appearing in various areas in which they do not belong.

Causes

1. Heat exchanger has burned out. Burnout may be no more than a small hole, but this is enough to contaminate adjacent areas.
2. Not enough stack draft.
3. Unit is being overfired.
4. Filter is dirty.
5. Burner spuds are oversized for rest of equipment.

Recommended Actions

1. Verify stack draft. Correct if necessary.
2. Check filters and clean or replace.
3. Check flue-gas temperature for overfiring. Verify proper sizes of the spuds.
4. Shut down unit and examine heat exchanger for possible burnout holes.

(i) ***Complaint.*** Burner short-cycles.

Causes

1. Setting of limit switch is too low for normal operating conditions.
2. Thermostat operation is defective.

Recommended Actions

1. Observe temperatures at which limit switch stops burner. Change setting to correct value.
2. Check operation of the thermostatic control, if such a control functions with the burner. Replace thermostat if found to be the cause of the short cycling.

(j) ***Complaint.*** Pilot light does not stay lighted.

Causes

1. Gas tube to pilot is dirty and partially clogged.
2. Gas pressure is fluctuating and at times drops to such a low value that the pilot goes out.

Recommended Actions

1. Check adequate delivery of gas to the pilot. Inspect the tube for foreign material and clean out.
2. If possible to be present at a pilot failure, note the gas pressure at

the time. If too low to maintain the pilot, consult gas supplier. If liquefied petroleum gas, check liquid in tanks.

3. Note if draft across pilot flame tends to blow out the flame. Correct as necessary. A pilot baffle is sometimes installed to protect the flame.

4. *Precautionary directive:* In situations involving "no pilot," *do not attempt to light the pilot until the combustion space is thoroughly vented.* Neglect of this warning can result in a serious explosion, injury, fire, and possibly death.

(k) ***Conversion Burners, Discussion.*** Conversion burners are used as the name implies, to replace some other method of firing in a given furnace. Three types of conversion burners are recognized:

1. *Inshot Burner* is fired horizontally and is designed for placement in the ash pit of a coal-fired furnace.

2. *Upshot Burner* is mounted in the furnace at the grate level of a coal-burning unit, with the gas flame taking a vertical position.

3. *Firing-door Burner* is mounted in the furnace firing door, with the burner at least 2 in. above the grate level. The flame is vertical.

In troubleshooting a conversion installation, verify the compliance of the design with the applicable standards and codes. This is important, since a unit not complying with the codes may be hazardous and should be replaced.

Conversion burners are subject to the same troubles of performance as are other gas burners, and the complaint made should be investigated in accordance with the suggestions given in Table 2-6.

Table 2-6. Troubleshooter's Field Guide for Gas Burners

Possible cause	How to check for cause	Recommended ways to correct
COMPLAINT: With aspirating burners, the burners lose ignition. REFERENCE: Sec. 2-8(c) and Sec. 2-8(j).		
1. Gas pressure is higher than burner is designed to handle.	*a.* Take a pressure reading of gas supply. Compare with burner design.	*a.* Adjust pressure as necessary.
2. Liquid in fuel supply line to burners. May-be water, tar, oil.	*b.* Run off sample of gas into closed container. Check for liquid. Inspect gas line. Disassemble as necessary to check by draining.	*b.* Install a filter or separator in gas line to remove any liquid that might be carried to the burner.

Table 2-6. Troubleshooter's Field Guide for Gas Burners (*Continued*)

Possible cause	How to check for cause	Recommended ways to correct
3. Pilot is out and automatic gas valve shuts off main supply.	*c.* Make sure combustion space is free of gas; then relight pilot, after cleaning pilot.	*c.* Refer to Sec. 2-8(j) for recommended action.
COMPLAINT: Flashback with aspirating burner. REFERENCE: Sec. 2-8(d).		
1. Gas pressure is too low for the burner design.	*d.* Check gas pressure and compare with minimum required by burner.	*d.* Increase gas pressure locally if possible. Inform gas supplier and ask for correction.
2. Fuel gas line is partially obstructed by foreign material, filter is clogged, or mixing orifice is dirty.	*e.* Blow out line. Examine filter and orifice.	*e.* Clean all components as may be necessary.
3. Venturi may be out of its proper position.	*f.* Check relationship of venturi with the gas nozzle. Obtain correct relationship from burner manufacturer if necessary.	*f.* Reset the venturi if found necessary.
COMPLAINT: Burners are not putting out enough heat. REFERENCE: Sec. 2-8(e).		
1. Burner is operating below rating due to low pressure of gas.	*g.* Compare existing gas pressure with burner design pressure.	*g.* See actions recommended under item (d).
2. Calorific value of gas is less than burner is designed to handle.	*h.* Obtain Btu per cubic foot from supplier.	*h.* Change gas spuds as necessary to suit Btu value.
3. Obstructions in the burner port or spuds.	*i.* Examine for foreign material. Check flow of gas in cfm by reference to meter.	*i.* Clean and restore to service.

Table 2-6. Troubleshooter's Field Guide for Gas Burners (*Continued*)

Possible cause	How to check for cause	Recommended ways to correct
COMPLAINT: Burners pulsate.		
REFERENCE: Sec. 2-8(f).		
1. Insufficient draft.	*j*. Check draft with draft gage.	*j*. Reduce firing rate until pulsing stops.
	k. Verify easy action and balance of stack damper, if in use.	*k*. Free up and open damper as needed.
	l. Check chimney for blockages.	*l*. Lower brick in chimney on rope and knock loose soot and adhering small objects.
	m. See that proper amount of primary air is reaching burner.	*m*. Set primary shutter to pass the correct amount. Open door or window if needed. Provide permanent opening for air.
COMPLAINT: Stack temperature is much too high.		
REFERENCE: Sec. 2-8(g).		
1. Inadequate combustion air.	*n*. See how to check in *j*, *k*, *l*, and *m* above.	*n*. See recommendations in *j*, *k*, *l*, and *m* above.
2. Overfiring.	*o*. Check Btu input to burner and compare with burner rating.	*o*. Measure gas flow by the gas meter. Convert to Btu per hour.
3. On warm air systems, exchange of heat from furnace to circulating air is too low.	*p*. Check flow of circulating air to system in cfm, using a velometer at the outlets. Check speed and capacity of circulating air blower.	*p*. Speed up blower as necessary, if this speedup does not exceed the design speed.
	q. Examine air filter for dirt and excessive clogging.	*q*. Clean or replace the filter as necessary.

Table 2-6. Troubleshooter's Field Guide for Gas Burners (*Continued*)

Possible cause	How to check for cause	Recommended ways to correct
COMPLAINT: Soot and smells are becoming an annoyance. REFERENCE: Sec. 2-8(h).		
1. Not enough stack draft.	*r.* Measure draft with a draft gage. Evaluate with the needed draft.	*r.* Increase the draft or correct conditions as recommended in Sec. 2-8(f).
2. Unit is being overfired.	*s.* Measure input of Btu as described in Sec. 2-8(g).	*s.* Correct overfiring condition as recommended in Sec. 2-8(g).
3. Air filter is dirty.	*t.* Inspect filter.	*t.* Clean or replace.
4. Heat exchanger has burned through in one or more areas.	*u.* Remove heat exchanger jacket and inspect. Use hammer test lightly in suspected areas.	*u.* If burnout is small and concentrated and exchanger is otherwise in good condition, patch and weld.
COMPLAINT: Burner short-cycles. REFERENCE: Sec. 2-8(i).		
1. Setting of limit switch is too low for normal operating conditions.	*v.* Observe temperatures at which limit switch functions.	*v.* Reset limit switch if new setting does not exceed safe temperature of the unit.
2. Thermostat is faulty in operation.	*w.* Operate thermostat as necessary to establish integrity or defect.	*w.* Replace the thermostat if the check shows any malfunction.
COMPLAINT: Pilot light does not stay lighted. REFERENCE: Sec. 2-8(j).		
1. Gas tube to the pilot is fouled with foreign material or is partly clogged.	*x.* Examine the gas supply lines to the pilot. Use a light to see through and pull a wire through.	*x.* Clean as necessary and replace in position.

Table 2-6. Troubleshooter's Field Guide for Gas Burners (*Continued*)

Possible cause	How to check for cause	Recommended ways to correct
2. Gas pressure is fluctuating and at times is so low that pilot goes out for lack of fuel.	*y*. Check gas pressure in line to pilot. Take reading with gage.	*y*. If pressure is too low, notify gas supplier. For LPG, replacement bottles may be all that is needed.

CAUTION: Never light a pilot until the absence of all gas in the combustion chamber has been determined beyond question. Purge thoroughly to the outdoor air. Failure to observe this precaution may result in property damage, serious injury, or death.

(1) ***Combination Burners for Coal-Oil-Gas.*** The coal in this arrangement is pulverized. Figure 2-8 illustrates a burner designed for these three fuels.

Troubles with oil and gas firing will be as previously discussed. For pulverized coal, refer to Sec. 2-10 and the appropriate subheading.

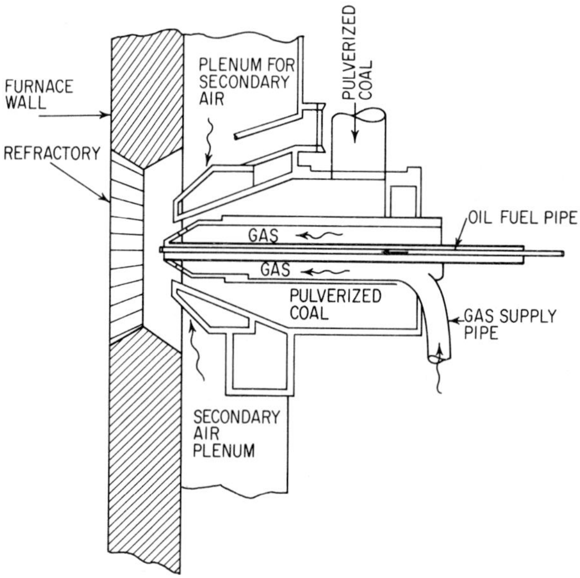

FIG. 2–8. Coal-oil-gas burner (no scale; schematic only).

2-9. FIRING EQUIPMENT, COAL, HAND

(a) ***Discussion.*** Hand-fired boilers and warm air furnaces are not the common makers of comfort they once were, yet there are enough of them in service to justify an examination of their troubles and how to correct them. Conversion to automatic firing with oil or gas may not be practical for reasons of cost or because the age of the hand-fired plant indicates the end of its career is within an early future. In general, hand-fired troubles originate in an inadequate chimney or draft, or in the use of a less than suitable fuel. These conditions will be noted and prescribed for in the following paragraphs.

(b) ***Complaint.*** In a hot water heating system, boiler fails to meet its rating, and the building is not satisfactorily heated in subfreezing weather.

Causes

1. If a high wind is blowing and the building is not weather-stripped, storm-windowed, or insulated, the chill may be coming directly from the extreme rate of cold infiltration, combined with an excessive loss of heat through single-pane windows and uninsulated walls.
2. Draft may be too little to carry the furnace at high ratings. This condition can result from a chimney partially choked with birds' nests, soot, tree branches, and anything else that can fly as high as the roof.
3. Calorific value of coal may be low, a cheap grade of coal of high ash content.

Recommended Actions

1. Obtain a proximate analysis of the coal being burned from the coal supplier. If calorific value is less than 13,000 Btu per lb, consider a better grade of coal. It will cost more but the results could justify the cost.
2. Observe draftiness of rooms. If noticeable, recommend the installation of storm windows, doors, weather stripping, and insulation.
3. Measure draft at boiler outlet. Calculate friction losses from chimney, breeching, and bends in breeching. If natural draft is insufficient to overcome friction losses, install a draft inducer. If draft should be sufficient but is not, clean chimney by means of a brick on a rope dropped down the chimney and jiggled against the sides to loosen soot accumulations and other foreign materials. Also note whether downdrafts are present as the result of overhanging trees or higher adjacent structures. These can be cured by a draft inducer or an extension of the chimney.

(c) ***Complaint.*** Warm air heating system is not adequate.

Causes

1. Check outdoor air intake of fresh air. The damper in the intake duct should be closed or nearly closed in severe weather.

2. Structure is drafty as noted in (b).

3. Damper positions in leaders to various rooms may be partially closed.

4. Heat exchanger may be fouled with soot.

5. Coal may be of poor grade as considered in (b).

6. Chimney condition may be inhibiting proper draft as in (b).

Recommended Actions

1. Verify the correctness of positions for each damper.

2. Obtain proximate analysis of coal and replace with better grade if necessary.

3. Make sure chimney is clean, operating at designed capacity. Correct as necessary, as suggested in (b).

4. Clean heat exchanger with vacuum cleaner.

5. Recommend ways to increase structural tightness and heat loss.

(d) *Complaint.* Fireplace smoke is heavy with wood fuel, less with coal, but undesirable in both instances.

Causes

1. Downdrafts from the presence of roof overhangs in the form of trees or an adjacent higher building.

2. Flue is undersized. Flue cross-sectional area should equal at least 10 percent of the area of the fireplace opening.

3. Flue has been constructed without a smoke shelf. See Fig. 2-9. Purpose of shelf is to divert downdrafts and reflect them back up the chimney.

Recommended Actions

1. Inspect chimney for presence of smoke shelf. If none is present, further use of fireplace may depend on the addition of a draft inducer to prevent smoke from blowing into the room.

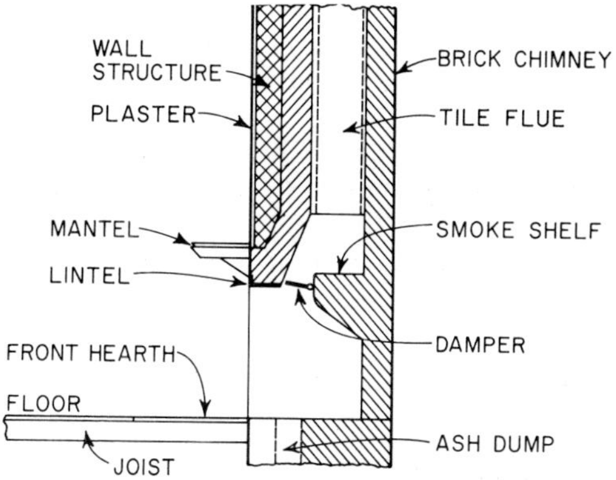

FIG. 2–9. Fireplace section (no scale).

2. Measure size of chimney versus area of fireplace opening. If flue is too small, again a draft inducer may be the only solution short of rebuilding the chimney. However, in some instances, restricting the fuel to coke, which requires relatively less air for combustion, may provide a satisfactory fire.

3. Consider extending or capping the chimney to eliminate downdrafts.

2-10. FIRING EQUIPMENT, COAL, STOKERS

(a) ***Discussion.*** Commonly encountered are three broad forms of stokers: those designed for single- or multiple-retort underfeed firing; the chain-grate or traveling stokers; and finally the spreader stokers.

Commercial and small to medium industrial steam plants are likely to operate the single-retort, chain-grate, or traveling-grate construction. The multiple-retort and spreader stokers appear in the larger plants, except when these larger units are burning pulverized coal.

Complaints with stokers generally take one or more of these forms:

1. The stoker parts are so exposed to the fire as to become overheated and ultimately burn out.

2. There is more or less smoke from the stack as the result of an excessively thick fuel bed. Tarry particles may be ejected.

3. Formation of clinkers on grates and water walls from burning a clinkering type of coal.

4. Fly ash gets into the neighborhood air.

Since these troubles are likely to happen with any of the stokers, their causes and treatments will be discussed as suitable for any of the mentioned designs.

(b) ***Complaint.*** Stoker parts burn out from fire exposure.

Causes

1. The use of preheated air produces a fire of maximum temperature. This temperature may be high enough to burn out stoker parts and castings.

2. Fire is too thin during a time of heavy overload. Green coal and a layer of ash are necessary to protect the metal. A thin fire does not produce this protection.

3. Segregation of coal by sizes in the feed hopper can produce holes in the fire and exposures of the stoker metal to radiant heat in these holes.

4. Not enough air is entering the fire through the stoker grates to offer any cooling of the parts.

5. Coal is of low ash content, thus denying the grates the protection of the ash layer that they need.

Recommended Actions

1. Check the temperature of the preheated air entering the stoker by upward passage through the grates. Reduce air temperature as necessary to prevent burnouts. A relatively low air temperature is needed for low-ash coal; higher temperatures must be restricted to high-ash fuel.

2. Increase the stoker feed to create a satisfactory fire bed on the grate layers of green coal and ash.

3. Replace the segregating coal feed with a nonsegregating type of hopper.

4. Increase the primary air through the fire bed but not enough to lower the combustion efficiency.

5. Switch to another coal, one with a high ash. This coal may be a little cheaper, but it offers needful protection for the stokers.

(c) ***Complaint.*** Smoke and some tarry particles are issuing from the stack and forming a nuisance.

Causes

1. Not enough air for efficient combustion.

2. Enough air but not enough turbulence, or air and fuel mixing, in the furnace.

3. Use of a high-volatile coal. These volatiles are driven from the coal at the first sense of heat as it enters the furnace. The result is smoke.

4. Stoker is being overloaded. Gas formations are more rapid than stack can carry them away.

Recommended Actions

1. Increase the air supply. Take CO_2 readings at close intervals until the optimum operating conditions are attained.

2. Increased furnace turbulence can be obtained by introducing secondary air from the *back* of the furnace, thus putting it into conflict with the air entering from the front. In some installations, instead of air, steam is blown into the furnace from nozzles mounted between the tubes above the bridge wall. This is effective but can be expensive in the use of steam which is not recovered.

3. Change the coal being burned from a high-volatile to a lower-volatile value.

4. Slow down the stoker, or stop shoveling additional coal on the grates by hand, until the stoker speed and stack capacity match.

(d) ***Complaint.*** Clinkers form on grates and walls of furnace.

Causes

1. Wrong type of coal. Ash content probably has a melting temperature of less than 2000 F.

2. Boiler is being pushed beyond its capacity so that furnace temperatures are higher than good practice permits.

Recommended Actions

1. Switch to a coal having ash that will not turn liquid and then glassy, at the normal operating temperature of the furnace.

2. Meet the overload with a new boiler.

(e) ***Complaint.*** Fly ash is a neighborhood nuisance.

Cause

Fly ash is a very light material, easily carried out by air movements. Its formation is a common problem with multiple-retort stokers, spreader stokers, and pulverized-coal systems.

Recommended Actions

1. The installation of a cinder catcher in the breeching before the flue gas reaches the stack can be expected to reduce the amount of fly ash but does not eliminate all of it. Two methods of treatment are presently in favor. One is to install an electrostatic precipitator, and the other is to increase the height of the stack. Both treatments for the same plant may be needed in some areas of important public opinion.

2. In addition to the two methods noted above, this writer has achieved satisfactory results with a flue-gas water washer, all of the flue gas being obliged to pass through water sprays prior to discharge to the atmosphere. However, because of acid conditions being created by the sulfur in the flue gas, spray piping and nozzles must be of stainless steel or lead in order to resist otherwise rapid corrosion and failure. This system is quite expensive to install and, because of water requirements, can be expensive to operate in some areas.

2-11. FIRING EQUIPMENT, COAL, PULVERIZED

(a) ***Discussion.*** Pulverizer mills are likely to be met in the two major forms of the bowl mill or the impact mill. In the former, the coal is ground up with rollers, and in the latter, it is pounded down to size by a set of hammers. Both designs feel the hard wear of grinding coal; the rollers can have an extended life by switching their positions in the mill, the hammers by reversal. But their life is limited at best.

Emergency troubles are likely to appear as some degree of furnace explosion or as a fire in the coal pipelines. Personally, I have had the embarrassment of a middle-pass explosion lifting the top of a three-drum boiler 3 weeks after I had reported the design as being good and most remotely susceptible to an explosion.

Because each particular installation has its own individualistic troubles, complaints will be similarly individualistic, and treatments must be prescribed accordingly. Therefore, the troubleshooter in a pulverized-coal installation will be advised to study the design carefully, to collect all the symptoms, and to develop corrective measures on this basis.

As for explosion protection, it is the responsibility of the designer to provide the proper safeguards, but the operator must observe them, not the troubleshooter. The troubleshooter comes in as an investigator—after the explosion. In a solid majority of these explosions, the cause can be charged against an operator's error. This leaves the question, How do you troubleshoot a man?

Table 2-7. Troubleshooter's Field Guide for Coal-firing Equipment

NOTE: Performance troubles with large public utility and industrial boilers that have constant, trained, technical attendance throughout a 24-hour day are not being examined in detail, since correction is a responsibility of the regular operators and not delegated to an outside troubleshooter as a usual procedure. The following Field Guide is intended primarily for the individual who is called in to correct a particular condition as a single item, not within the scope of the regular attendants.

Possible cause	How to check for cause	Recommended ways to correct
COMPLAINT: Hand-fired boiler does not heat building in subfreezing weather.		
REFERENCE: Sec. 2-9(b).		
1. High winds; building is not tight.	*a.* Use smoke test near doors and windows.	*a.* Weather-strip, add storm doors, etc.
2. Furnace chimney draft is being inhibited by soot or birds' nests in chimney.	*b.* Check draft with draft gage.	*b.* Clean out the chimney. Consider draft inducer.
3. Poor grade of coal is being used.	*c.* Get proximate analysis of coal from supplier. Check calorific value.	*c.* Switch to a better coal with more Btu per lb.
COMPLAINT: Warm air system is inadequate.		
REFERENCE: Sec. 2-9(c).		
1. Outside air connection for fresh air is letting in too much.	*d.* Find open air connection. Note whether air is blowing in by use of draft gage.	*d.* Close up opening as needed.

Table 2-7. Troubleshooter's Field Guide for Coal-firing Equipment (*Continued*)

Possible cause	How to check for cause	Recommended ways to correct
2. Structure is drafty.	*e.* Check with smoke movement.	*e.* Weather-strip, add storm doors, etc.
3. Dampers in leaders are partially closed.	*f.* Examine each damper in leaders to cold rooms.	*f.* Free up dampers if necessary and set properly.
4. Heat exchanger in furnace is dirty.	*g.* Inspect heat exchanger.	*g.* Clean as needed.
5. Chimney is dirty and partially blocked by soot and other materials.	*h.* Inspect chimney visually. Check draft with gage.	*h.* Clean chimney.
6. Poor grade of coal is being used.	*i.* Obtain a proximate analysis of coal.	*i.* Provide better coal.
COMPLAINT: Fireplace smokes. REFERENCE: Sec. 2-9(d).		
1. Downdrafts from trees, buildings, or absence of smoke shelf.	*j.* Check for draft with gage. Inspect flue for smoke shelf.	*j.* Correct draft conditions as necessary.
2. Flue is undersized.	*k.* Compare flue cross-sectional area with required size.	*k.* Consider draft inducer for undersize flue.

COMPLAINTS WITH STOKERS: These are varied and may involve technical operating problems. For discussions and recommended actions, see the appropriate sections.
REFERENCE: Sec. 2-10(a)–(e).

COMPLAINTS WITH PULVERIZER INSTALLATIONS: For discussion and recommended actions, see Sec. 2-11.

2-12. QUESTIONS AND ANSWERS

QUESTION 1. Are cast-iron boilers presently being built for pressures greater than 15 psig steam and 30 psig water?
ANSWER 1. There appears to be just one manufacturer, Hydrotherm,

Inc., who builds a gas-fired boiler for steam pressures up to 100 psig. Several manufacturers build units designed for operating water pressures of 80 psig, but do not exceed the 15 psig for steam.

QUESTION 2. What is the effect of elevation on combustion efficiency?
ANSWER 2. Because of a decrease in density of the air that accompanies an increase in elevation above sea level, the capacity of the firing equipment is likewise decreased. As a rule of thumb, it is safe to accept nameplate ratings anywhere between sea level and 2,000 ft. Above 2,000 ft, reduce the rating by 4 percent for every additional 1,000 ft of elevation.

Elevation changes similarly affect the combustion in diesel and other internal combustion engines.

QUESTION 3. How do you test the performance of a draft hood?
ANSWER 3. The simplest way is to run the flame of a lighted candle around the edges of the draft hood's relief opening. If the flame is extinguished rather than drawn into the opening, draft conditions are faulty, and the gas burner should not be lighted until the cause of the downdraft is identified and corrected.

QUESTION 4. How can fires be extinguished in coal storage piles or bunkers?
ANSWER 4. Experience indicates that digging out the burning coal is the surest way. Blanketing a bunker or silo with carbon dioxide has been a success in some instances, a failure in others. Flooding with water will do the job if the burning material can be kept under water for a period of about 24 hours. If the water is permitted to drain off as applied, the results are likely to be unsatisfactory.

Layers of sand poured on top of a burning coal bed achieve nothing but an undesirable mixing of the sand with the coal; the fire continues to burn.

Foam treatment should not be undertaken except on the approval of the foam manufacturer. In some instances, the fire continues to burn under the foam, living on the oxygen in the air that fills the voids between the particles of coal. In these circumstances, the cost of the foam treatment is wasted.

QUESTION 5. How long is a water-tube boiler out of service after a tube failure?
ANSWER 5. Straight-tube boilers, as in Fig. 2-3, may be expected to lose 1 hour's operating time while the handhole cover over the bad tube is removed and the tube end is plugged. For multidrum bent-tube boilers, the end of the defective tube cannot be reached without all but dumping the water, and from 1 to 1½ days will be needed to plug or cut out the bad tube.

QUESTION 6. What is the life of a fusible plug?
ANSWER 6. For safety's sake, a fusible plug should not be kept in service longer than 1 year. Replacement of a plug, even apparently in good condition after a year in the boiler, is required under the ASME Boiler Code.

QUESTION 7. In lighting off a pulverized fuel furnace, what safety precautions should a troubleshooter observe?
ANSWER 7. Purge the furnace with air for at least 5 minutes, using a volume at least equal to one-quarter of the full load air. Then, if ignition does not take place after 5 seconds of fuel feed, the fuel feed must be stopped immediately and the furnace again purged for 5 minutes before attempting a relight.

For full details, refer to the NFPA No. 60 "Standard for the Installation and Operation of Pulverized Fuel Systems," latest edition.

DIVISION 3

Hydronic Heating Systems

3-1. GRAVITY HOT WATER HEATING SYSTEMS

(a) *Discussion.* The circulation of hot water by gravity, or as it is sometimes called, "natural convection," results from the differences in density of water at different temperatures in the same circuit. These differences are shown on the curves in Fig. 3-1.

Hot water is pushed to the top by the cooler, denser water seeking the bottom, and the speed of this movement is determined by the relative difference in temperatures, the frictional resistance of the piping, and difference of elevation, top to bottom, of the containing system.

A customary difference is 20 deg F; for example, water entering a radiator at 190 F and leaving that radiator at 170 F generates a motive force calculated from Fig. 3-1 as follows:

At 190 F, 2.3827 ft of water = 1 lb pressure
At 170 F, 2.3683 ft of water = 1 lb pressure

Difference 0.0144 ft of water × 12 = 0.1728 in.

Obviously this is an extremely small force, and to simplify the use of such small values, the trade has adopted a term called the "milinch." A milinch is $1/1000$ of an in. or $1/12000$ of a ft. Since a pound pressure is exerted by 2.30 ft of cold water (Fig. 3-1), one pound pressure is also equal to 2.30 × 12,000 = 27,600 milinches.

Applied to the difference of 0.0144 ft as given above, 27,600 milinches × 0.0144 = 397 milinches, approximately. This is the motive power resulting from the differences in temperature only, but there is also a difference in altitude between the point where the cooler return-

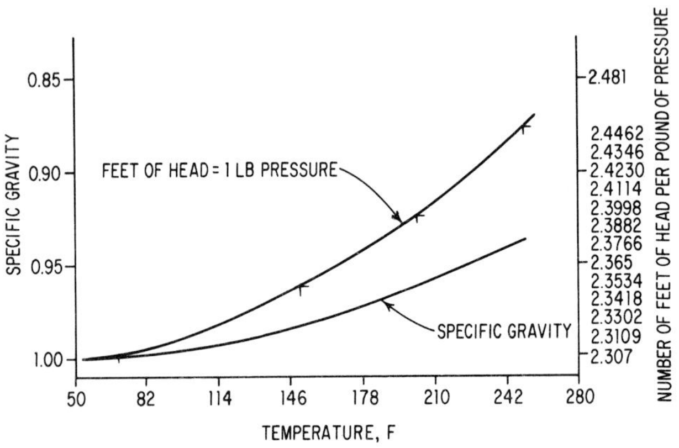

FIG. 3–1. Temperature-head water data.

ing water enters the boiler and the radiator in which the 20-degree temperature drop takes place.

For a single-story house with the boiler in the basement, this difference in elevation might amount to 10 ft; for a two-story house, 20 ft; and for a three-story building, 30 ft. These differences of elevation give the motive power an overall boost, the total for the single-story building being 10 × 397 = 3,970 milinches; for the two-story building, 20 × 397 = 7,940 milinches; and for the three-story building, 11,910 milinches.

This effect of elevation on the motive power in a gravity system explains why the upper radiators in a building get the hottest water and why underheating on lower floors is a common complaint to the troubleshooter. The total represents also the power limit within which pipe friction must be overcome and circulation maintained. If the pipes are too small or too corrosion-choked to pass the necessary volume of water under the milinches of head available, cold radiators can be expected.

Gravity systems for new installations have become generally obsolete for both economy and performance reasons. Relatively large pipes are needed to stay within the motive power and are items of undesirable expense. Moreover, the tendency of pipes to sag, and thereby upset circulation to a degree, and to accumulate corrosive materials in pockets and obstruct circulation are ailments likely to occur with time. However, enough gravity systems remain in service to require a knowledge of their characteristics by the complete troubleshooter.

Figure 3-2 illustrates two common piping arrangements for gravity systems, and each has its own potential for trouble. Each of these will be examined strictly as piping situations; the expansion tanks and their troubles are analyzed in Sec. 3-1(c).

(b) ***Complaint.*** Radiator no. 4 in Fig. 3-2(a) does not heat.

Causes

1. Air pocket in radiator.
2. Water has been so cooled and diluted by the preceding radiators that the temperature difference is inadequate to overcome the frictional resistance of the circuit.
3. Frictional resistance has built up with time in the branch to the radiator, often in combination with cause no. 2 as given above.
4. A pipe sag somewhere in the connections is obstructing circulation to the cold radiator, while producing a condition of overheating in the lower radiation.

Recommended Actions

1. Vent air from the radiator through the manual vent valve.
2. If radiator remains cold, run hand over pipes in an attempt to possibly locate the point at which circulation is being stopped. An abrupt change in pipe temperature indicates a point of stoppage. If cooling is gradual, not abrupt, in supply piping, step up the boiler water temperature in 5-deg increments until the radiator heats.
3. If a temperature of 200 F or more is needed in mild weather, check all preceding radiation for conditions of overheating. If overheating is found, throttle supply to these radiators by means of inlet valves.
4. Use bubble level to check proper pitch of piping. A reverse pitch slows circulation and might even stop it entirely.
5. If internal pipe obstructions are indicated, open the system piping at the suspected points and either clean out or replace.

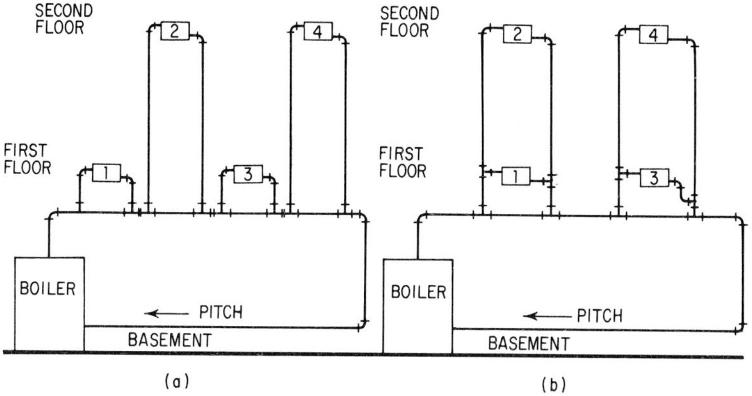

FIG. 3–2. Gravity circulation piping.

Economically, replacement of the system by a forced circulation arrangement may be found best. The conversion of a gravity to forced circulation system is covered in Sec. 3-2.

(c) ***Complaint.*** Radiator no. 4 in Fig. 3-2(b) does not heat. Or nos. 1 and 3 do not.

Causes

For radiator no. 4, causes to be checked are the same as for no. 4 nonheating in Fig. 3-2(a).

For radiators nos. 1 and 3, consider that circulation is taking the path of least resistance, and this is leading right past the connections to these radiators in favor of those at the higher levels. As noted previously, the higher radiators have a tendency to "hog" the hot water circulation to the detriment of the lower. Also check each of these radiators for air pockets by opening the manual vent valves. Check condition of inlet valves. Sometimes these become fouled with muck or lint and block normal entry of the hot water.

Recommended Actions

For radiator no. 4, same as for no. 4 unit in Fig. 3-2(a).

For radiators nos. 1 and 3, throttle inlet valves to upper-level radiators and thereby produce a more positive flow pressure at the inlets to the nonheating units. Check and clean, as necessary, the valves and connecting piping.

In planning corrective actions, remember that radiators are cold simply because the hot water is not being supplied to them in proper volumes or the temperature of the water going through the radiator is too low to be effective. Once the unsatisfactory condition is identified, the causes of that condition can be found by the logical comparison of each possibility against the facts as known.

(d) ***Expansion Tanks.*** Since the increase in volume of water heated from 50 to 200 F is a little more than 4 percent, facilities for accommodating an increase of at least 6 percent are essential. Many designers go to the extra prudent sizing of 10 percent in selecting an expansion or compression tank to suit system conditions.

Figure 3-3(a) shows an expansion tank commonly observed and serving a gravity system. This tank is open to the atmosphere; consequently, it must be sufficiently high above the boiler to avoid steaming any time the temperature of the water in the boiler exceeds 212 F. To illustrate what happens, the absolute pressure at 220 F is 17.186 psi or 2.49 psi above atmospheric (17.186 − 14.696). Since a pound pressure is created by 2.416 ft of water at 220 F, as noted on the curve in Fig. 3-1, the open-to-atmosphere system will need 2.49 lb × 2.416 ft/lb = 6.016 ft minimum head above the topmost point of the hot water system.

Figure 3-3(a) shows a minimum head of 4 ft, which is the usual allowance, but the extra 2 ft are desirable if the climate in the area is severe and the water temperature likely to be higher than 212 F.

Troubles with this open tank system may be expected as follows:

(e) ***Complaint.*** Overflow pipe spills water excessively.

Causes

1. Cold water fill valve for the system water supply is leaking and admitting water when it should be tightly closed.

2. Water temperature in the boiler is substantially above 212 F, and the extra volume is spilling out through the overflow.

3. Position of the tank is too low relative to the high point of the system.

Recommended Actions

1. Check elevation of tank above the high point of the system and equate this with the temperature and pressure as shown on the boiler gages. If pressure and/or temperature are too high for the tank elevation, reduce the latter values.

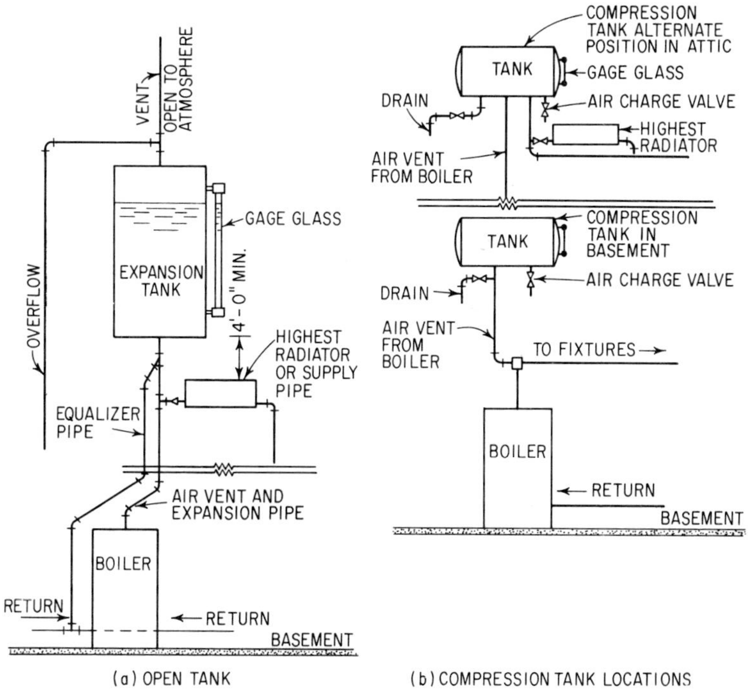

FIG. 3–3. Expansion tanks and locations.

2. If the house generally fails to heat satisfactorily unless the expansion tank is overflowing, calculate the heat loss for the structure and compare this with the amount of radiation presently in place. Also check carefully the pipe sizes. Adequate radiation surface can be starved by connecting piping that is too small; this hinders circulation at normal operating temperatures. Radiation surface can be increased in cold spots at reasonable cost, but a repiping with larger pipes becomes so expensive a project that conversion of the system to one of forced circulation must be seriously considered.

Procedures for a gravity-to-forced-circulation conversion are given in Sec. 3-2.

3. Check fill valve for tightness.

Referring to Fig. 3-3(b), we have the closed or compression tank in which the expanded volume of water makes room for itself in the tank by compressing whatever air happens to be in there. This tank provides several advantages over the open tank. There is no spillage, the tank is readily located in the basement, and the connecting piping is of limited quantity. Also, circulation is improved by the pressure exerted on the water by the air in the tank, a desirable condition most noted in the instances of radiators relatively distant from the boiler.

Here are the usual troubles with this closed-tank design, when trouble develops:

(f) ***Complaint.*** Boiler safety valve spills when demand for heat is strong.

Causes

1. Compression tank is too small to accommodate the temperature expansion of the water in the system; therefore excess goes out through the boiler safety.

2. Tank is waterlogged, meaning that the necessary air cushion for compressing is not there, and the excess expansion escapes through the safety valve.

Recommended Actions

1. With boiler in operation, check level of water in tank by gage glass. If tank is not fitted with a gage glass, place hand on tank at various elevations from the bottom, in order to determine the level of water in the tank. (The air space, if any, will be relatively cool.)

2. On the basis of water level in tank, determine if tank is too small or waterlogged, the latter condition being indicated by a tank almost entirely filled with water and the former by a tank showing about three-fourths water and the balance air, while simultaneously the boiler safety valve is spilling.

3. For the tank suspected of being too small, trip the tank safety valve

by hand, releasing some of the air and note whether this release of air pressure allows the boiler valve to close. This action, if continued, puts the small tank into a waterlogged condition, but one that cannot be relieved by the injection of air. If the air bleed does in fact permit the boiler valve to close, this is a borderline tank with replacement desirable for future dependable operation. A fast estimate of the tank size needed can be reached by dividing the total square feet of radiation by 40. Thus 2,000 sq ft/40 = 50-gal tank. Another fast figure can be obtained by calculating the total water in the system in gallons, on the basis of 50 F water. Then provide a tank equal to at least 10 percent of this for one-story structures, 15 percent for two stories, 24 percent for three stories, the tank in all cases being in the basement or elsewhere below the radiation. If the tank is mounted *above* the radiation, as in an attic, a somewhat smaller tank is satisfactory since the air space can be smaller.

In any event, for an accurate calculation of compression tank size, refer to the latest edition of the ASHRAE Guide.

4. For a tank that has become waterlogged by the protracted absorption of air by the water in contact with the air, correct the condition by a combination of air injection and water bleed to restore the proper water level in the tank. This level will range from one-third to one-half the total volume of the tank, depending on the characteristics of the system. The greater the temperature range of the water from cold to hot, the greater must be the allowance for volumetric water expansion at the maximum temperature. The troubleshooter must consider this aspect of performance when evaluating the size and condition of the tank.

Air for injection to a compression tank may be available in industrial buildings from a compressed-air source that has been provided for other operations of the plant. Otherwise, a simple hand pump may be used of the kind for inflating bicycle or automobile tires.

(g) ***Series Return Arrangement.*** Figure 3-2 illustrates a flow arrangement in which both the supply and return connections to each radiator are served by a common header. Generically this is called a series arrangement, since each radiator is connected in turn and there is a continual dilution of the hot supply water by the cooler returns. Such systems are not suited to large buildings, since radiators become cooler and cooler as their locations take them farther and farther from the boiler, and a great deal of extra heating surface becomes essential to compensate for the cooler water.

(h) ***Complaint.*** Series radiators do not emit sufficient heat.

Causes

1. Water temperature is too low to start the circuit.

2. Radiator surface is insufficient for the water temperature.

3. Individual radiators show abnormally low surface temperatures with hotter radiators farther along the line.

Recommended Actions

1. Increase water temperature in increments of 5 deg until last radiator shows an acceptable heat emission for the surface provided. If preceding radiators in the series are overheating the areas in which they are located, throttle the flow to each radiator by the inlet valve until a normal heat output is realized.

2. Instead of throttling the inlet valves, install in the main header at each radiator offtake a special diversion fitting, sized to divert the correct quantity of water to that radiator and no more. Where the radiator is below the main, install a diversion fitting in the return to the main, as well as in the supply branch to that radiator.

(i) ***Direct and Reversed Returns.*** The direct return, in which the distance of water travel for each radiator circuit is different, is illustrated by Fig. 3-4(a). Since the total length of each individual circuit produces its own level of frictional resistance as a reflection of that

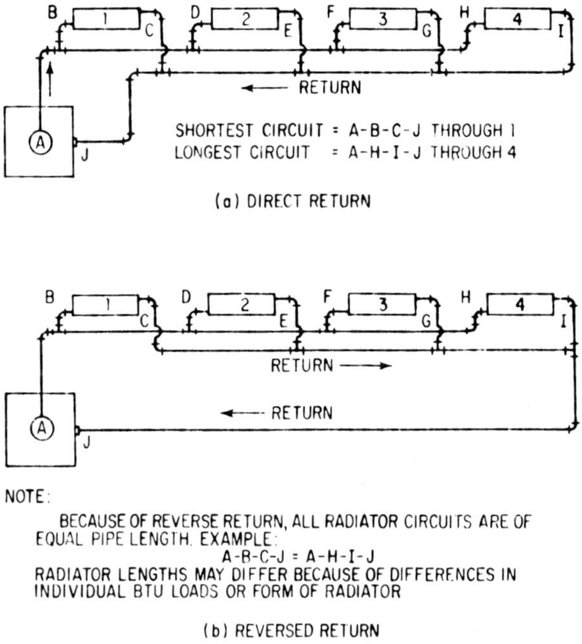

FIG. 3-4. Direct and reversed returns.

length, circulation with an unmodified direct return is notably nonuniform. The result is uneven heating, with the long circuits being relatively starved and the short ones overheated.

(j) ***Complaint.*** As noted above.

Cause

The laws of nature constrain the water to flow in the direction of least resistance.

Recommended Actions

1. Install a balance cock in the return from each radiator and set it to pass the proper volume of water for that radiator and no more. In a gravity system, satisfactory arrangement of these cocks might be difficult or even impossible, because of the frictional resistance they introduce. The motive power in milinches might well be unable to overcome this resistance.

2. Provide the system with a reversed return arrangement, as suggested by Fig. 3-4(b). The purpose of the reversed return is to approximate a uniform resistance level for every radiator circuit in the building, thereby eliminating, or mostly eliminating, any invitation to the flow to skip circuits because of relatively high resistance. By substituting pipe friction for the usually greater friction of balance cocks, the reversed return may work well where a direct return with balance cocks does not work at all.

Incidentally, we note that with forced circulation, the additional friction of the cocks is of no controlling significance, and for these systems, cocks are preferred to the reversed return with its comparatively clumsy piping layout.

(k) ***Complaint.*** Some of the radiators do not heat adequately.

Recommended Actions

1. For a single cold radiator, check for air, and vent the radiator by means of the manual vent.

2. For several radiators standing cold or cool, while the water temperature is normal for heating, check the pitch of all lines by means of the bubble level.

3. Look for an internal obstruction, usually corrosion, by feeling the pipe with the bare hand, noting where the hot surface suddenly turns cool. The hand test should be applied to both the supply and return.

4. Correct sag or pitch as necessary. Open pipe at point of obstruction and clean out or replace.

5. We assume that the system at one time worked satisfactorily, but if there has been faulty circulation from the start, check all piping for size. There is a chance it has been undersized, and in this circumstance, conversion to forced flow may be the most economical cure.

As a guide in the appraisal of pipe size, velocity flow in excess of 6 ft per second increases frictional resistance and for forced circulation promotes noise.

3-2. CONVERTING GRAVITY SYSTEMS TO FORCED CIRCULATION

(a) *Converting a Gravity Series Arrangement to a Series with Forced Circulation*

STEP 1 Examine the existing header to determine its condition for continued service, looking inside as well as outside. If found satisfactory, install pump on return side near boiler. If corroded with crusts or pits, replace the entire supply-return loop with new piping. This new piping may be smaller, and its size calculated on a flow velocity of 4 ft per second.

STEP 2 Install a diversion fitting at every place in the supply line where a branch is taken off to a radiator. For a radiator *below* the header, mount a diversion fitting on the return connection to the header as well as on the supply. Each fitting is to be sized to divert the correct volume of hot water to the radiator it serves. Data on the correct sizes can be obtained from the fitting manufacturer for the temperature and capacity conditions selected.

STEP 3 If gravity system has open expansion tank, replace this with a closed or compression tank, putting the new tank in the same location (usually the attic) and retaining the direct air vent from the boiler in the basement to the tank. Size the tank by data given in Sec. 3-1(c) or from the latest edition of the ASHRAE Guide. If system already operates with a closed tank, verify the size of the existing tank and replace with a larger one if necessary.

STEP 4 Provide and install an automatic air vent valve on the highest point of the system, if such a valve is not already there.

STEP 5 Provide all circulator controls and wiring.

STEP 6 If boiler is fired automatically, check and verify the safe and satisfactory operation of all automatic firing controls.

STEP 7 Examine and clean boiler inside and outside. Look for internal corrosion and cracks. If boiler is significantly deteriorated, replace with new unit.

STEP 8 If building tenants desire sharp control of the heat, install a flow-control valve to prevent gravity circulation when the circulator is not running.

General note: As stated previously, a series arrangement with forced circulation is somewhat better than a series operating by gravity but is not practical for any installation but very small ones. The limit of rooms served satisfactorily by a series layout can seldom exceed five or six, depending on the individual sizes and exposures.

(b) ***Converting Gravity Two-pipe Direct Return to a Two-pipe Forced Circulation Direct Return.***

STEP 1 Examine all piping for condition as previously described.

STEP 2 Provide and install pump circulator in the return piping near the boiler.

STEP 3 Cut into the return branch from each radiator and insert a balance cock for use in restricting flow, in Btu, through the branch to the design load of the radiator.

STEP 4 Provide and install all circulator automatic controls, such as control thermostat and wiring.

STEP 5 Examine the boiler in detail as previously described and clean, repair, or replace as necessary.

STEP 6 Evaluate the expansion or compression tank as previously indicated, resizing or replacing if necessary.

STEP 7 Install automatic air vents at high point of system if not already there.

(c) ***Converting Gravity Reversed Return to Forced Circulation Reversed Return.***

STEP 1 Verify condition of all piping.

STEP 2 Provide and install a pump circulator in the return piping near the boiler.

STEP 3 Provide and install all circulator automatic controls, such as thermostat and wiring.

STEP 4 Evaluate the compression tank for size. Replace if necessary.

STEP 5 Mount automatic air vent valves at high point of system if not already there.

STEP 6 Verify automatic safety controls on boiler.

STEP 7 Install a flow-control valve on the boiler outlet piping to prevent natural circulation when the circulator is shut down, if the building tenants so desire. In some systems, natural circulation may suffice during the fall and spring, with the circulator coming into action only during severe weather. Flow control will not be wanted for that sort of arrangement.

STEP 8 Examine the boiler for condition, repairing, cleaning, or replacing as necessary.

(d) ***Converting a Gravity One-pipe System to a Forced Circulation Two-pipe***

STEP 1 Examine all piping to determine its fitness for continued service. If acceptable, disconnect all radiator *return* branches from the main. If badly deteriorated, replace.

STEP 2 Provide and install a separate return header as shown on

Fig. 3-4(a) for a direct return and Fig. 3-4(b) for a reversed return.

STEP 3 Connect each radiator return branch to the new return header.

STEP 4 If a direct return is selected, provide and install a balancing cock in the return branch from each radiator. These are not needed for the reversed return.

STEP 5 Provide and install a circulator in the new return pipe near the boiler.

STEP 6 Provide and install all automatic controls for the circulator, such as thermostat and wiring.

STEP 7 Verify condition of boiler and its safety controls.

(e) ***About Pipe Sizes.*** For gravity piping, the allowable frictional resistance in milinches *per foot* is determined from the formula:

$$\text{Maximum allowable friction} = \frac{\text{total generated head, milinches}}{\text{total equivalent pipe length, ft}}$$

The total equivalent length is determined by adding together the feet of pipe plus the losses in valves, fittings, and radiators, these latter losses being expressed in equivalent feet of straight pipe.

The use of existing gravity piping with a circulator offers no problem in circulator selection, since the equivalent pump discharge head in feet is obtained by dividing the total generated head in milinches by 27,600, which is the equivalent of 2.30 ft of water.

However, if the gravity piping is being replaced in whole or in part for service with a circulator, the resistance of the longest equivalent circuit must be calculated using the established friction loss tables for the kinds and sizes of pipe involved.

3-3. FORCED CIRCULATION HOT WATER HEATING SYSTEMS

(a) ***Discussion.*** Forced circulation has made gravity water heating obsolete because forced circulation is positive circulation, it is readily and comparatively exactly controlled, and the smaller pipes, fittings, and valves conserve space and materials.

The observance of just two rules by the designer and installing contractor can be expected to produce a satisfactory installation. The first rule is to deliver enough hot water to properly sized radiation in the space to be heated. The second rule is to do it quietly.

While forced circulation is used with both direct- and reversed-return piping arrangements, the former is the more frequent because of the presence in most installations of zone control plus a considerable variation of heat-emitting forms. We can have baseboard, cabinet convectors, unit heaters, fan-coil combinations, and even radiant surfaces as part of the whole coordinated heating program.

In some instances, operating troubles do not develop from the fact that the heating medium is pumped hot water, but from certain characteristics of the heat-emitting surface. To illustrate this point, crackling sounds from the expansion and contraction of fin-tube baseboard can be observed in both hot water and steam installations.

These equipment-born complaints and their corrections will be identified and treated in detail in subsequent sections. In each case of trouble from whatever cause, the first step is to determine which of the two rules has been broken and thereafter to proceed accordingly.

(b) ***Complaint.*** One or more radiation units do not heat.

Causes When Only One Unit Is Involved

There are six possible causes to consider.

1. Radiator is air logged, and hot water is unable to enter the upper areas.

2. An air pocket is present in the branch pipe supply or return. Two ways these pockets occur are illustrated by Figs. 3-5 and 3-6.

3. Look for a lack of balance resulting in the water flow bypassing the cold radiator. Evidence of this condition is a fully heated and perhaps overheated, condition of other units on the system while this particular one is an orphan.

4. If the cold radiation is below the supply main and is *not* supplied through a diversion fitting in the main, flow may bypass it. If more than 7 ft below, it will need a diversion fitting on the return as well as on the supply (see Fig. 3-7).

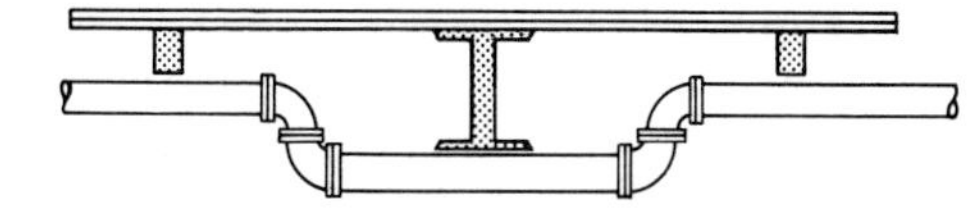

(a) AIR POCKET CREATED BY DROPPED SECTION

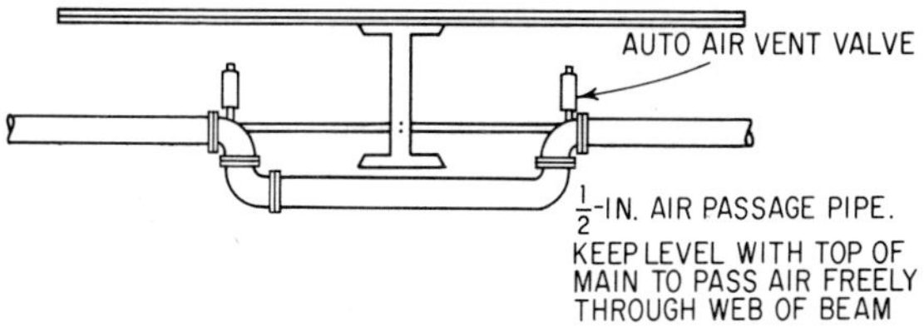

(b) TWO (2) WAYS OF HANDLING AIR POCKET

FIG. 3–5. Air pockets and corrections.

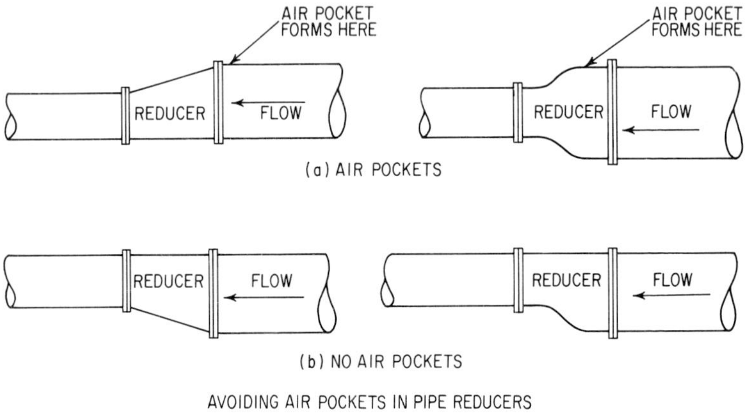

FIG. 3–6. Air pockets in pipe runs.

5. Water temperature may be too low for the square feet of heat-emitting surface that has been provided. Check temperature with a contact thermometer, or alternatively, judge the adequacy of temperature by laying the hand on top of the radiation surface.

6. Piping connections are too small for the pump head available.

Recommended Actions

1. Vent radiator by means of manual air vent.
2. Correct under-beam pocket as shown in Figs. 3-5(a) and 3-6(a).
3. Correct balance by resetting balance cocks on overheating sections of system to augment circulation through the main.
4. Insert diversion fittings as necessary.
5. Increase water temperature from boiler in 5-deg increments until radiator heats satisfactorily or water temperature reaches 210 F. If still too cool at top temperature, check surface for possible inadequacy of area and add surface as necessary. *Note:* Sometimes an auxiliary electric heating device, such as a length of baseboard or a fan-coil unit, is the simplest and cheapest way to make up the lack.
6. Examine connecting piping for size. If calculated to be too small, either replace with properly sized pipe or increase the discharge head of the circulator. This may require speeding up of the circulator with the possibility of noise or replacement of the whole unit by one with higher head characteristics.

Causes When a Section of Units Is Involved

1. The system is short-circuiting. Taking the path of least resistance, the water is returning to the boiler through unmonitored or incorrectly balanced returns.

2. Insufficient hot water is being pumped into the system.

3. An air pocket in the main supplying that section or group of radiation units is interrupting circulation. Run the bare hand along the main and note where it turns cool, giving particular attention to relative elevations of the cool area and adjacent hot piping. Consider a fitting trap as shown in the reducer in Fig. 3-6.

4. If cool section constitutes a zone, check thermostat and zone valve for correct operation. A valve that does not pass the calculated necessary flow must be repaired or replaced. Thermostat troubles are rare but should not be ignored as a possibility.

5. Check each radiator for a choke connection. By this is meant a supply branch that is substantially smaller than the radiator inlet, for example, a 3/4-in. pipe trying to serve a 1 1/4-in. fin-tube baseboard.

6. Circulator capacity, head, or both are too low. Check load against model and size pump being used.

7. If any part of the piping runs through unheated or exposed spaces, check for an ice plug. Test suspected area of plug with hand; ice can be felt.

Recommended Actions

1. If trouble is determined to be short-circuiting, insert balance cocks and set to suit conditions.

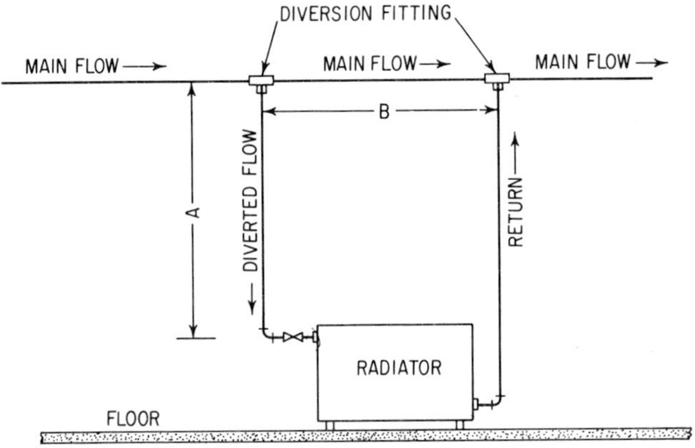

NOTES:
1. DIVERSION FITTING ON RETURN IS ESSENTIAL WHEN DISTANCE A IS 7'-0" OR MORE
2. DISTANCE B MUST BE AT LEAST EQUAL TO LENGTH OF RADIATOR
3. CONNECTIONS TO RADIATOR TO BE AS DIRECT AS POSSIBLE
4. FOR BEST RESULTS CONNECT SUPPLY AND RETURN AS SHOWN

FIG. 3–7. Connections to below the header radiation.

2. Build up the system pressure, starting with no less than 12 psig on the boiler when cold.

3. Eliminate air pockets as indicated previously in Figs. 3-5(a) and 3-6(a).

4. Correct thermostat and zone valves if found defective in performance.

5. Replace any choke connection with piping of the correct size.

6. Establish proper pump capacity and head, installing a new replacement unit if necessary.

7. After thawing out frozen section of pipe, improve the insulation on the vulnerable sections. If exposure is too great for reasonable insulation or weather is extremely severe in the area, consider the circulation of an antifreeze solution rather than water. Because of different thermal characteristics, adoption of the antifreeze may call for a new circulator of larger capacity, more heating surface, or both. Obtain characteristics of the antifreeze from the manufacturer before adopting.

(c) ***Complaint.*** Radiator is hot; room is cold.

Causes

1. Check surface temperature by a contact thermometer. The total heating surface might be such that water temperature of 200 F or more will be needed for comfort.

2. The building construction is leaky. There is too much clearance around window sash and doors, or the subfloor is full of cracks or knots, and there is no insulation in the ceiling. If the heating system has been designed for 72 F on the premise of a normally constructed building, the best that can be expected is a very drafty 72 F and discomfort for everybody there.

3. Room exposure has been increased after the heating system was in place. For example, a large picture window has been cut into a heretofore solid wall, or a standard window has been expanded into a large glass area.

4. A feeling of chill results from the radiant loss of body heat to walls and floors that are substantially lower in temperature than the surface temperature of the human body. This is called "cold 70" or "cold 72," depending on the space temperature being provided.

5. Radiation is low; ceiling is high. All the heat heads enthusiastically upward where, other than warming the ceiling, it has little effect. A difference of 20 deg between living levels and the air under a 14-ft ceiling is not unusual. The difference in any given complaint should be ascertained in order to best prescribe relief.

Recommended Actions

1. Increase the water temperature at the boiler by 5-deg increments until a maximum of 210 F is reached. If room remains cold, consider other appropriate correctives as described hereinafter.

2. If trouble is observed to be generated by leaky building construction, recommend the installation of storm windows and doors, weather stripping, and insulation in the walls, in the ceiling, and under the floor. Adding to the radiation will not be economically desirable so long as the leaky condition continues.

3. Compensate for changes in room exposure by increasing the radiating surface, raising the water temperature, or providing auxiliary heating units to be used for temporary extreme conditions.

4. For the "cold 70" condition, elevate the water temperature enough to add 1 or 2 deg to the room space temperature. Usually this is enough; otherwise, insulation shot into the walls may be necessary.

5. For high ceilings, a fan-coil type of auxiliary heater is effective, since the fan sets up a certain amount of air circulation and provides better mixing. In industrial buildings with high ceilings or the exposed underside of the roof serving as a ceiling several expedients are possible. These include the overhead mounting of infrared heaters, the installation of downdraft nullifying coils under the sills of clerestory windows, the setting up of fan-coil overhead units, and, occasionally, the spreading of a dropped ceiling, using suitable stiff materials.

(**d**) ***Complaint.*** The circulator rumbles or crackles.

Causes

1. Rumbling generally is the sign of a pump that does not match the load, being either too big or too small.

2. Rumbling has been observed also where the pump discharge has been bushed down to fit a smaller pipe.

3. Crackles occur when the pump is cavitating, meaning that the suction is being fretted by slugs of air or vapor that are banging around in the attempt to escape.

4. Vibration to some extent is observable in all pumps, but when excessive, sometimes travels throughout the building. This comes from a lack of balance in the pump or a stressed connection to the pump foundation. Also it may be the result of a wrong speed.

Recommended Actions

1. Weigh carefully the capacity, head, and speed of the pump against the optimum requirements of the system. If the pump is too big or too small, replace it with the proper size.

2. Check the speed using a shaft tachometer in rpm. High speed is considered to be 3,600 rpm, and if this is the speed found and the pump is also noisy, replace it with a pump of lower speed, say 1,800 rpm.

3. A change in speed may change the vibration for the better, but if not, remount the pump on vibration absorbers.

4. To eliminate the crackling of cavitation, rearrange the piping on the suction side to make sure the pump receives solid water and not a

mixture of water and vapor. If the pump is too large, replacing it with a smaller unit is likely to eliminate the crackling.

5. If the rumbling pump has a bushed discharge, remove the bushing, extend the discharge full size for at least 1 ft, and there reduce the discharge to the pipe size by means of an eccentric reducer, the eccentric being chosen to maintain the top of the discharge level with the top of the piping. This avoids a high pocket at that point. See Fig. 3-6(a).

(e) ***Complaint.*** The baseboard is noisy.

Causes

There are at least four conditions that give rise to noises in baseboards.

1. For squealing sounds, examine the point at which the supply and return connections to the length of baseboard transit through a floor or wall. If the opening is too small, the expansion and contraction movements of the pipe set up rubbing and binding sounds where it contacts with the floor. These sounds are most noticeable at night after the system has been given its night setback and are quite disconcerting to some persons.

2. Fin-tube baseboard is not beyond making crackling sounds as the

Table 3-1. Troubleshooter's Field Guide for Gravity Hydronic Heating Systems

Possible cause	How to check for cause	Recommended ways to correct
COMPLAINT: A given radiator does not heat. REFERENCE: Fig. 3-2(a).		
1. Air pocket in radiator.	*a.* Open manual vent.	*a.* Keep vent open until water appears.
2. Water has cooled so much that differential cannot maintain circulation through the radiator.	*b.* Take temperatures of water at inlet and outlet. Difference should be at least 20 deg.	*b.* Step up the water temperature at boiler in 5-deg increments. *Caution:* Watch out for overflow, if open expansion tank.
3. Circulation through radiator is blocked by corrosive deposits or other material in inlet valve or supply and return branches.	*c.* Run bare hand over piping; feel for sudden change of hot to cool or cold.	*c.* Open and clean inlet valve and piping as indicated.

Table 3-1. Troubleshooter's Field Guide for Gravity Hydronic Heating Systems (*Continued*)

Possible cause	How to check for cause	Recommended ways to correct
4. Air pocket in supply or return piping.	*d.* Look for sag in piping. If sag is also cold, compared with adjacent piping, consider evidence. Usually an old system.	*d.* Reestablish correct uniform pitch of piping, using bubble level.
5. Short-circuiting of hot water flow causes bypass of cold radiator.	*e.* Check adjacent radiators for overheating, particularly their returns.	*e.* Balance the circuits by resetting inlet valves, or install balance cocks. *Caution:* High resistance of cocks may affect circulation. Get resistance from manufacturer in milinches.

COMPLAINT: Expansion tank overflows in cold weather.
REFERENCE: Open expansion tank, Fig. 3-3(a).

6. Cold water makeup valve is leaking.	*f.* Open piping downstream from fill valve and note any leakage through valve.	*f.* Repair or replace leaky valve.
7. Water is being heated above 212 F and is boiling over and out through expansion tank overflow.	*g.* Note temperature of water in boiler and height of tank above boiler in feet.	*g.* Reduce water temperature or raise tank enough to compensate for the plus-212 F temperature.
8. Expansion tank is not sufficiently high above topmost radiator to contain the expansion at elevated temperatures.	*h.* Note temperature of water in boiler. Equate height of tank above top radiator in feet, for the noted temperature in boiler.	*h.* Reduce the temperature to suit height or elevate tank to suit temperature.
9. House is cold although tank is over-	*i.* Calculate the required radiation and	*i.* Increase the amount of radiation or call

Table 3-1. Troubleshooter's Field Guide for Gravity Hydronic Heating Systems (*Continued*)

Possible cause	How to check for cause	Recommended ways to correct
flowing, because of either insufficient radiation or building leaks around windows, doors, etc.	equate this with square feet in place. Check points of draft and crack leakage.	for weather stripping, more insulation, etc. For economy, suggest use of auxiliary heat device.
10. The pipe sizes in branches to radiation are too small, making high temperatures necessary to obtain any circulation.	*j.* Check square feet in each radiator against the supply and return piping sizes.	*j.* Increase pipe sizes or convert to forced circulation.

COMPLAINT: With closed compression tank, boiler safety valve spills when boiler is fired to capacity.

REFERENCE: Compression tank, Fig. 3-3(b).

11. Tank is too small.	*k.* Calculate the correct size and compare with what is being used.	*k.* Replace with properly sized tank.
12. Tank is waterlogged.	*l.* Feel tank while system is operating. If hot all over or nearly all over, there is not enough air in tank.	*l.* Bleed water until tank is no more than two-thirds full, injecting air to provide the needed cushion.

COMPLAINT: Series radiation is putting out too little heat.

REFERENCE: Supply and return header is provided by a single main. Fig. 3-2.

13. Water temperature is too low leaving boiler.	*m.* Check temperatures leaving and returning to boiler, using a contact thermometer if insertion thermometers are not in place.	*m.* Increase the water temperature leaving boiler.

COMPLAINT: With a reversed-return arrangement, one or more radiators fail to heat.

REFERENCE: Fig. 3-4(b).

14. If only one radiator is not heating, check for air binding.	*n.* Open manual vent.	*n.* Keep vent open until water appears.

Table 3-1. Troubleshooter's Field Guide for Gravity Hydronic Heating Systems (*Continued*)

Possible cause	How to check for cause	Recommended ways to correct
15. For several radiators staying cool, check pitch of piping.	*o*. Use bubble level.	*o*. Correct pitch as necessary.
16. Obstruction in the pipeline, either supply or return.	*p*. Use bare hand to find point of temperature change.	*p*. If not due to sag or wrong pitch, open piping and clear out obstructing material.
17. Wrong pipe sizes.	*q*. Recompute the flow through each cold pipe run, based on heating surface connected to that run. Calculate the motive force in milinches and equate against friction in affected piping.	*q*. Replace with right size pipe or convert to forced circulation. See Sec. 3-2(e).

result of expansion setting up strains in the relationship of the fin tubing to its cover or backboard.

3. Creaking and knocking result from the same forces that produce crackling and should be treated the same way.

4. Humming and vibration can be irritating occurrences. Sometimes these sounds are projections from the circulating pump and should be corrected there, as previously described. Vibration and humming also can appear as a harmonic from outside sources. This is particularly familiar in areas traversed by airplanes and is usually expressed by unreinforced sheet-metal surfaces, such as the covers for fin-tubing baseboard.

Recommended Actions

1. To correct squealing noises, provide at least ¼-in. clearance around the pipe in all directions, when the pipe is *hot*.

2. Locate the points of stress and relieve them by tightening and loosening connections as appropriate. To find the points of trouble, paint the suspected surfaces and connections with Prussian blue, give the system a chance to put on its performance, and then examine for smearing. Where the blue has smeared, movement has occurred and is likely to be the point requiring correction.

3. Stiffen the vibrating surfaces by tack welding a length of light angle iron on the underside of the surface where it cannot be seen. The added weight throws it out of synchronism with the external vibrations. Temporary corrections have been reported by applying strips of masking tape rather than the angle. This idea is worth trying where physical limitations disallow the more permanent iron.

(f) ***Complaint.*** The radiant floor, with hot water coils embedded in concrete, takes too long to warm up when heat is needed and too long to cool off when heat is not needed.

Cause

Because the concrete floor acts as a heat reservoir, what this system needs is an anticipating control by which outside cooling or warming trends are sensed before their effects are felt inside. These anticipating units react not only to the outside temperature, but also to the influence of sun and wind, and by reducing or increasing the flow of hot water in reflection of the outside conditions, compensate for the heat storage characteristics of the concrete.

To be sure that slab cooling is never too rapid for room comfort, the inside thermostat is given overriding authority to maintain the water flow necessary to provide whatever inside temperature is desired.

Alternatively, temperatures can be controlled to reflect outside trends by installing a water temperature–modulating valve, rather than by a flow-modulating valve. The temperature modulation is accomplished by mixing a percentage of the returns into the hot supply.

Recommended Action

Install an anticipating type of controller and have it function through either a flow-modulating valve at constant temperature or a temperature-modulating valve at a constant rate of flow.

(g) ***Complaint.*** The radiant floor is warm, but the room remains cold.

Causes

1. The operation of a radiant floor at surface temperatures higher than 85 F is not practical, since effects of the warmer floor on some persons is to promote swelling of the feet and ankles and an uncomfortable tight feeling of the shoes. In view of this limitation, the total output of the radiant area at 85 F may simply be insufficient to satisfy the heating demands of the space.

2. Drafts across a radiant floor can largely nullify the floor's output of heat. These drafts may be downdrafts from windows, exposed walls, or both, or they may come from loose-fitting sashes and leaky doors. When the presence of drafts is uncertain, check by means of the smoke-maker listed in the troubleshooters' inventory of needed instruments and apparatus.

Recommended Actions

1. For insufficient heat availability, set up some auxiliary source of heat, such as an electric fan coil unit, a length of hot water baseboard, or even an infrared overhead unit, gas-fired or electric as suitable to conditions.

2. For chilliness resulting from drafts, set up a program of weather stripping for doors and windows and possibly the placement of insulation in walls or ceilings.

(h) ***Complaint.*** The unit heaters fail to put out enough heat.

Causes

1. If they have *ever* put out enough heat and have been in their present positions for several years, chances are the air passages have become more or less clogged with airborne lint. The hot water is getting to them, but the heat exchange to air is reduced substantially from the original rating.

2. Air pockets have developed in the supply piping or in the branch return from the ailing heater. As described previously, the presence of air can be detected by the simple examination of the piping with the bare hands, by checking the pitch with bubble level, and by the opening of the air-vent valve. If this last is automatic, there is a likelihood of its being clogged and in need of cleaning.

3. The possibility of one or more unit heaters being on an unbalanced circuit is as valid as it is for any radiator being heated by circulating hot water. The tests for short-circuiting are as described previously, including the observance of overheating in adjacent unit heaters, radiators, or whatever type of equipment is functioning in the same zone.

Recommended Actions

1. For recently inadequate unit heaters, clean the heat exchange surfaces thoroughly by an alternation of vacuuming and blowing out with compressed air. As lint and threads tend to wrap themselves around the tubes and fins, a final cleaning by hand picking may become an unhappy necessity.

2. For air pockets, open vent valves and fittings, even lengths of pipes in some instances, and release the trapped air. Then correct the pitch or the air-trapped fitting [Fig. 3-6(a)] and make sure that the automatic vent valves, if in service, are clean and functioning properly.

3. For short circuits, install balancing valves in the overloading circuits and set them to the correct flow of hot water for each of the overloading circuits. This should force hot water in proper volumes into the heretofore short-circuited units. If there is still an insufficient supply, check the circulator for capacity and head. One or the other or both may be too small for the job and require replacement with a new and properly sized pump. In this connection, we must remember that

the passage of years tends to build up frictional resistance from corrosive and other materials in the piping, and a pump that was barely adequate originally with good clean piping may presently be unequal to the enlarged task.

(i) ***Complaint.*** Fine rust particles are being carried around the building and smudging walls and ceilings. The rust is believed to be originating in the metal baseboard.

Causes

1. If the baseboard is of cast iron, lift out of position in contact with the wall and clean thoroughly. Paint the surface with a rust inhibitor, such as some form of zinc chromate in a glycerol phthalate alkyd vehicle modified with linseed oil. Or obtain a recommendation from any anticorrosive paint manufacturer.

2. Check the ambient atmosphere for the possibility of the rust dust originating elsewhere.

Recommended Actions

1. After treating the baseboard as above, replace with a thin strip of asbestos board between the baseboard and the wall. The purpose of the asbestos is to prevent moisture in the wall from making direct contact with the iron.

2. If the structure is over a crawl space, examine carefully under the floor and directly below the baseboard for evidences of moisture. Moisture can promote the rusting, and evidence of its presence suggests that more ventilation of the crawl space is needed. In addition, a layer of roofing material on the ground, with 4-in. laps, will inhibit the migration of moisture from the ground into the building structure.

3. The wall immediately above the baseboard should be checked at least biweekly for evidence of smudging, and the wall wiped carefully whether or not smudging is found. Some of the rust particles that take to the air and eventually appear as smudges cannot be seen with the naked eye until they conglomerate into measurable smears.

(j) ***Complaint.*** The boiler shows signs of excessive corrosion.

Causes

1. The boiler stands in a basement or machinery room that becomes both damp and cool in the nonheating season. As a result, atmospheric moisture condenses from time to time on the boiler, and corrosion develops under the droplets.

2. Atmospheric contaminants attack the metal, notably sulfur in combination with a high humidity, producing sulfuric acid. If the corroding boiler is located within 2 or 3 miles of an industrial or chemical plant, an analysis of the atmosphere is desirable.

3. Internal corrosion of the boiler results from an excessive presence of oxygen in the boiler water or from acids that attack the metal.

Recommended Actions

1. For nonoperating boilers in cool, damp locations, provide ventilation to dry out the atmosphere and to keep it dry.

2. For corrosion in a suitably dry atmosphere, arrange for an analysis of atmospheric contaminants.

3. For internal corrosion, do not add fresh water to the boiler any oftener than necessary. Makeup water is charged with dissolved oxygen that is freed when the water is heated and attacks the iron. For persistent corrosion that obviously does not come from frequent makeup, analyze the water for acids and add a suitable alkalizing agent.

4. For the gas passages through the boiler to the breeching, protect with trays of lime to absorb the atmospheric moisture that enters these passages.

5. See that all suspected corrosion spots are wire brushed to remove corroded material. Then touch up with an anticorrosive such as red lead. Do not fire the boiler until all the anticorrosive paint is completely dry.

(k) ***Complaint.*** The hydronic piping is perforating in various spots.

Cause

The evidence points to electrolytic action. This produces craters and perforations in the anodic areas and tumors in the cathodic.

Recommended Actions

1. Remove all sections of corroded pipe and replace these with piping that has been coated with a good grade of bitumastic paint. In selecting the paint, be sure that it is suitable for the high temperatures of a hydronic system. This treatment has worked satisfactorily, both internally and externally, in a number of instances.

2. If corrosion continues at an unacceptable rate, a corrosion specialist should be consulted, since the environment might require particular forms of treatment.

3. If there is reason to suspect the presence of stray currents being collected by the pipeline, the insertion of dielectric unions at intervals of about 50 ft will have beneficial results.

Cause

There is oxygen in the system. The corrosive deposits are often magnetic, and if the material is powdered and spread out on paper and a magnet is held underneath the paper, magnetic lines of force can be expected to appear in the dust. This is a reasonable indication that oxygen attack is present.

Recommended Actions

1. Check automatic air-vent valves at high points of system and pressure in system at these points. If positive pressure is not kept on these points at all times, probability is that air is being sucked into the piping

through these vents. Correction may involve shifting of pump from one side of the boiler to the other, depending on location of compression tank, in order to assure adequate pressure at the top of the system. See the latest edition of the ASHRAE Guide for explanations of the compression tank location and how it affects the pumping.

2. A chemical inhibitor offers protection. The inhibitor chosen should be the outcome of a chemical analysis of the water, this analysis determining the acidity or alkalinity of the water and the amount of oxygen present. The inhibitor interposes a protective surface between the metal and the water.

(l) ***Complaint.*** In a multiple-zone system, heating one or more zones causes one or more shutdown zones to heat simultaneously. Controls seem to be futile.

REFERENCE: Fig. 3-8.

Cause

Hot water backflows through return connections to zones not intended to be heating, because of lack of check valves, balance valves, and cutoff valve for each zone supply.

Recommended Actions

1. The least expensive correction is to install check valves in the returns from each zone to prevent back circulation.
2. Provide segregated supply and return circuits for each zone.

3-4. QUESTIONS AND ANSWERS

QUESTION 1. In the American Southwest, some of the rural churches have no floor under the pews but the ground. Does a ground surface

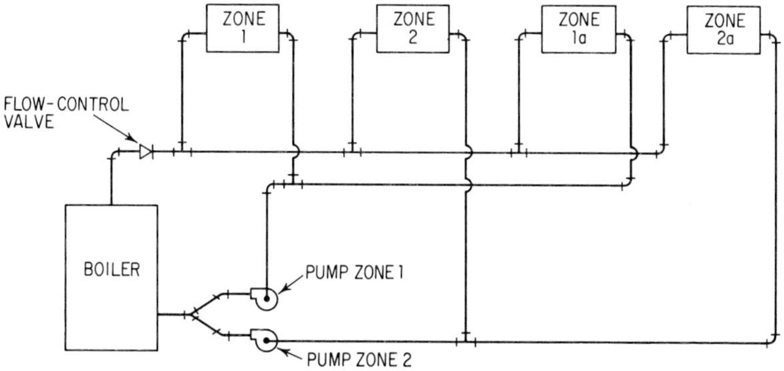

ASSUME ZONE 1 IS HEATING; CIRCULATOR ON ZONE 2 IS STOPPED. THEN RETURN FROM RADIATOR 2 REVERSE FLOWS THROUGH 2a AND RADIATOR 1a TO CIRCULATOR ON ZONE 1

FIG. 3–8. Incorrect piping for 2-zone system.

Table 3-2. Troubleshooter's Field Guide for Forced Circulation Hydronic Heating Systems

Possible cause	How to check for cause	Recommended ways to correct
COMPLAINT: One or more radiation units do not heat.		
REFERENCE: Figs. 3-5 and 3-6.		
1. Radiation is air-logged.	*a*. Open vents.	*a*. Keep vents open until water appears.
2. Air pockets in the piping.	*b*. Feel the pipe for changes in temperature.	*b*. As suggested in Fig. 3-5(a) and 3-6(a).
3. Unbalanced water flow.	*c*. Adjacent areas may be overheated. Check volume of flow through adjacent hot radiation.	*c*. Correct balance with balancing cocks.
4. Radiation is under the main and has no diversion fittings.	*d*. Inspect for location and presence of fittings.	*d*. Insert fittings as shown in Fig. 3-7.
5. Water temperature is too low.	*e*. Elevate temperature by 5-deg increments at boiler. Check water temperature at boiler and at radiator.	*e*. When satisfactory temperature is established, set controls accordingly.
6. Pipe too small for pump head available.	*f*. Calculate proper size of pipe for volume of hot water required for existing radiator and equate with pump head available in milinches per foot.	*f*. Increase pipe size or increase discharge head of pump to overcome friction. Increase cold pressure.
7. Zone valve may be malfunctioning.	*g*. Check operation of valve by manipulating thermostat. Observe speed and extent of valve response.	*g*. Repair or replace zone valve as necessary.

Table 3-2. Troubleshooter's Field Guide for Forced Circulation Hydronic Heating Systems (*Continued*)

Possible cause	How to check for cause	Recommended ways to correct
8. Ice plug has formed in radiator circuit.	*h.* Feel pipe for ice-cold area.	*h.* Thaw out ice plug. Insulate piping, if none, or increase thickness.
COMPLAINT: Radiation is hot; room is cold.		
9. Water temperature is too low.	*i.* Take surface temperature of radiation.	*i.* Raise temperature to a maximum of 210 F in 5-deg increments.
10. Building construction is air-leaky and full of drafts.	*j.* Note air movement in room by means of smoke-maker. Hold lighted match near sash cracks.	*j.* Weather-strip or provide storm doors, storm windows, or both. See that insulation is put into walls and ceilings.
11. Exposure has been increased since heating system was installed by addition of windows, doors, or skylights.	*k.* Calculate heat loss and compare with amount of heating surface available.	*k.* Add heating surface. For peak conditions use a portable type of auxiliary heater.
12. A condition of "cold 70" or "cold 72" exists.	*l.* Note thermometer. If these temperatures show, but room is cold, this is the reason.	*l.* Increase temperature of room by 2 deg. If heating capacity cannot do this, set up some form of auxiliary heater. Insulate walls.
13. Ceilings are high; radiation is low.	*m.* Observe the ceiling height and positions of radiating surface.	*m.* Provide auxiliary heating units. Consider infrared. Stop the downdrafts from clerestory windows.

Table 3-2. Troubleshooter's Field Guide for Forced Circulation Hydronic Heating Systems (*Continued*)

Possible cause	How to check for causes	Recommended ways to correct
COMPLAINT: Circulator rumbles or crackles. REFERENCE: Fig. 3-6(a).		
14. Pump too big or too small.	*n.* Compare the pump nameplate data of head and capacity with calculated requirements.	*n.* Replace pump with proper size or change pump speed.
15. Pump discharge connection is bushed down.	*o.* Observe pump opening and size of connecting pipe.	*o.* Extend discharge full size for 12-in. minimum; then bush down with eccentric bushing.
16. Pump is cavitating.	*p.* Check suction lift and temperature. Inspect impeller for cavitation pits.	*p.* Modify suction to make sure pump gets solid water and not a mixture of vapor and water.
17. Pump is vibrating.	*q.* Vibration shows up in sound and touch. Pump is out of balance.	*q.* Correct balance so that operation is not close to critical speed. Mount on vibration pads.
COMPLAINT: Baseboard is noisy, squealing, and crackling.		
18. Baseboard piping connections are rubbing on sides of holes in floor through which they pass.	*r.* Inspect holes. They must have a minimum clearance of 1/4 in. all around when pipe is hot.	*r.* Provide the needed clearance.
19. Fin tube, when expanding and contracting, is putting stress on cover or backboard.	*s.* Paint suspected areas with Prussian blue. Smeared paint indicates contact friction.	*s.* Tighten or loosen holding bolts and saddles as needed to release stress.

Table 3-2. Troubleshooter's Field Guide for Forced Circulation Hydronic Heating Systems (*Continued*)

Possible cause	How to check for causes	Recommended ways to correct
20. Baseboard hums and vibrates in harmony with certain external sounds, such as those of a passing airplane.	*t*. Note coincidence of baseboard sounds with external noises.	*t*. Stiffen the baseboard cover by means of a length of angle iron welded to underside of cover.
COMPLAINT: Radiant floor heats and cools too slowly.		
21. Concrete floor acts as a heat reservoir.	*u*. Note time of heating and cooling by measuring surface temperature of floor at 15-minute intervals.	*u*. Install an anticipating type of control of the coils in the floor.
COMPLAINT: Radiant floor is warm, but room remains cold.		
22. Area of warm floor is too small to deliver comfort with a surface temperature of 85 F.	*v*. Calculate heating needs and compare with heat emission of floor.	*v*. Provide some form of auxiliary heat or reduce heat losses by use of extra insulation, weather stripping, etc.
23. Floor is swept by cool-to-cold drafts.	*w*. Check location and amount of drafts with the smoke-maker.	*w*. Stop drafts with weather stripping, extra insulation, etc.
COMPLAINT: Unit heaters have lost capacity to meet heat requirements.		
24. The air passages through heaters have become clogged with dust and lint.	*x*. Inspect heaters for fouling of passages.	*x*. Clean thoroughly.
25. Air pockets exist in the connecting piping.	*y*. Locate pockets by change in pipe temperature.	*y*. Vent air-bound sections. Establish free flow of water.
26. Unit heater circuit is unbalanced or is being bypassed by a short circuit.	*z*. Inspect system for unbalance or short circuit as previously explained. See Sec. 3-3(b).	*z*. Balance and install balancing cocks in all overloaded circuits.

Table 3-2. Troubleshooter's Field Guide for Forced Circulation Hydronic Heating Systems (*Continued*)

Possible cause	How to check for causes	Recommended ways to correct
COMPLAINT: Fine rust particles from baseboard are being airborne throughout building.		
27. The cast-iron baseboard, closely fitted to wall, is rusting.	*aa*. Lift out baseboard and inspect. If rust dust is present, it will rub off on finger. Analyze area atmosphere for some foreign origin of the rust dust.	*aa*. Paint the baseboard with an anticorrosive paint on side nearest the wall. Remount, using an asbestos board between wall and baseboard. Wipe off wall above baseboard thoroughly at least once every 2 weeks.
COMPLAINT: Boiler is corroding excessively.		
28. During nonheating time, boiler stands in a cool, damp environment. Condensate occurs and produces pits on iron surface.	*bb*. Scrape away scale. If pits are round, they can come from water drops. Check atmosphere for sulfur content. This is positive contributor to corrosion. Paint pits with anticorrosive.	*bb*. Provide plenty of summer ventilation for basement or boiler room. Do not change boiler water if possible, to keep down oxygen. Set dishes of lime in gas passages.
COMPLAINT: Piping is perforating in various spots.		
29. Electrolytic action.	*cc*. Holes and craters are in anodic areas; tumors, in cathodic.	*cc*. Replace with new pipe which has been painted with a bitumatic paint.
30. Oxygen in water.	*dd*. Appearance of iron oxide in corroded areas. Check with magnet.	*dd*. Eliminate points of air in-leakage. Treat water with inhibitor.
COMPLAINT: Unable to effectively separate individual zone controls. REFERENCE: Fig. 3-8.		
31. Zones cross-connected through returns.	*ee*. Check by surface temperature of return pipes for reversed flow.	*ee*. Insert check valves in each return.

perform satisfactorily as a radiant heating surface if hot water pipes are arranged in the ground as a heating grid?

ANSWER 1. The idea is considered practical, but as with any other radiant floor, a maximum surface temperature of 84 F should not be exceeded to avoid undesirable physical reactions in people using the floor.

Where much kneeling takes place, as in a church, an even lower temperature may be required. The optimum temperature must be determined by experience in the individual church.

QUESTION 2. Are new installations of gravity heating being made anywhere at present?

ANSWER 2. In farmhouses, the installation of hot water pipes in the back wall of the open fireplace is often used to generate hot water that serves radiation on upper floors by gravity circulation. When contemplating such an installation, be sure to include an expansion tank.

QUESTION 3. Can antifreeze be used instead of water in a hot water heating system?

ANSWER 3. Yes. Note the specific heat of the proposed antifreeze since it may be less than that of water and thus require increased circulation to equal the heat-bearing capacity of water. This may lead to the need for a larger circulator.

QUESTION 4. If hot water piping is run between the joists under a wooden floor, will the floor act as a radiant heating surface?

ANSWER 4. Yes, until it warps and buckles all out of shape. There is no better way to ruin a good floor.

QUESTION 5. Is there any danger in the placement of electric cables and radiant heating pipes in the same concrete floor?

ANSWER 5. It has happened that stray currents go to work on the hot water piping and cause leaks. This can occur in less than a year after the floor is finished, and of course it must be dug up for repairs. It is best to keep them separated.

DIVISION 4

Steam Heating Systems

4-1. STEAM GENERATED ON THE PREMISES

(a) ***Discussion.*** The recipe for a successful steam heating performance is to deliver enough steam at a high enough temperature to the heat-emitting surfaces to meet their ratings, while simultaneously, and without disturbing the steam movements, to remove the water of condensation.

Failure of a steam system to deliver the heat it should results from the failure of something or somebody to follow the recipe. The somebody could be the original designer; the something, the defective operation of a valve or perhaps a trap. Finding the source of the trouble is the job for the troubleshooter.

First step in the troubleshooter's program is to identify the basic design of the system, such as one-pipe, two-pipe, vacuum return, gravity or pumped return, upfeed, downfeed, live steam or exhaust steam, pressure, in psig, with a particular notation for those systems that present a mixture of design, usually the result of structural additions. Since each of these arrangements has certain common susceptibilities for trouble, knowing about these weaknesses ahead of the complaint may well save the troubleshooter some time. Here are common ailments.

1. *One-pipe System:* In vertical risers, the riser pipes are too small to let the steam rise quietly and effectively while the condensate is flowing downward in the same riser. This is especially true for buildings more than two stories high. In this and other one-pipe installations, inadequate drainage of the condensate in the horizontal distribution header is another common source of trouble.

2. *Two-pipe Systems:* Condensate flow through a radiator is hindered

by air trapped in the upper sections of the radiator, rather than passing out through either the automatic vent valve or the thermostatic trap.

3. *Upflow, Headers in the Basement:* Risers should take off from the header at 45° and in no case pitch downward after leaving the header. These headers should be drained adequately and should pitch in the direction of steam flow.

4. *Downfeed, or Header in the Attic:* Attic header tends to accumulate air pockets and thus interfere with free circulation. This header should pitch in the direction of flow and carry an automatic air-vent valve at the highest point, usually atop the supply riser or risers.

5. *Gravity Returns:* For very small structures frequently of the one-pipe arrangement, return may back up higher than the low point of the steam header and thus obstruct or even stop steam entry into one or more risers.

6. *Tank and Pump Return:* Operation of pump is controlled by water level in boiler. If boiler is hard-pushed for a brief period and returns are unable to meet the steam delivery, the automatic feeder may flood the system unless the excess, when returns do catch up, is run off to waste.

7. *Exhaust Steam:* In an engine exhaust, the steam may carry oil which coats the interior surfaces of the heating equipment and checks the rate of heat transfer. For zoned systems having fluctuating zone demands, back pressure on engine or turbine may not be uniform unless control means are provided.

8. *Live Steam:* When steam is being generated solely for premises heating plus perhaps domestic hot water, a common trouble under heavy-load conditions is a priming of water out of the boiler. The reason is frequently too high velocity in the steam outlet pipe; the water is carried out with the steam. The equalizing pipe from the outlet header to the return connection in a Hartford return tends to relieve this condition, but the header connection must be in the right place. An unexpected development may be encountered from steam that reaches the heating system by way of a pressure-reducing valve, a common occurrence in mills and factories. The steam is superheated by the sudden drop in pressure through the valve, and heat transfer from superheated steam is substantially slower than with saturated steam. The effects may take the form of inadequate heating, despite the fact that radiating equipment is fully charged with steam, or if the heating steam is metered for any reason, steam consumption seems to be excessive.

(b) ***Complaint.*** In a one-pipe system, radiators do not heat.

Causes

1. Individual radiators are air-bound.

2. Individual radiators *do not* pitch back to the entry valve, causing water block.

3. Water pockets in radiator connection prevent steam from reaching the radiator.

4. Supply pipe is too small to accommodate steam and water simultaneously.

5. Pitch of connecting pipe is wrong as the connection leaves the supply header.

Recommended Actions

1. Remove air-vent valve. If radiator does not heat now, hold flame of match near to vent opening. If flame is sucked in, steam is condensing before it reaches radiator and flow, pressure, or both must be increased. If flame blows away from vent, line is air-bound and may be expected to clear. Keep vent open until steam appears. *Caution:* If emission of air from the open vent is strong, do not strike a flame until the pressure is relieved and the blowing action subsides. In any event, when holding the flame near the vent, stand as far back as possible and extend match from a position that provides some protection if a flash or explosion should occur. Cases have been reported in which disassociation of the steam or an in-leakage of some flammable gas has been the fuel for radiator explosions via the vent valve. Clear the area of other persons when making the flame test.

2. For a water pocket at the steam entry valve, block up the far end of the radiator or convector, leaving the entry where it is. Check the radiator pitch by bubble level. Generally a lifting of the far end of the unit of 1 in. or less is enough to establish satisfactory drainage.

3. By use of the level, verify the pitch of all connecting supply piping, making sure that the condensate flows down and back to the main header. Rearrange piping as necessary.

4. A supply connection that is too small frequently tells on itself by gurglings and thumpings within itself. Such noises are sure evidence that water and steam are in a traffic conflict. Check the vent valve for the automatic admission of air when the steam is turned off by the controls. If the valve fails to do this, a partial vacuum is developed within the radiator and prevents the condensate from draining freely. This means that steam, when again turned on, meets a lot of water that should have been out of there long before.

If the vent valve is performing normally but there is still grumbling in the pipes and a relatively cold radiator, replace the connection with a larger pipe. For the minimum right size, refer to the tables provided in the latest edition of the ASHRAE Guide.

5. For noises and water hammering in the main supply header,

rather than in the branches, one or more additional drainage connections must be provided. As a rule of thumb, add together the internal cross-sectional areas of all the branch takeoff pipes, and at any point where the sum of these areas exceeds the internal area of the main install a new drain connection from the bottom of the main. If a drain is already there, make it bigger.

(c) ***Complaint.*** In a two-pipe system, indicated radiators do not heat.

Causes

1. Check the pitch of the radiator from inlet valve to thermostatic trap on the outlet. Condensate must drain toward the trap. In old buildings, settlement of the structure often disturbs the correct pitch and may even reverse it.
2. Verify the satisfactory operation of the automatic vent valve. The top third of radiator may be air-bound if vent is not performing.
3. Bellows of thermostatic trap, if cracked, will fail to act as it should and may leave the valve in a closed position.
4. Water pockets in various places along the steam supply lines may be preventing steam from reaching the radiator.

Recommended Actions

1. Correct pitch of radiator.
2. Check vent valve and replace if necessary.
3. Lift bellows and needle valve from thermostatic trap and inspect the bellows for cracks. Clean point of needle and the valve seat. Replace trap if found defective.

(d) ***Complaint.*** In a two-pipe system, system is noisy. Piping frequently presents an anvil chorus.

Cause

Water and steam are meeting head on in one or more reaches of the system.

Recommended Actions

1. If not already present, install a trap at the end of each long run of 50 ft or more. If end traps are in service, make sure of their effective operation. If found satisfactory, new extra drainage connections to the headers must be provided. As suggested previously, install a drip point wherever the areas of the preceding offtakes exceed the internal area of the steam main.
2. Make sure the mains are all pitched in the direction of steam flow. Correct as necessary.
3. For buildings more than two stories in height, trap the base of every supply riser, as shown in Fig. 4-1.

Caution: Do not permit water hammer in a steam line to go on indefinitely. Neglecting the correction of such a condition can cost lives at worst and injuries at best, if a fitting fails under the hammering. This writer has been tagged as an expert witness in a court suit that may cost the owners of such a heating system $250,000. The suit was instituted by the family of the dead.

(e) ***Complaint.*** In a two-pipe system, all radiation is hot, but the building remains cold.

Causes

1. Insufficient heat-emitting surfaces to satisfy the structural demand.

2. Surface is adequate, but the building is drafty with excessive leakage around doors and windows. In some instances, high ceilings, of perhaps 10 ft or more, allow the heat to stagnate under the ceiling with lower levels remaining chilly.

3. Steam is wet and operating at a pressure that is too low to supply enough Btu to the radiation to meet the design values.

Recommended Actions

Check possibilities in the following order:

1. Take the temperature of the steam at the radiator most distant

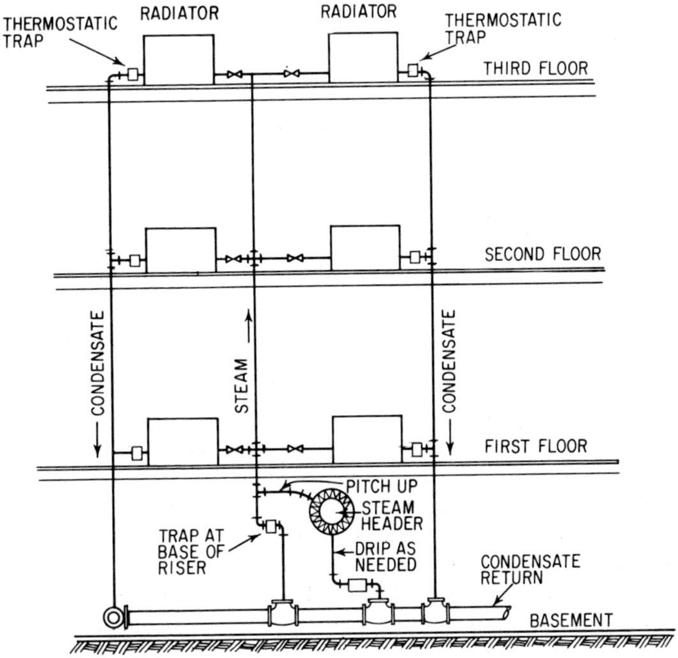

FIG. 4–1. Riser trapping.

from the boiler and at the radiator nearest to the boiler. If obviously low at the far radiator, step up pressure 2 or 3 lb at the boiler.

2. If heating remains unsatisfactory at increased pressure, use the smoke-maker to ascertain the presence of drafts and their direction of movement. Recommend the installation of weather stripping, storm doors, and additional insulation in the walls and ceiling.

3. As a temporary expedient, set up small propeller type fans to blow room air across the radiator. This increases the rate of convective heat exchange and may be sufficient for meeting occasional extreme weather conditions.

4. For high ceilings, placement of a fan that moves the warm pocket under the ceiling toward the outside walls will help to nullify the cold downdrafts over windows and exterior walls. Do not use a fan so large that it creates noticeable drafts at the living levels.

5. Calculate the necessary heating surface in square feet and compare with surface presently in place. Install additional radiation or replace undersized existing radiation units with ones of correct size.

(f) ***Complaint.*** In a two-pipe system, the unit heaters do not put out enough heat.

Causes

1. Units are undersized and cannot meet the load.

2. Heat exchange surfaces are fouled with lint and dust.

3. Water or air pockets in the connecting piping obstruct circulation.

4. Units have been suspended too far inboard from the outside walls, so that the heaters offer no neutralization to the cold downdrafts.

5. Insulation has fallen from some section or sections of the steam supply lines, so that only very wet steam is reaching the heaters.

Recommended Actions

1. Check for absent insulation on the supply piping.

2. Clean the exchange surfaces by blowing through them *in reverse* with compressed air.

3. Check for water and air pockets as directed by the presence of cold areas in the piping or unit. Test vents.

4. Verify thermostatic controls.

5. Verify capacity of the heaters by comparison with the calculated heating requirements.

6. If strong floor-level drafts are diluting the output of the heaters, set up barricades to block the drafts. This is a common trouble in shops and warehouses.

7. Step up steam pressure to deliver more Btu per hour to the unit heaters.

8. For factories and plants, provide storm doors.

9. If exhaust steam is being supplied to the unit heaters, open thermostatic trap and examine for the presence of oil. Clean if oil is found and install an oil separator in the main supply pipe carrying the exhaust steam.

10. For a vacuum return system with either live or exhaust steam, check level of vacuum being pulled at the pump. If insufficient, steam will be found in return piping below the thermostatic trap where its presence tends to keep the valve closed and to lock up condensate in the heater. Check over the entire return system, looking for a lift fitting that exceeds the practical value for such arrangements. The maximum "hump" over which the vacuum pump can be asked to pull the condensate in the form of a lift fitting is 5 ft.

Table 4-1. Troubleshooter's Field Guide for Steam Heating Systems Using Premises Steam

Possible cause	How to check for cause	Recommended ways to correct
COMPLAINT: One or more radiators fail to heat. REFERENCE: One-pipe Steam System, Sec. 4-1(b).		
1. Individual radiators are air-bound.	*a.* Remove vent valve.	*a.* If radiator now fills with steam, replace vent valve.
2. Radiator pitches away from inlet valve instead of toward it.	*b.* Check pitch of radiator with bubble level.	*b.* Block up far end of radiator to drain condensate through inlet valve.
3. Connecting piping is blocked with water pocket.	*c.* Feel along supply branch. Where pipe turns cold, test pitch with level.	*c.* Remove pocket by repitching pipe.
4. Steam supply pipe is too small to handle steam and condensate simultaneously.	*d.* Pipe gurgles and pounds when steam is turned on.	*d.* Replace with pipe of correct size as indicated by ASHRAE Guide.
5. Main steam header is noisy with hammerings.	*e.* Noise is evidence of inadequate header drainage.	*e.* Provide new drain connection to header wherever total areas of offtake pipes exceed area of the header.

Table 4-1. Troubleshooter's Field Guide for Steam Heating Systems Using Premises Steam (*Continued*)

Possible cause	How to check for cause	Recommended ways to correct
COMPLAINT: One or more radiators fail to heat.		
REFERENCE: Two-pipe Steam System, Sec. 4-1(c).		
1. Radiator does not pitch from entry valve to thermostatic trap on outlet.	*f.* Check pitch with bubble level.	*f.* Block up inlet end or lower outlet end to suit.
2. Air vent valve not operating properly.	*g.* Remove valve. Note if radiator now fills with steam.	*g.* Install new valve if necessary.
3. Thermostatic trap is inoperative from sludge or a split bellows.	*h.* Take out thermostatic element and inspect.	*h.* Clean or renew bellows element as necessary.
4. Water pockets in supply piping are blocking steam to radiator.	*i.* Test for pipe cooling by hand contact. Check pocket by level.	*i.* Eliminate pocket by repitching or rerouting of pipe.
COMPLAINT: System is noisy.		
REFERENCE: Two-pipe System, Fig. 4-1, Sec. 4-1(d).		
1. Water and steam are in conflict within the piping.	*j.* Hammering is evidence of unacceptable conditions.	*j.* See that steam and condensate travel in the same direction in the same pipe. Increase the drainage connections. Always have one at end of header.
COMPLAINT: All radiation is hot but building remains cold.		
REFERENCE: Two-pipe System, Sec. 4-1(e).		
1. An insufficient heat-emitting surface.	*k.* Calculate the surface needed and compare with installed radiation.	*k.* Add new radiation or replace some units with larger ones.
2. Building is drafty.	*l.* With smoke-maker, check presence and direction of drafts.	*l.* Stop drafts with weather stripping and storm doors. Insulate walls and under floor.

Table 4-1. Troubleshooter's Field Guide for Steam Heating Systems Using Premises Steam (*Continued*)

Possible cause	How to check for cause	Recommended ways to correct
3. Steam pressure and temperature are too low.	*m.* Take temperature at boiler outlet and farthest radiator.	*m.* Raise pressure as necessary and practical. Set fans to blow across heating surfaces.
4. High ceilings causing upflow of heated air.	*n.* Compare temperature just below ceiling and at 5 ft above floor.	*n.* Arrange fans to establish a ceiling-to-floor circulation if practical.
COMPLAINT: Unit heaters do not put out enough heat.		
REFERENCE: Two-pipe Steam System, Sec. 4-1(f).		
1. Units are undersized for the load.	*o.* Compare capacity of heaters with calculated load.	*o.* Replace with units of right size.
2. Heat exchange surfaces are fouled with lint and dust.	*p.* Inspect for dirt.	*p.* Blow through in *reverse* with compressed air. Wipe out with cloth, not with waste.
3. Water or air pockets obstruct steam circulation.	*q.* Locate pockets with hand touch and level.	*q.* Eliminate pockets by repitching or replacing that section of pipe.
4. Unit heaters are too far from exterior walls.	*r.* Measure blow pattern and compare with distance in feet between unit center line and wall.	*r.* Rearrange blow, or add new units to obtain warm air contact with wall.
5. Supply pipe is being cooled on way to unit by a loss of insulation.	*s.* Check full distance of supply piping for fallen insulation.	*s.* Reinsulate as necessary.
6. Thermostats are not functioning correctly.	*t.* Check thermostat operation against admission of steam.	*t.* Repair or replace defective instruments.

Table 4-1. Troubleshooter's Field Guide for Steam Heating Systems Using Premises Steam (*Continued*)

Possible cause	How to check for cause	Recommended ways to correct
7. Cold drafts from doors or windows along floor are nullifying warm blows.	*u*. With smoke-maker, locate drafts and directions they take.	*u*. Set up draft barriers along main walkways. Barrier can be portable and no more than 3 ft. high.
8. Steam pressure is so low that wet steam is the rule.	*v*. Check pressure and temperature at boiler outlet and at most distant heater.	*v*. Set up pressure as practical.
9. Exhaust steam is coating exchange surfaces with oil.	*w*. Open thermostatic trap and inspect for oil.	*w*. Clean out as practical and install oil separator in steam line from engine.
10. Vacuum-return vacuum is too low.	*x*. Check the vacuum against pump rating. Look for lift fitting that is too high.	*x*. Increase the vacuum. Break lift fitting down into two or more steps so vacuum can handle it.

4-2. STEAM GENERATED FROM HIGH-TEMPERATURE–HIGH-PRESSURE WATER

(a) ***Discussion.*** High-temperature–high-pressure (ht–hp) water can be pumped over hill and dale with a minimum of trouble as compared with steam. In a steam line, condensate must be removed periodically, and the disposal is not always easy. For example, in city areas, hot condensate may not be dumped into either the sanitary or storm-sewer systems, since the heat and moisture may create organic reactions that produce foul gases. Therefore it must be pumped back to the boilers or to some process usage or be sufficiently cooled to permit its waste somewhere along the line.

Hot water also has the advantage of allowing the design to employ piping of uniform size. Economy is possible because the water is neither wasted nor used up, but is pumped in a closed circuit where the heating and cooling continue to alternate as long as desired.

Since the water temperature is generally in the area of 400 F, its passage through a heat exchanger can be expected to generate steam in the exchanger (or converter as it is sometimes called) at a pressure of 200 psig or more.

This is much too much for a heating system; consequently, the installation of a pressure-reducing valve on the steam outlet is common practice. Alternatively, steam production can be monitored by limiting the volume and velocity of hot water admitted to the exchanger.

High-temperature–high-pressure water systems are best suited for application to large projects and those which involve considerable territory. For example, army camps with numerous buildings are readily served in each building by diverting the required volume of hot water to on-the-premises heat exchangers and domestic water heaters. The minimum plant size justifying ht–hp water appears to be about 5,000,000 Btu per hour.

Once the steam leaves the exchanger, it can run into any of the troubles already examined in one-pipe and two-pipe steam systems, and the complaints should be handled accordingly.

The heat exchanger itself and the troubles to which it is susceptible are examined in Division 2. This is because the heat exchanger as a steam generator is comparable to other nonfired boilers and, in many respects, to the fired boilers as well.

Mixed applications are common with the high-temperature–high-pressure heat exchanger installations. To illustrate: in a given building steam at at least 110 psig is needed for tailor shop activities; heating steam at less than 5 psig; steam for steam tables, kettles, and other restaurant equipment at from 5 psig up; finally, of course, there is domestic hot water which can be produced at any desirable steam pressure.

Any one of these uses can and often does present a complaint. Here is a common one:

(b) ***Complaint.*** Steam kettles do not produce the desired temperature in restaurant cooking.

Causes

1. The discharge of condensate from the kettle traps is being made to a condensate header that is also accepting the discharge from traps at substantially higher pressure. This means the kettle traps cannot develop enough pressure differential relative to the return piping to expell the condensate.

2. Return piping is too small to handle the total volume of condensate from this and other traps, resulting in the creation of back pressure.

3. Temperature of steam is too low to produce the desired kettle temperature.

Recommended Actions

1. Install a check valve in the discharge from *every* trap currently discharging into the return header.

2. Install a new low-pressure return line to handle low-pressure traps exclusively.

3. Adjust pressure-reducing valve *upward* in the line that is delivering steam to the kettles.

(c) ***Complaint.*** In the tailor shop, presses are being affected adversely by wet steam. Steam gage shows 110 psig, but slugs of water are recurrent.

Causes

1. Velocity of steam leaving heat exchanger is too high.

2. Pressure differential across the reducing valve, if steam is being generated at a pressure above that needed for the tailor shop equipment, is not wide enough to produce any superheating, or erratic superheating.

Recommended Actions

1. Calculate steam velocity at exchanger outlet under full load and, if so high as to produce priming, provide an expanded steam outlet by at least one pipe size.

2. Mount a separator steam drum on the exchanger, arranged to drain carry-over back to the steam space of the exchanger.

3. Step up the steam pressure as generated to obtain a wider pressure reduction across the reducing valve. Check minimum acceptable differential as recommended by the reducing-valve manufacturer.

Table 4-2. Troubleshooter's Field Guide for Steam Systems and HT–HP Water

Possible cause	How to check for cause	Recommended ways to correct
COMPLAINT: Various units of steam heating system fail to function.		
REFERENCE: See Table 4-1 for details of heating complaints and recommended actions.		
COMPLAINT: Certain restaurant equipment, such as steam kettles, does not maintain necessary temperatures.		
REFERENCE: Premises steam generated from a heat exchanger using high-temperature–high-pressure water as the source of heat. See Sec. 4-2(b).		
1. Back pressure is present on discharge	*a.* Measure pressure in return line. Look for	*a.* Eliminate back pressure. Install a check

Table 4-2. Troubleshooter's Field Guide for Steam Systems and HT–HP Water (*Continued*)

Possible cause	How to check for cause	Recommend ways to correct
from the condensate traps.	high-pressure traps discharging into the return from low-pressure traps.	valve in discharge of each trap. Provide a separate high- or low-pressure condensate line to suit.
2. Return line is too small to handle total condensate of all traps discharging into this line.	*b*. Calculate size of line needed and compare with line in service.	*b*. Provide an extra condensate return line or replace existing line with a larger one.
3. Steam temperature is too low to produce desired equipment temperature.	*c*. Check temperature with thermometer.	*c*. Set up a reducing valve to deliver higher pressure to equipment. Do not exceed setting of relief valves.
COMPLAINT: Water slugs reaching various items of equipment. REFERENCE: Steam from HT–HP water by means of heat exchanger. See Sec. 4-2(c).		
4. Steam velocity from heat exchanger is too high.	*d*. Calculate velocity versus steam outlet size.	*d*. Increase size of outlet pipe by one pipe size minimum.
5. Low-pressure differential across reducing valve allows water to pass through unchanged.	*e*. Check pressures on both sides of reducing valve. Compare differential with minimum recommended by valve manufacturer.	*e*. Increase pressure differential by stepping up pressure at discharge of heat exchanger. Alternatively, mount a separator drum on heat exchanger to remove water.

NOTE: Similar difficulties in restaurants, tailor shops, and any place that steam is being provided at multiple pressures can be expected when the steam is generated in a direct-fired boiler. For discussions of the boilers and heat exchangers, refer to the appropriate sections of Division 2.

4-3. STEAM FROM A DISTRICT STEAM SYSTEM

(a) ***Discussion.*** The nature of steam supplied from a district utility network usually differs from that generated on the premises in pressure, total heat, and uniformity.

Whereas premises generation takes place at the pressure needed for its ultimate use, district steam is almost universally delivered at high pressure, 100 psig or more, in order to make the delivery through pipes of economic size. Pressure-reducing valves, which often have no place in low-pressure heating systems, become part of the control system in nearly every district steam installation.

Figure 4-2 shows a typical large city district steam control and meter section on the customer's premises. Note that two reducing valves are needed; the step from 200 psig down to 5 psig is too great for a single valve.

Special problems appear as direct results of the wide differentials between delivery and use pressures and of the four possible methods

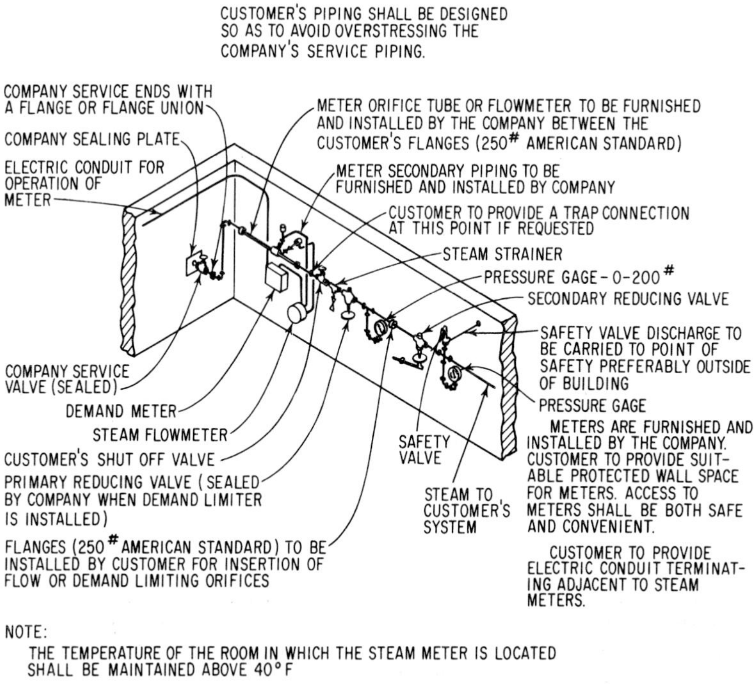

Fig. 4–2. District steam service connection.

of metering. These methods are (1) metering steam delivery by a steam flowmeter; (2) metering the hourly or half-hourly demands by demand meter; (3) metering the total consumption of steam by condensate meter; and (4) metering by a flow-limiting orifice that is installed in the steam supply pipe and for which the customer pays a flat fee.

The presence or absence of steam in the radiators of a structure can be evaluated in the same way as the troubles are evaluated in Sec. 4-1 and Table 4-1. The troubles peculiar to a district steam supply and contract will be considered next.

(b) ***Complaint.*** "We are paying for steam and getting hot water."

Causes

1. The service trap, located immediately inside the premises, is blowing because of a defect.

2. The service trap is blowing because the water leg of the trap has been insulated like the rest of the piping and no water seal is being maintained.

3. The street main is insulated ineffectively, and the steam is condensing in this main faster than the traps and line pumps can remove the water.

Recommended Actions

1. Inspect the service trap and strip off the insulation on the inlet connection for a distance of at least 2 ft ahead of the trap. This bare pipe allows sufficient loss of heat from the steam in the inlet leg to establish a water seal and keep it there. This is a fairly common trouble.

2. If inlet leg is properly bare of insulation, shut the trap valves, open the trap, and inspect it for malfunction. Repair or replace as necessary.

3. Water in the street main is a problem for the steam utility company to solve. It can be checked by making a calorimeter test of the steam quality entering the premises. A test that reads satisfactory in dry weather may show unacceptable moisture during periods of heavy rains as the result of rain percolation through the soil and into contact with an inadequately protected steam main. Several tests under various conditions must be run in order to reach a reasonable conclusion.

(c) ***Complaint.*** "The meter goes round and round whether we are taking steam or not."

Causes

1. Refer to Fig. 4-3, which illustrates a typical steam flowmeter that operates by registering the differential pressure across an orifice. On each side of the orifice is a condenser with water or another

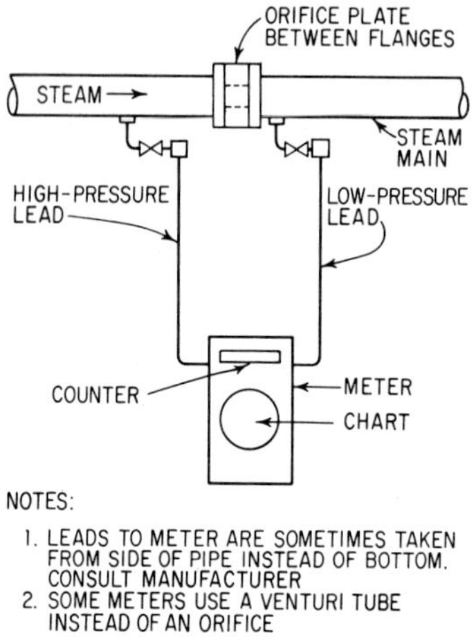

FIG. 4–3. Flowmeter connections.

liquid filling the piping between the condenser and the integrating mechanism in the instrument case.

Depending on the flow across the orifice (or through a flow nozzle in some instances), the levels of the water in the condensers depart from a condition of equilibrium, and the difference between them reflects the flow of steam. Calibration is usually in pounds of steam per hour and is totalled and chart-recorded by the instrument.

From time to time, the meter must be calibrated for accuracy, which usually means that water levels in the condensers are not returning to equilibrium after the steam is turned off, and thus the meter continues to register a flow although none is taking place. When this occurs, the utility company is likely to develop a credit for the customer against his next bill.

2. Steam actually is still being delivered to some remote area or forgotten equipment.

Recommended Action

In all cases, check into the possible presence of steam at points in the main from 20 to 100 ft downstream from the orifice plate and the condensers. Take the pipe surface temperature either by contact thermometer or by hand, being careful in the latter action to avoid burns from hot piping.

If no flow is established, check water levels in condensers and equalize.

(d) *Complaint.* "Our system has an orifice-type demand limiter, and during severe weather, it does not pass enough steam to maintain comfort. The next size limiter jumps the fixed charge nearly 50 percent per month."

Cause

The ability of an orifice to pass steam increases with an increase of pressure on the inlet side, but is *not* increased by a drop of pressure on the outlet side once this pressure has reached a level of about 58 percent of the supply.

This being true, the steam utility company is likely to have a locked pressure-reducing valve ahead of the limiting orifice, thus maintaining constant pressure at all times to the orifice. In this condition of supply, an increased steam flow is not possible as the result of the pressure in the building system falling below that 58 percent of supply, even if the piping is discharging to the atmosphere.

Recommended Action

If the customer feels unable to afford the fixed billings of a larger limiter, some form of auxiliary heating should be considered for severe weather. This emergency heat can be in the form of electric fan-coil units, infrared electric or gas heaters, or possibly direct-fired unit heaters.

(e) ***Complaint.*** Hot condensate is being dumped into the building sewer system and is creating odors.

Cause

The hot water promotes organic decay which in turn generates gases.

Recommended Action

Run the condensate through a cooling radiator ahead of the discharge to the sewer. From the temperature of the condensate, subtract the temperature allowed for the sewer and provide enough *hot water radiation* to dispose of the difference.

(f) ***Complaint.*** Ever since district steam was substituted for an on-the-premises boiler supply, the system becomes cold periodically and must be bled of water.

Cause

Condensate is not returning rapidly enough from the radiation. This can result from faulty air valves that are permitting a vacuum to develop in the radiation. In one case of record, the changeover of return-piping connections from the boiler to the sewer discharge involved relocating a check valve, and this valve was reinstalled backward.

Recommended Actions

1. Follow all return pipes until the points at which the piping becomes cold are located. Examine each fitting and valve thereafter, looking for possible obstructions. Remove when found.

2. Check one or two of the air valves on cold radiation and note whether steam enters and condensate drains with the air valves removed. Repair or replace as necessary.

(g) ***Complaint.*** Several radiators, floor-set and on the same floor that carries the boiler, heat only across the top. Figure 4-4 illustrates.

Cause

Since this is a gravity return arrangement, most of the radiating

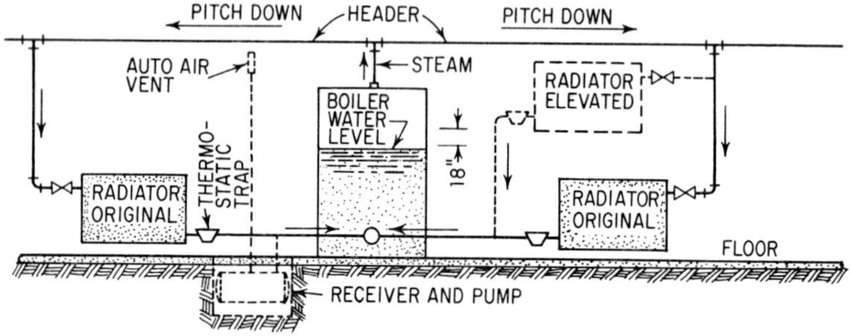

NOTES:

1. DOTTED LINES SHOW SUGGESTED TREATMENTS FOR RADIATION BELOW BOILER WATER LINE
2. SHADING INDICATES WATER IN BOILER AND RADIATION

FIG. 4–4. Radiation below supply main.

surface is located below the waterline of the boiler; consequently, this area of the radiator is always flooded. The result is heat emission from a hot water radiator rather than steam.

Recommended Actions

1. Provide a receiver in a floor pit, so that condensate flows into the receiver by force of gravity. From the receiver, it is moved into the boiler by a pump that is controlled by the water level in the boiler. This is a fairly expensive but effective cure for the trouble.

2. Elevate the radiators sufficiently so that at least an 18-in. difference is provided between the maximum water level in the boiler and the outlet of the radiator. This works, but the output of a high-set radiator never heats the space with the uniformity of the floor-set, steam-filled, unit.

3. Disconnect the radiator from the steam supply and install a small circulating pump that moves hot water from the boiler, through the radiator, and thence back to the boiler. This is simply converting the steam heated radiation to hot water radiation using forced circulation. Since the output of heat from the hot water may be somewhat less than the output from steam, some additional surface may become necessary. This will have to be determined by actual performance.

(h) ***Complaint.*** Several radiators spit water more or less continuously from the vent valves.

Causes

1. Valves are defective. Remove, open, and inspect.

2. Condensate is not leaving radiation freely. Refer to Complaints 4-1(b)–(d) and to the appropriate tables.

Recommended Actions

1. Replace valves with new ones, properly adjusted to pass air but not water.
2. Other steps are noted in the tables.

Table 4-3. Troubleshooter's Field Guide for Heating Systems with District Steam

Possible cause	How to check for cause	Recommended ways to correct
COMPLAINT: Various complaints concerning piping and radiation not specifically chargeable to district steam characteristics. REFERENCE: See Table 4-1 for premises steam.		
COMPLAINT: District steam supply pipe is delivering hot water rather than dry steam. REFERENCE: Sec. 4-3(b).		
1. Service trap, first inside of premises, is blowing.	*a.* Pour a little water on trap discharge pipe. If it boils, trap is blowing.	*a.* Open trap and repair or replace defective part.
2. Drip line from main is insulated right up to the trap.	*b.* By observation.	*b.* Strip off insulation and leave pipe bare for at least 2 ft ahead of trap.
3. Water is coming in from street main faster than trap can handle it.	*c.* Listen for slugs of water. Make calorimeter test of steam quality.	*c.* Report findings to steam company for correction.
COMPLAINT: Flowmeter registers steam flow when no steam is flowing. REFERENCE: Sec. 4-3(c).		
1. Water level in meter condensers has become out of equilibrium.	*d.* Check levels by observation.	*d.* Bring levels into equilibrium when no steam flow is occurring.
2. Steam is actually being supplied to some remote area or forgotten units.	*e.* Follow hot steam pipe to destinations. Note whether temperature does or does not fall.	*e.* If use is legitimate, so report. If leaking valves are causing flow, repair the valves.

Table 4-3. Troubleshooter's Field Guide for Heating Systems with District Steam (*Continued*)

Possible cause	How to check for cause	Recommended ways to correct
COMPLAINT: Orifice-type demand limiter prevents adequate steam flow in cold weather. REFERENCE: Sec. 4-3(d).		
3. The orifice limiter is designed to limit flow.	*f.* If pressure-reducing valve ahead of limiter is locked and cannot be adjusted except by the steam utility, the orifice is doing exactly what it is designed to do.	*f.* Ask utility to unlock valve and step up pressure to orifice, pay utility for bigger orifice, or set up some form of auxiliary heating for severe weather.
COMPLAINT: Hot condensate to sewer is causing obnoxious odors. REFERENCE: Sec. 4-3(e).		
1. Hot water is promoting organic decay and bacterial action in sewer system.	*g.* Note temperature of condensate by thermometer. Query local health department on temperature at which decay begins and compare with present discharge temperature.	*g.* Reduce temperature by running condensate through a cooling radiator. Do not exceed the allowable maximum.
COMPLAINT: System waterlogs and must be bled periodically. REFERENCE: Sec. 4-3(f).		
1. Vacuum in radiation is holding up returns.	*h.* Check radiation for hot radiators having cold outlets.	*h.* Replace air valves with ones that permit the entry of air when steam is cut off.
2. Physical obstruction in piping return system is holding up condensate.	*i.* Travel all return lines until cold pipe is located.	*i.* If caused by a bad valve or a valve wrongly installed, replace or reinstall correctly.
COMPLAINT: Radiator valves spit water more or less continually. REFERENCE: Sec. 4-3(h).		
1. Radiator drainage is faulty.	*j.* Check valve and repair. Look for back pressure or pocket in returns.	*j.* Put in new valve if repair is not practical.

Table 4-3. Troubleshooter's Field Guide for Heating Systems with District Steam (*Continued*)

Possible cause	How to check for cause	Recommended ways to correct
2. Wrong kind of vent valve is being used.	*k*. Try another type of valve.	*k*. Install proper kind of vent valve.

4-4. QUESTIONS AND ANSWERS

QUESTION 1. Does the presence of superheat in steam increase or decrease heat-exchange capacities in the units to which the steam is supplied?

ANSWER 1. It decreases the rate of exchange. Tests over a period of years by various laboratories have shown a reduction of up to 20 percent in the rate of heat transfer from superheated steam to other media. In some situations, superheated steam is made saturated by the injection of water sprays prior to its reaching the point of usage.

QUESTION 2. Is the changeover of steam heating to hot water heating considered economically feasible?

ANSWER 2. Each case has its own particular features. In general, the changeover is best justified when extensive alterations are being made to a building or the source of steam supply is lost. An example of the latter situation occurs when a district steam utility of the neighborhood category goes out of business.

QUESTION 3. Why is galvanized piping not used more often for steam service?

ANSWER 3. Primarily it is a matter of cost for benefits received. While corrosion will occur in some steam lines, most piping carrying steam does so for years without significant deterioration. This is because the steam is generated from carefully treated water in most industrial and some commercial boilers of size. Therefore the extra cost for galvanizing usually is not justified.

DIVISION 5

Electric Space Heating Systems

5-1. RESISTANCE SYSTEMS

(a) *Discussion.* At present, 19 different forms of electric space heating may be encountered by the troubleshooter. The identification, capacities, and applications of these 19 forms are summarized in Table 5-1.

Eight of these forms depend on a resistance element which delivers its heat by convection, radiation, or both, without help from a fan or any other auxiliary assistant. Five resistance applications are coupled with a fan to expedite the air movement over the hot surfaces. Four arrangements incorporate water or steam in the heat-emitting service. And the final two are highly individualistic arrangements of heating with infrared elements and heat pumps.

The relative frequency of one arrangement compared with another does not indicate that one is better than the other. What it does indicate is that one is simpler to install for that particular requirement than the others, that it is cheaper, or that it possesses qualities of performance singularly appropriate to the observed environment. For example, infrared concentrates its effects on selected areas, a very desirable characteristic for warehouses and some shops; but if a draft is blowing through the area, some form of air heating is preferred, either as the whole system or as an auxiliary to infrared.

Complaints centered around electric heating of whatever type will be discussed in the following subsections.

(b) *Complaint.* Electric bills are much higher than had been predicted when the electric heating system was installed.

Causes

1. Economy has been oversold. The total comparative operat-

ing costs of electric heating include the savings on the original investment, and the owner may not have appreciated this and expected to realize a gratifying thriftiness in his electric bills.

2. The building is not effectively insulated, there is cold inleakage around doors and windows, or the ceilings are high and the heat is going to the upper reaches of the rooms.

3. A number of rooms are being heated simultaneously, although some of these are not occupied.

4. Interior doors are being left open, so that heat migrates from heated rooms to unheated hallways and other spaces.

Recommended Actions

1. Recalculate the electric bills and show the owner how these bills are evaluated by the preinstallation estimator, taking full advantage of the installation savings and counting on adequate insulation and tight weather joints.

2. Point out the economy of keeping doors closed and of stepping back the thermostat for rooms that are not in use.

3. If insulation is incomplete (for instance, none under the floor although this space is unheated), compute the savings possible by adding insulation in such areas.

4. For high ceilings, consider the possibility of replacing part of the convection heating equipment with electric radiant or infrared surfaces. This expedient will maintain the installed overall wattage but will counteract the rapid rising of heated air by providing radiant heat at the living level.

5. Make sure the installation is enjoying the best electric rate available. Some utility companies offer special low rates for buildings that are all-electric, including heating, hot water, and ventilation.

(c) ***Complaint.*** The heat pump supplies relatively little heat during cold weather, and the needs of the building must be made up by resistance units, the operation of which runs up the overall heating cost.

Cause

The heat pump compressor has a fixed volumetric capacity for handling refrigerant gas, and in cold weather, the amount of heat available per cubic foot of gas decreases. This means the output of heat from the heat pump also decreases at these times.

Recommended Action

Explain that the economy of the heat pump is realized most when the climatic temperature ranges from 35 to 65 F, since no auxiliary is needed over that range and the whole comfort load can be provided by the pump. By referring to the data published in the ASHRAE

Guide for various locations in the country, the number of times a local temperature falls below 35 F can be compared with the times and hours it stays above 35 F during a normal heating season.

(d) ***Complaint.*** Electric heating cables laid in a concrete floor cause swelling of the feet in some individuals who must spend much time on floors so heated. The room barely reaches comfortable temperature in severe weather.

Causes

1. The temperature of the floor is carried too high. To avoid foot swelling, the surface of the concrete should never exceed 84 F.

2. The room heat requirements are such that an extra high temperature of the floor is needed to satisfy them.

Recommended Actions

1. Control the energizing of the floor cables by means of an outdoor, anticipating type of thermostat. This thermostat will cut off the cable energy soon enough to keep the concrete slab from overheating when the thermometer is rising outside.

2. As a basic control, use an inside thermostat that limits the surface temperature to a maximum of 84 F at any time. Since this limitation will probably result in a chilly room, install some form of auxiliary heat that comes into action at peak times and when the temperature of 84 F is insufficient to meet the demands of comfort. This auxiliary heater could be a length of resistance baseboard, a radiant panel in the wall, or another suitable unit.

5-2. HEATING BY LIGHT

(a) ***Discussion.*** Heating by light is a method of recovering the heat that is emitted by every electric light bulb or fluorescent tube and has come into recent favor with the designers of several office or commercial structures.

The design is concerned with the heat-collecting medium and its handling. In those installations that work solely with air, the air from the spaces below the ceiling lights is taken into the light troughs by means of a fan; thence it passes over the light bulbs or tubes and rises in temperature as the result of the picked-up heat. This warmed air may or may not be heat-boosted before being returned to the spaces needing heat, depending on the weather and the structural insulation and tightness. In general, we can expect heat recovered this way to need help when the outside thermometer approaches freezing or goes below that level. As with the heat pump, heat from the lighting does its best work during times of mild temperatures.

Instead of using air as the heat-collecting medium, some designs circulate water around the lighting elements. Water at these higher temperatures is pumped through heat exchange coils where the heat is transferred to forced circulation air and delivered to the spaces needing it.

Both designs require heat "dumps" during the summer weather when the active spaces call for cooling rather than heating. These dumps can take the form of cooling towers mounted where convenient. Water is usually too valuable to dump into the sewer and must be reused. There are future possibilities of using this heat in the summer in absorption refrigeration systems.

(b) ***Complaint.*** The burning out of bulbs and tubes can be expected to temporarily reduce the heat recovery until new elements can be mounted in their places. Also air and water circulating troubles involving fan, pump, and damper performances are likely to develop over a period of time.

Recommended Actions

1. Determine source of trouble. Do this for an air system by taking temperatures of the air entering and leaving the light exchange areas. This will indicate whether the system is heat-starved or normal.
2. Check speed of fan with a shaft tachometer and measure output of air by means of velometer or anemometer.
3. Take temperature of the air at the delivery points. This will indicate whether percentage of outside air is too great or too little. Adjust dampers to preserve the proper proportion.
4. For water circulation, check water temperatures in and out of lighting troffers.
5. Record pump speed, in rpm, using the shaft tachometer. Read suction and discharge pressures and compare with design requirements.
6. Measure flow of air over the heat exchange coil, in cfm, and set down temperatures in and out of coil, thus obtaining temperature rise of the air.
7. Determine if the trouble occurs about the same time each day; this might indicate that a basic design error is present.
8. On the basis of knowledge acquired by these tests, decisions can be made on the cure for the situation, such as adjustments to fan, pump and dampers or the addition of some auxiliary heat-emitting equipment to meet a basic deficiency. Some relief may be obtained in certain instances by simply changing the time control, such as turning on the lights earlier and keeping them on longer each day. This action will help store heat in the structure during the night and holiday hours and thus reduce the morning warm-up time.

For auxiliary heat emitters, consider the installation of a duct heater in the discharge of the fan and on the way to the diffusers or grilles. This is readily reserved for peak demands by a thermostatic control.

5-3. INFRARED SYSTEMS

(a) ***Discussion.*** On the electromagnetic spectrum, the region of infrared occupies wavelengths of from 10,000 to 15,000 Å. It is just above the visible spectrum, as "black light" is just below it. We do not see the infrared energy, although the filament of the lamp producing it is quite clear. This filament is operated at a relatively low temperature, since the objective is infrared energy and not light. Theoretically, this low operating temperature should extend the life of an infrared lamp into the thousands of hours, but the troubleshooter should remember that this seldom happens, for the filament is readily destroyed by shock or vibration. Å = abbreviation for Ångström units.

Infrared energy travels through air without warming the air in the least; its peculiar characteristic is that it immediately is converted into heat on impact with a solid material, such as the human body, furniture in a room, floors, walls, or whatever it hits. Thus it is expected to travel through the air of a frosty day and warm up any workman whom it hits instead of warming the air through which it passes.

However, another peculiarity of infrared is its restriction to traveling in straight lines; it cannot go around corners. This means the man to be warmed must be in direct line with the infrared heater; if he cannot see the heater, he will not be warmed by it.

This ignoring of air as a medium to be warmed is the basis of many complaints arising in drafty areas. The infrared is warming but the drafts are chilling, and some form of air heater usually becomes essential in windy locations despite a plenitude of infrared.

Infrared lamps used outdoors, on loading platforms, at bank drive-in windows, or at other exposed mountings must be selected for this type of environment; just any bulb will not do. Some of them are susceptible to cold drafts and will fail early in life unless their service is restricted to a quiet part of the indoors.

Complaints usually arise from one of these conditions: either draft discomfort or bulb short life. Ways to correct these complaints are given in Table 5-2, the troubleshooter's field guide for electric space heating.

5-4. QUESTIONS AND ANSWERS

QUESTION 1. How can summer air conditioning be incorporated into electric heating installations?

ANSWER 1. For heat pumps, it is already there, since the compressor is the heart of both the cooling and heating cycles.

For resistance baseboard, some arrangements include duct connections to the baseboard, and these ducts deliver air from a fan to expedite convective heating in winter. This same air is cooled by other equipment for delivery to room spaces in summer. The electric baseboard elements are not energized of course in summer, except if needed by an unseasonably cold day, and the chilled air passes over them.

For electric ceilings or floor cables, the summer air conditioning must be provided as a totally separate system. This is true of radiant panel heating and infrared systems, also.

QUESTION 2. Are the surface temperatures of radiant panels and baseboards high enough to start a fire on contact with furniture, curtains, or rugs?
ANSWER 2. That depends on the ignition temperatures of the materials making contact with the electric units. As a rule, there is no danger, but surface temperatures should be obtained from the electric equipment manufacturer, and steps should be taken to isolate dangerous surfaces if found to be present.

QUESTION 3. Are electric cables under driveways satisfactory in keeping the driveway clear of snow?
ANSWER 3. This is an established practice and has been used successfully on the sloping and therefore dangerous approaches to many bridges.

QUESTION 4. When old buildings are equipped with electric space heating systems, is rewiring of the building usually required?
ANSWER 4. Generally not the whole building, just the heating service. This is likely to call for a new panel and lead-in wires from the public utility company supplying the energy.

Table 5-1. Electric Heating Forms

	Form	Output, watts	Usual applications °
1	Ceiling cables	2.75 per ft	Residential, commercial
2	Floor cables	2.75 per ft	Residential, commercial
3	Wall heater, fan	2,500–5,000	Residential, commercial
4	Wall heater, small, natural-draft	1,500–3,000	Residential, commercial
5	Wall heater, large, natural-draft	3,000–8,000	Residential, commercial
6	Resistance baseboard	150–300 per ft	Residential, commercial
7	Glass baseboard	135 per ft	Residential, commercial
8	Glass panels	750–3,000	Residential, commercial
9	Electric steam radiators	950–1,200	Residential, commercial, industrial
10	Electric boiler for hydronic heating	5,000–72,000	Residential, commercial, industrial
11	Electric boiler, steam and process	5,000–400,000	Industrial
12	Electric furnaces, warm air circulating	10,000–30,000	Residential, commercial
13	Electric furnaces, gravity floor type	3,000–8,000	Residential, commercial
14	Duct heaters and blast coils	500 and up	Commercial, industrial
15	Unit heaters	3,000–60,000	Commercial, industrial
16	Ventilating cabinet heaters	6,000–18,000	Institutional, residential
17	Infrared, reflective	500–6,000	Commercial, industrial
18	Heating by light	Depends on light load	Commercial
19	Heat pumps	As designed	Commercial, residential

° "Usual applications" is not to be interpreted as meaning exclusively. "Residential" includes single homes, apartment houses, clubs, and motels. "Institutions" refers to hospitals and orphanages. "Commercial" means stores, restaurants, banks, and wherever commerce takes place. "Industrial" applies to shops, factories, warehouses, etc.

Table 5-2. Troubleshooter's Field Guide for Electric Space Heating Systems

Possible cause	How to check for cause	Recommended ways to correct
COMPLAINT: The electric bills are much too high. REFERENCE: Sec. 5-1(a) and (b).		
1. Economy of system was oversold.	*a*. Compare the incurred bills with the estimated bills.	*a*. Correct owner's picture of where and how much saving should be anticipated.
2. Building not well insulated and drafty.	*b*. If draft is too gentle to be readily noticeable, put the smokemaker in operation to find source and direction of air movement.	*b*. Recommend additional insulation, plus storm doors, weather stripping.
3. Ceilings so high that bulk of heat rises to elevations beyond the comfort level.	*c*. Observe height; if over 9 ft, loss to ceiling will be considerable.	*c*. Install one or more radiant type of heater or infrared to directly warm living levels.
4. Building is usually open throughout its parts, with all room doors open.	*d*. Inspect the building. Note whether halls and unoccupied rooms are warmer than necessary.	*d*. Point out economy of closing doors and dropping thermostat setting in unoccupied rooms.
5. Customer is operating on an unnecessarily high energy rate.	*e*. Examine bills to see if a special electric heat rate is being used or not and if such a rate is available.	*e*. Ask utility company for best rate.
COMPLAINT: Heat pump does very little good in subfreezing weather. REFERENCE: Sec. 5-1(c).		
1. Heat pump compressor has fixed volumetric capacity.	*f*. Examine the performance curves of the machine, and note the balance point at	*f*. No correction is needed. This is characteristic of the machine.

Table 5-2. Troubleshooter's Field Guide for Electric Space Heating Systems (*Continued*)

Possible cause	How to check for cause	Recommended ways to correct
	which auxiliary electric heaters are energized.	
COMPLAINT: Floor is heated by electric cables in concrete. Feet of individuals working on the floor swell uncomfortably.		
1. Floor temperature exceeds the 84 F that is physiological comfort limit for many individuals.	*g*. With contact thermometer, note surface temperature of the floor.	*g*. Install an anticipating outdoor thermostat to keep lag out of concrete slab. Provide inside thermometer to interrupt energy to cable when surface temperature moves above 84 F. Install auxiliary heat if 84 F slab does not heat room.
COMPLAINT: Heat from a heat-from-light installation is not stable and is sometimes inadequate under normally satisfactory conditions. REFERENCE: Sec. 5-2(a) and (b).		
1. Light tubes or bulbs have burned out.	*h*. Inspect light troffers. Note burned-out elements.	*h*. Replace with new, good units.
2. Circulating air is not picking up heat called for by design as a result of volume deficiency.	*i*. Record temperature of circulating air into light troffers and emerging temperature. If less than design, check speed of fan in rpm with shaft tachometer.	*i*. Adjust fan speed up to rating.
3. Circulating water is not picking up heat called for by design as a result of volume deficiency.	*j*. Record temperature of water into troffers and exit temperature. If less than design, take speed check and delivery volume of pump.	*j*. Correct pump operation as necessary

Table 5-2. Troubleshooter's Field Guide for Electric Space Heating Systems (*Continued*)

Possible cause	How to check for cause	Recommended ways to correct
4. Percentage of outdoor air is too high.	*k*. Compare temperature of air being sent to diffusers and grilles with temperature of air leaving light troffers.	*k*. Adjust fresh air damper as necessary.
5. System was overrated from design error.	*l*. Compare with design performance.	*l*. Add auxiliary heaters as needed.
COMPLAINT: Infrared lamps appear to be putting out heat as designed, but people working under them say they are cold. REFERENCE: Sec. 5-3(a) and (b).		
1. The wind or draft through area heated by infrared lamps is chilling workers.	*m*. Take ambient air temperature. Should not be more than 4 deg below comfort level. Check for drafts with smokemaker.	*m*. Provide an air heater for tempering drafts. Block drafts by barricades across corridors.
2. The infrared lamps should have larger Btu output or be closer together.	*n*. Check original calculations with units installed. Observe if effects cones overlap adjacent cones.	*n*. Increase number or size of infrared units. Effects cones must overlap if working area is to be kept comfortable.
COMPLAINT: Infrared lamps have very short lives. REFERENCE: Sec. 5-3(a) and (b).		
1. Infrared bulb is not suited to environmental conditions. Particularly, choose outdoor installation bulbs only as recommended by the manufacturer.	*o*. Check records for average life of lamp under the existing conditions. Compare with manufacturer's estimate of bulb life.	*o*. Change bulb models as may be indicated to increase life of bulbs.

DIVISION 6

Domestic Hot Water Systems

6-1. DIRECT-FIRED HEATERS

(a) *Discussion.* The majority of direct-fired water heaters are burning gas: natural, artificial, or liquefied petroleum. Oil-burning heaters are available but from relatively few manufacturers. The reason is largely economic, the initial overall investment favoring the self-contained gas-fired unit when compared with the oil burner that requires an oil storage tank and interconnecting piping.

Direct-fired water heaters may be chosen to function either with or without a hot water storage tank. The storage tank levels out the peaks on the burner and permits use of equipment having less than peak capacity, since the stored hot water provides the difference on demand.

Certain complaints are common to the direct-fired heater, whether the fuel is gas or oil. These are examined in the following paragraphs.

(b) *Complaint.* Following the opening of a faucet, a loud pounding or knocking seems to occur in the heater or in the hot water supply piping.

Causes

1. There are patches of scale inside the heater, causing the metal under the scale to overheat. The opening of the faucet produces a sudden brief reduction in pressure inside the heater, and this pressure drop is enough to allow water in the scale area to flash into steam at the existing temperature. These steam bubbles then collapse, setting up minute vacuums that fill with solid water to the tune of miniature thunder claps.

2. The reduction of pressure allows the formation of steam in the

pipeline in those instances of an overheated water supply. Generally, if steam as well as water emerges from a lavatory faucet, the water temperature is too high for the safety of users. Temperature should be dropped until the steam disappears. The travels of the steam bubbles inside the piping and the booming of their collapse are common phenomena where overheating is present.

3. Opening and closing of a faucet disturbs the pressure in the supply pipe, and this disturbance often signifies its presence by the familiar banging of water hammer. The meaning of this is that the supply pipe is too small and produces an excessive water velocity. A velocity of 4 ft per second is generally considered suitable, but there may be a greater speed than this if the flow was calculated on iron pipe dimensions and then served by copper tubing. To illustrate, the cross-sectional internal area of a $\frac{3}{4}$-in. iron pipe is 0.5335 sq in. and that of $\frac{3}{4}$-in. copper tubing, 0.4841 sq in., or a reduction of not quite 10 percent.

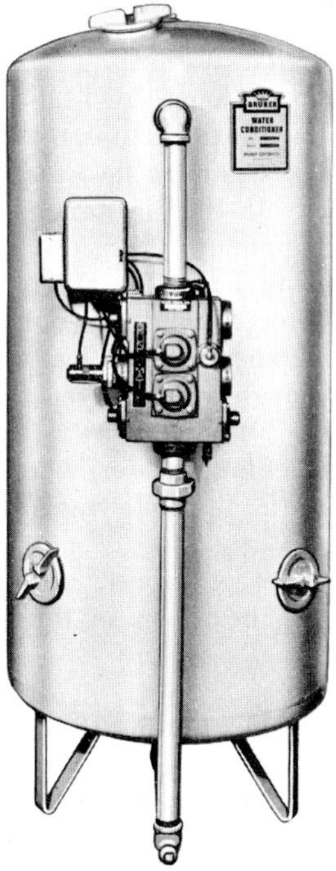

Fig. 6–1. Neutralizer for iron in water. (*Courtesy Bruner Corp.*)

Recommended Actions

1. If hammer is from overheating, reduce temperature.

2. If noise arises from excessive velocity, increase the pipe size or install a flow-limiting orifice or valve in the line.

3. If trouble is steam formations in the heater, take heater off the line and clean thoroughly. Cleaning will involve the use of acid, frequently a 50-percent solution of muriatic, to remove the scale.

(c) **Complaint.** The water is rusty.

Causes

1. Iron is in the water supply to the heater and is being carried along in solution.

2. Stray electric currents are pitting the inside surface of the heater or the storage tank.

3. The water is excessively charged with air, and when heated, the air is separated and adheres to the inner tank surfaces, causing oxygen corrosion.

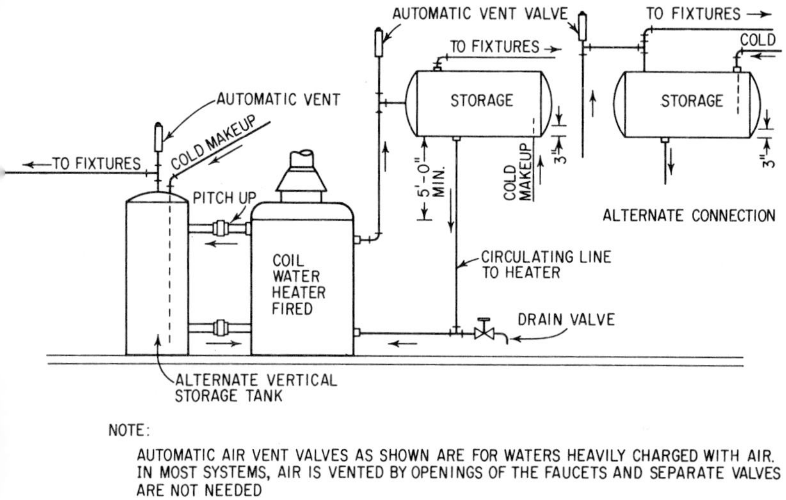

FIG. 6–2. Heater connections and vents.

Recommended Actions

1. Install an iron neutralizer in the cold water supply line to the heater, as illustrated in Fig. 6-1. This is a chemical iron coagulator.

2. For electrolytic pitting, install dielectric unions in the piping both ahead of and after the water heater or water storage tank. Alternatively, install magnesium rods in the heater or storage tank. Destruction of the rods preserves the heater or tank.

3. Air accumulates at high points; consequently, install an automatic air-vent valve where appropriate. Figure 6-2 suggests one possibility for the heater, another for the storage tank. Be sure the vent valve is suitable for the maximum pressure likely to be present.

(d) *Complaint.* The water is too hard for the lavatories, dishwashing, or laundry.

Cause

Water contains calcium and magnesium.

Recommended Action

Obtain a standard soap solution and test the water for hardness. On basis of test findings, size and install a water softener on the cold side of the heater. Include automatic control for the softener.

(e) *Complaint.* Hot water is very slow appearing at the faucet after the faucet is opened.

Causes

1. Supply piping is substantially distant from the heater or storage

tank and is a one-way connection. Water standing in pipe becomes cold between draws. If pipe is not insulated, this condition is aggravated.

2. Water pressure is too low in view of the frictional resistance to flow that is being imposed by the pipe. Also, pipe may be too small for a normal pressure.

3. Water temperature, measured at the heater or the storage tank outlet, is low to start with.

4. Demand for hot water at certain times exceeds the capacity of the heater or the storage available. There is not enough hot water to go around.

Recommended Actions

1. Check discharge desired from the faucet in gallons per minute and relate this to the size of the supply pipe by computing the normal time for water to reach the faucet from the heater or tank. Consider whether velocity is introducing excessive pressure drop and slowing down the flow.

2. If pipe size is correct and piping is bare, install insulation.

3. Convert single pipe to a two-pipe circuit, so that hot water moves through the pipe at all times by natural convection. If natural convection calculates to be unsatisfactory, install a circulating pump. This will keep hot water at the faucets at all times.

4. Try raising temperature of water at heater or in tank 10 deg at a time if presently under 130 F.

5. Increase size of storage tank or heater, in this order. Larger tank may meet requirements by carrying a relatively small heater over the peak.

(f) ***Complaint.*** Restaurant dishwasher gets cold water.

Causes

1. Heater is not generating water at the proper temperature for dishwashing. This should be at least 180 F.

2. Referring to Fig. 6-3, there is a possibility of cold water to the mixing valve finding a way to join the 180 F water to the dishwasher.

3. Heater without a storage tank does not have enough reserve generating capacity to meet the peak loads on the dishwasher.

Recommended Actions

1. Step up temperature of water leaving heater as necessary.

2. Referring to Fig. 6-3, install a check valve as shown. In addition, a flow regulator mounted in the cold water supply piping to the mixing valve limits the quantity of cold water that can be taken into the system for any temperature.

3. Calculate flow of 180 F water through the existing pipeline and determine whether this piping is large enough to accommodate the volume of hot water required. Amount of water needed per hour is

obtained from the manufacturer of the dishwasher or from his catalog. If friction loss is so high as to restrict the amount of water reaching the machine, either replace the piping with the correct size or install a hot water pump in the line.

4. If peak capacity of the heater alone is unable to meet the simultaneous peaks on the dishwasher and on the 140 F lavatory circuits, either install a hot water storage tank or provide an additional heater to handle one or the other of the water temperatures. This additional heater might be simply a booster unit *if* the heater that is unable to supply both temperatures at peak times is able to meet the total volume requirement at a common temperature of 140 F. Available space and the comparative economics may decide the form of correction to be taken.

(g) ***Complaint.*** Relief valve on water heater spills frequently.

Causes

1. Valve has accumulated lime crusts on the seat and cannot close tightly. Lime deposits form in hard-water areas when the temperature of the heater is carried above 150 F.

2. Pressure in the supply line to the heater fluctuates and at times exceeds the setting of the valve.

3. Valve also contains a thermostatic element and opens when water temperature exceeds the setting.

4. Valve is defective.

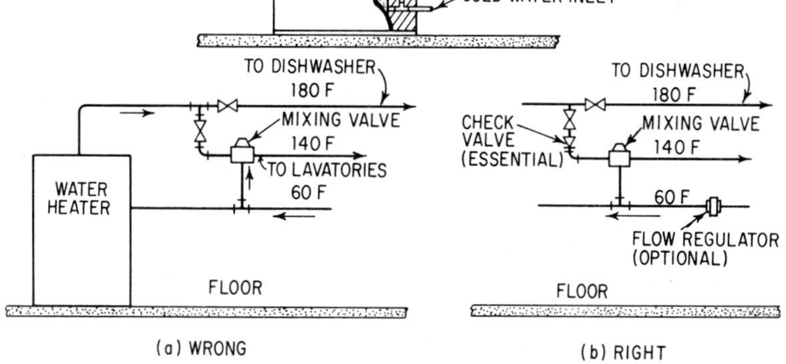

FIG. 6–3. Mixing-valve connections.

Recommended Actions

1. Check pressure in supply line. If possible, take pressure at a time when the valve is spilling. Install a pressure-regulating valve in the supply line if the pressure is not controlled prior to the entrance of the water service to the building.

2. Open valve and inspect seat and disk for liming. Clean and verify working condition of valve.

3. If thermostatic element is opening the valve, reset the thermostat to a higher temperature.

4. Replace a valve if found defective or designed for the wrong opening pressure.

(h) ***Complaint.*** Water carries a foreign material that accumulates in the bottom of a glass or bottle as a sediment.

Causes

1. Water supply has not been adequately filtered by the source of supply.

2. If water is being softened before entering the heater, the filter element in the softener may have become fouled and needs cleaning.

3. Sediment is a metallic oxide being generated in the heater by oxygen corrosion.

Recommended Actions

1. Request a filtering improvement from the water supply authorities. If unsuccessful, install a filter in the main supply line.

2. Backwash the softener. Establish a clear effluent.

3. Identify the nature of the sediment. If from corrosion, heater bottom probably is covered by a flocculent. Wash out heater after draining off by way of bottom drain valve. Set up a schedule of regular washdown, such as every 60 or 90 days. Install, or consider renewal of, magnesium rods.

(i) ***Complaint.*** Large hot water storage tank, as shown in Fig. 6-4, is served by direct-fired water heater, but tank does not fill with hot water.

Cause

This tank is served by natural circulation, and there is insufficient head to move the water from the heater to the tank. The minimum head (distance) between the bottom of the storage tank and the hot water outlet on the heater, in order to establish natural circulation, is 5 ft.

Recommended Actions

1. Relocate the tank by lifting it enough to provide the 5 ft of head.

2. Install a motor-driven circulator pump in the circuit between the bottom of the tank and the inlet of the heater.

3. Calculate the necessary flow of water in gallons per minute from the heater to the tank and compare this with the friction loss in the piping presently in place, expressed in milinches. If this friction loss per lineal foot is greater than the milinches per foot of head generated by the difference in temperatures between water leaving the heater and the water returning to the heater from the tank, the system will not work. Sometimes this situation can be remedied by increasing the size of the piping connections between the heater and storage tank, thus reducing the friction.

(j) ***Complaint.*** Hot water discharge from faucets turns cool almost immediately, as though storage tank is too small.

Causes

1. Cold water makeup to domestic water system is getting directly, or almost directly, into the hot line to the fixtures. See Fig. 6-4 for wrong locations of cold water connection and Fig. 6-2 for proper connections.

2. If this is a two-temperature system, the setting of the mixing valve could be incorrect.

Recommended Actions

1. Reconnect the cold water makeup piping to the tank as shown in Fig. 6-2, after verifying this as the cause of the trouble.

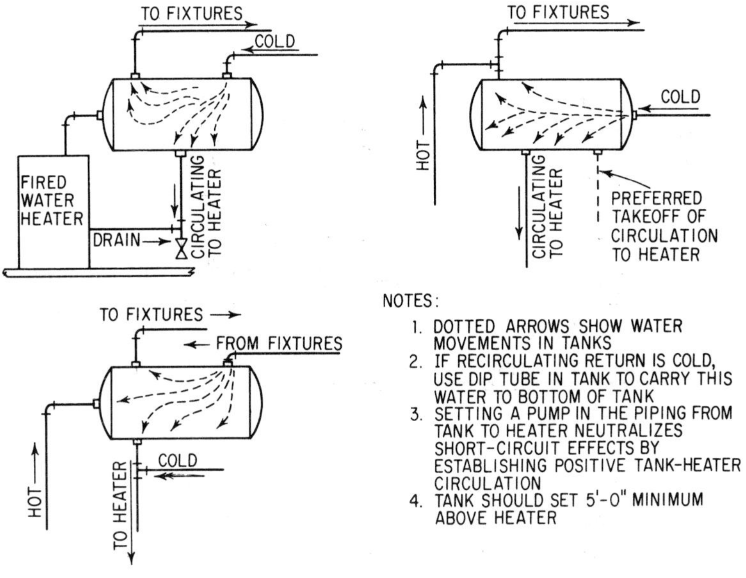

FIG. 6–4. Tank connections.

2. Check the temperature of the water leaving the mixing valve and adjust valve setting if the cold water dilution is excessive.

(k) ***Complaint.*** Pilot light goes out too frequently, resulting in a condition of inadequate hot water supply.

Causes

1. Pilot flame may be set too low, so that faint floor drafts are enough to extinguish it.

2. Draft hood is not suitable to the heater. See Fig. 6-5 for correct arrangements.

3. Vent pipe may be subject to downdrafts as the effect of higher structures nearby.

Recommended Actions

1. Adjust pilot flame.

2. Install correct form of draft hood.

3. Extend vent pipe to an elevation of at least 2 ft above any structure within 10 ft of the outlet.

(l) ***Complaint.*** In oil-fired heater, oil burner lights off with a loud pop and sometimes shows signs of a flareback.

Causes

1. Stack or chimney draft is inadequate, and the heater is not purged of combustion gases between firing periods.

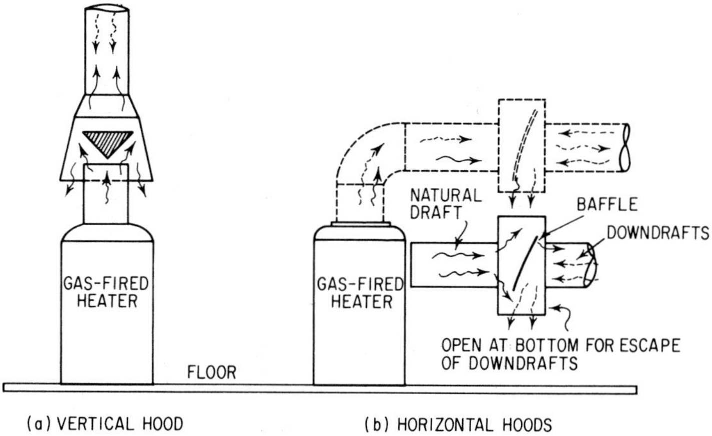

NOTES:
1. FOR MINIMUM DIMENSIONS OF HOODS SEE "AMERICAN STANDARDS" AS SPONSORED BY AMERICAN GAS ASSOCIATION
2. ALL HORIZONTAL BREECHINGS HEATER TO CHIMNEY, MUST SLOPE UP AT LEAST $\frac{1}{4}$ IN. PER FT
3. DOTTED LINES SHOW ALTERNATE CONNECTION

FIG. 6–5. Draft hoods.

2. Chimney terminal is too small or otherwise restrictive to the free venting of the combustion gases.
3. Oil burner is set with a too-rich mixture.
4. Gas passages and chimney flue are heavy with soot.
5. Ignition transformer is not furnishing a suitable spark.

Recommended Actions

1. Measure draft at outlet of heater with a draft gage. Compare measured draft with the calculated friction loss from burner to chimney outlet.
2. Check for draft loss as a result of excessive sooting, too short a chimney, or too small a flue. If soot is the source of trouble, clean chimney by dropping brick through on end of line, jiggling the brick to dislodge soot from sides. Then finish up by vacuuming the flue. If chimney is too short, extend it as necessary, when surroundings and appearance will allow. If too small in cross-sectional area, install a draft inducer in the breeching between the heater and the chimney.
3. For smoking, adjust burner to produce a clean fire. It may be necessary to install a new and smaller burner.
4. Check spark from ignition transformer. If inadequate, replace the transformer. Make sure that spacing of electrodes is correct and has not been opened too far.

Table 6-1. Troubleshooter's Field Guide for Direct-fired Domestic Hot Water, General

Possible cause	How to check for cause	Recommended ways to correct
COMPLAINT: A loud knocking follows the opening of a faucet.		
REFERENCE: Sec. 6-1(b).		
1. Scale patches in the heater overheat.	*a*. Examine water from heater drain for bits of scale.	*a*. Clean with a 50 % solution of muriatic acid.
2. Steam bubbles are present in the pipeline.	*b*. Reduce temperature of water and note whether noise stops.	*b*. Leave system operating at a nonsteam temperature.
3. Water hammer from variations in water pressure follows the opening and closing of faucet.	*c*. Operate the faucets slowly. Note whether this eliminates noise. Check flow velocity for possible excess	*c*. If excess velocity, increase pipe size or install flow limiters. Install new shock absorbers or recharge

Table 6-1. Troubleshooter's Field Guide for Direct-fired Domestic Hot Water, General (*Continued*)

Possible cause	How to check for cause	Recommended ways to correct
	above 4 ft per second and causing hammer.	waterlogged existing absorbers.
COMPLAINT: Water is rusty. REFERENCE: Sec. 6-1(c).		
1. Cold water supply to heater is carrying iron in solution.	*d.* Draw sample and make chemical test for iron. Refer test to laboratory.	*d.* Set up removal station (chemical filter) ahead of heater.
2. Stray electric currents are pitting the inner shell of the heater or storage tank.	*e.* Scale and pits are similar to oxygen corrosion. Look for dissimilar metals in contact with water. Also grounds.	*e.* Install dielectric unions on both sides of heater. Also install sacrifice rods.
3. Water is full of air which causes oxygen corrosion and forms iron oxide (rust).	*f.* Listen for possible sounds of air at high points. Draw water sample, cold, and heat up in a beaker to heater temperature. Note separation of air, if any.	*f.* Vent heater, tank, or pipeline, as appropriate. See Fig. 6-2.
COMPLAINT: Water is too hard. REFERENCE: Sec. 6-1(d).		
1. Water contains substantial amounts of calcium and magnesium.	*g.* Check with standard soap solution which indicates hardness. (Solution can be mixed by druggist.) *	*g.* Install zeolite or other suitable softener ahead of heater.
COMPLAINT: Hot water is very slow in appearing at a turned-on faucet. REFERENCE: Sec. 6-1(e).		
1. Faucet is served by single supply pipe	*h.* Inspect piping arrangement and mea-	*h.* Insulate bare pipe. Consider converting

* For details of procedure in making the soap test, see the U.S. Department of the Interior publication "Questions and Answers on Boiler Feed Water Conditioning." This is Q & A Handbook No. 3.

Table 6-1. Troubleshooter's Field Guide for Direct-fired Domestic Hot Water, General (*Continued*)

Possible cause	How to check for cause	Recommended ways to correct
and is a considerable distance from source of hot water.	sure length of line. Note whether pipe is bare.	to return pipe system.
2. Same as above and pipe is not insulated.		
3. Water pressure is too low.	*i.* If pressure cannot be increased, calculate flow to determine if supply pipe is too small.	*i.* Consider return pipe plus a circulating pump. Increase pipe size.
4. Water temperature is so low that line heat losses hold up hot water delivery to faucet.	*j.* Increase water temperature but not above 140F at the source. Note whether this improves conditions.	*j.* If test is satisfactory, leave system set at higher temperature.
5. Total water demand exceeds capacity of heater, or heater plus storage tank.	*k.* Check flow of hot water at nonpeak times. Time appearance of hot water at faucet after faucet is turned on.	*k.* Increase capacity by addition of booster heater or new peak storage tank.
COMPLAINT: Restaurant dishwasher gets cold water.		
REFERENCE: Sec. 6-1(f).		
1. Heater is not generating hot water at 180 F.	*l.* Check temperature at source and at dishwasher.	*l.* Step up temperature as needed.
2. On a two-temperature system, cold tempering water is getting into the 180 F supply through the mixing valve.	*m.* Take the temperature of the dishwasher water before and after the mixing valve.	*m.* See Fig. 6-3. Install check valves as shown.
3. There is insufficient generating capacity	*n.* Check hot water needs of dishwasher	*n.* Install a booster heater on the 180-F

Table 6-1. Troubleshooter's Field Guide for Direct-fired Domestic Hot Water, General (*Continued*)

Possible cause	How to check for cause	Recommended ways to correct
to provide 180 F at peak periods.	as called for by the machine manufacturer. Compare with available volume at 180 F.	line after the mixing valve.

COMPLAINT: Relief valve on water heater (or storage tank) spills frequently.
REFERENCE: Sec. 6-1(g).

Possible cause	How to check for cause	Recommended ways to correct
1. Lime deposits on seat of valve prevent tight closure.	*o*. Open valve and inspect condition.	*o*. Clean or replace to suit.
2. Pressure in water supply line to heater fluctuates in excess of design operating pressure of valve.	*p*. Check the pressure. Check operating design of the valve.	*p*. Install pressure control valve in water makeup piping.
3. Thermostatic element in valve is operating.	*q*. Check water temperature against setting of thermostatic element.	*q*. Reduce water temperature to stay within limits. Step up thermostatic setting if temperature is not excessive.
4. Relief valve is defective.	*r*. Examine and test.	*r*. Replace if found defective.

COMPLAINT: Water carries a sediment.
REFERENCE: Sec. 6-1(h).

Possible cause	How to check for cause	Recommended ways to correct
1. Water entering the heater has not been adequately filtered.	*s*. Take sample and let stand for 24 hours.	*s*. Take steps to ensure effective filtering.
2. Sediment is a metallic, corrosive oxide or scale.	*t*. Identify the material using a laboratory's services if necessary.	*t*. If corrosion, correct as previously directed.
3. Sediment is soft.	*u*. Identify. If organic, look for a polluted water supply, such as a crossed sewage—	*u*. Correct on premises or on incoming service as appropriate.

Table 6-1. Troubleshooter's Field Guide for Direct-fired Domestic Hot Water, General (*Continued*)

Possible cause	How to check for cause	Recommended ways to correct
	potable water seepage.	
COMPLAINT: Storage tank in Fig. 6-2 does not fill with hot water as intended. REFERENCE: Sec. 6-1(i).		
1. Arrangement of tank and heater denies tank the necessary head of at least 5 ft needed to generate natural circulation of hot water to the tank.	*v*. Relative positions of tank and heater show this to be a defective arrangement.	*v*. Elevate the tank to provide at least 5 ft of distance vertically between bottom of tank and top of heater. Alternatively, leave tank as is, but install a circulating pump between tank and heater.
COMPLAINT: Storage tank appears to have sufficient head, but natural circulation is inadequate to fill the tank. REFERENCE: Sec. 6-1(i).		
1. Piping between the heater and tank is too small, thus introducing excessive friction in the circuit.	*w*. Calculate flow of hot water in gpm, and compare with friction loss in existing piping, expressed in milinches.	*w*. Either increase size of piping or install a circulating pump.
COMPLAINT: Storage tank gets hot, but water quickly turns cool at faucets. REFERENCE: Sec. 6-1(j). Also Figs. 6-2 and 6-4.		
1. Cold water makeup to system enters tank at a location that results in short-circuiting of cold water to hot line to fixture.	*x*. Take temperature of hot pipe at tank outlet when faucets are closed and note speed with which the temperature falls when faucets are opened. If mixing valve is involved, verify correct valve setting using a surface-type thermometer.	*x*. Relocate cold water connection as shown in Fig. 6-2. Adjust mixing valve if necessary.

Table 6-1. Troubleshooter's Field Guide for Direct-fired Domestic Hot Water, General (*Continued*)

Possible cause	How to check for cause	Recommended ways to correct
COMPLAINT: Pilot light on gas burner goes out too often. REFERENCE: Sec. 6-1(k).		
1. Pilot flame is set too low.	*y*. Check by observation.	*y*. Increase flame.
2. Unsuitable draft hood is being used.	*z*. Inspect hood and compare with Fig. 6-5.	*z*. Replace with correct hood.
3. Vent pipe is too short and is promoting downdrafts.	*aa*. Inspect mouth of vent pipe with respect to nearby structures.	*aa*. Extend vent pipe 2 ft above surroundings. Install a draft inducer, if nothing else works.
COMPLAINT: Oil-fired heater burner is explosive in lighting periodically under control of the thermostat. REFERENCE: Sec. 6-1(l).		
1. Natural draft is inadequate.	*bb*. Measure with draft gage and compare with minimum draft needed. Determine cause as too short a stack, too small cross-section of flue, or flue choked with soot. Ignition spark too thin.	*bb*. Clean flue with brick on cord. Use vacuum. Increase chimney height. Install a draft inducer. Replace or repair ignition transformer.

(m) ***Special Caution for Gas-fired Heaters.*** In troubleshooting gas-fired heaters that are burning liquefied petroleum gases, always remember that propane and butane gases are heavier than air and will sink to the lowest ambient level when escaping from a leak. Therefore, in working on a liquefied petroleum gas installation, make sure the area is thoroughly ventilated, paying particular attention to the atmosphere immediately above the floor. Do not under any circumstances introduce an open flame where any gas, liquefied petrolem, natural, or artificial, *could possibly be present.* For natural and artificial gases, which are lighter in density than air, the upper strata of the atmosphere, where work is to be undertaken, becomes the space in which the most dangerous concentration will be found.

With any gas-fired appliance, the second important step is to verify the correctness of the installation under the applicable codes. These will be, as appropriate, local codes for gas fuel, the National Board of Fire Underwriters, the Interstate Commerce Commission, the ASME Code for Unfired Pressure Vessels, the American Gas Association, and the National Fire Protection Association.

In purging the system, always carry the purge pipe to the outdoors. Follow the purging procedures set up by the American Gas Association in "Purging Principles and Practice." This publication should be in the library of every troubleshooter.

Do not leave the immediate vicinity of an operating heater at any time if, during the troubleshooting procedure, the safety and limit controls have been temporarily disconnected or otherwise made inoperative. Failure to observe this rule has resulted in fires and explosions.

6-2. UNFIRED WATER HEATERS

(a) ***Discussion.*** Unfired water heaters are simply heat exchangers in which the source of heat is steam, hot gas, or hot water generated elsewhere. In form, they are available as instantaneous hot water producers complete in themselves, as indirect units closely connected to a steam or hot water boiler, and as storage heaters, in which the heating coil is commonly submerged in the storage tank.

The instantaneous, complete-in-itself heater is shown in Fig. 6-6; the indirect, in Fig. 6-7; and the storage type with submerged coil, in Fig. 6-8.

Troubles commonly reported with the unfired units include lack of capacity, corrosion, and noise. How to deal with these troubles will be described in the following subsections.

(b) ***Complaint.*** Steam heat exchanger, sized correctly, fails to deliver rated output of hot water from time to time.

Causes

1. Condensate trap finds itself attempting to discharge against a back pressure for which the trap was not designed. This back pressure is of fluctuating value and is caused by other traps of equally high pressure condensate, or possibly higher, discharging into the common condensate system.

2. Steam admission control valve is operating erratically because of foreign material in the valve, diaphragm trouble, or thermostat bulb in a wrong location.

3. Steam supply carries too much water in form of slugs.

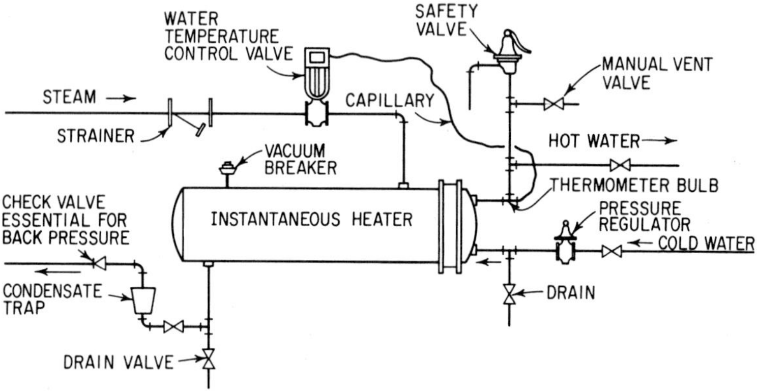

FIG. 6–6. Instantaneous heater.

Recommended Actions

1. Examine the condensate line from the exchanger all the way to the final receptacle of the condensate. Note whether other traps are discharging into this line and check the various pressures of the discharge. If this is the situation, install a check valve in the discharge branch of each trap before the branch ties into the common line.

2. Open steam admission control valve, inspect, check, and repair or replace as necessary.

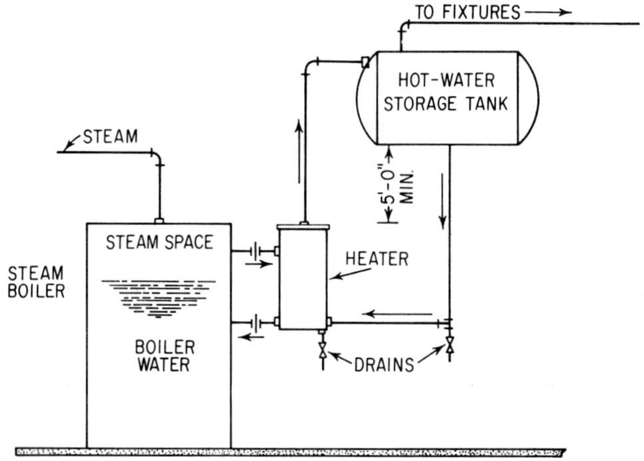

NOTE:

IF 5'-0" HEAD CANNOT BE PROVIDED FOR STORAGE TANK, INSTALL PUMP IN PIPING FROM TANK TO HEATER

FIG. 6–7. Indirect water heater.

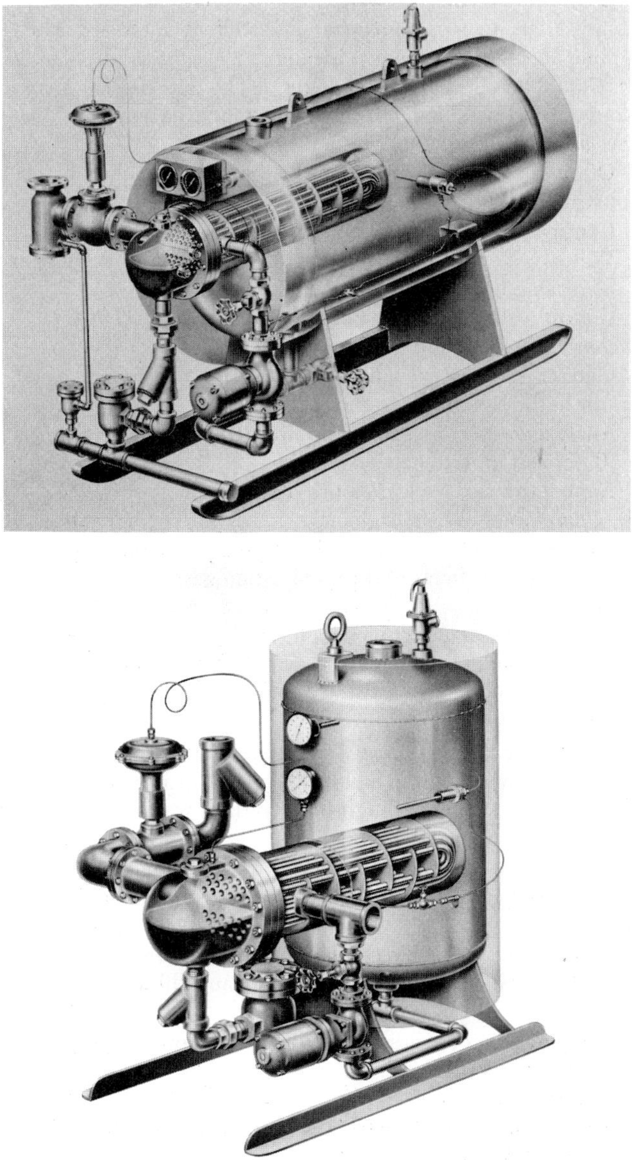

FIG. 6–8. (a) Phantom view of horizontal storage water heater. (b) Steam coils as fitted into a vertical storage water heater. (*Patterson-Kelley Co., Inc.*)

3. Examine steam generator for priming and foaming. Take necessary steps to stop carryover. Install a water separator in the steam line ahead of the heat exchanger.

(c) ***Complaint.*** Steam exchanger as in Sec. 6-2(b) *never* meets its rated output. Exchanger is operating on exhaust steam.

Causes

1. Oil in exhaust steam is gradually coating tubes of exchanger and obstructing efficient heat exchange.

2. Continual back pressure on the condensate discharge trap.

3. Water is depositing scale on the inside of the exchanger tubes, thus decreasing heat exchange.

4. Control valve defects as noted in the previous section.

Recommended Actions

1. Check inside of exchanger shell for oil. If found, install an oil separator in the steam line to the exchanger.

2. Analyze water for scale-forming character. Wash out tubes with 50-percent muriatic acid solution.

3. For control and back-pressure troubles, see Sec. 6-2(b).

(d) ***Complaint.*** Heat exchanger is served by hot gas originating in a waste-heat boiler. Tubes of exchanger corrode rapidly and expensively from the effects of the combustion gases.

Cause

Gases are carrying sulfur which, in contact with moisture from atmospheric condensation or other sources, becomes sulfuric acid and attacks the tubes.

Recommended Actions

1. Retube the heat exchanger using Monel tubes.

2. Switch to a sulfur-free fuel for the waste-heat boiler.

3. Substitute steam or high-temperature water as the source of heat in the exchanger.

(e) ***Complaint.*** Heat exchanger is a coil submerged in the boiler water. Output of hot water at peak periods is insufficient.

Causes

1. Coils of this type have relatively little storage capacity; consequently, the available hot water is used and diluted rapidly at peak times.

2. Temperature of water in boiler is too low to produce the full rating of the submerged coil.

Recommended Actions

1. Add a hot water storage tank to the system which the coil can fill between peak demands.

2. Step up temperature of water in boiler if basic purpose of the boiler allows operation at higher temperatures.

3. Install an auxiliary heater that can make up the temperature needs during peaks. This can be a fairly expensive treatment.

Table 6-2. Troubleshooter's Field Guide for Unfired Domestic Hot Water Heaters

Possible cause	How to check for cause	Recommended ways to correct
COMPLAINT: Heater fails to deliver rated output from time to time. REFERENCE: Sec. 6-2(b).		
1. There is back pressure on the condensate trap from the heat exchanger.	*a.* Open trap and inspect. Examine piping for other trap discharges.	*a.* Install check valves in each trap discharge before connection to common condensate line.
2. The steam admission valve is misperforming because of scale, puncture in diaphragm, etc.	*b.* Open valve and inspect.	*b.* Repair or replace valve as necessary.
3. Boiler is priming and sending over slugs of water.	*c.* Listen to steam line with stethoscope for water moving in pipe. Open pipe and check steam quality with calorimeter.	*c.* Eliminate cause of boiler priming. Install separator in steam line to exchanger.
COMPLAINT: Same as preceding complaint except that failure to meet rating is constant and steam is exhaust steam. REFERENCE: Sec. 6-2(c).		
1. Oil in steam is coating and heat insulating the tubes.	*d.* Check inside of exchanger for oil film.	*d.* Install oil separator.
2. There is continuous back pressure on condensate trap.	*e.* Open trap; check pressure on condensate line.	*e.* Install check valves as in 1(a) above.
3. Water is leaving scale on inside of exchanger tubes.	*f.* Open head of exchanger and inspect tubes. Analyze water.	*f.* Wash tubes with 50 % muriatic acid. Soften water if not too costly.

Table 6-2. Troubleshooter's Field Guide for Unfired Domestic Hot Water Heaters (*Continued*)

Possible cause	How to check for cause	Recommended ways to correct
COMPLAINT: Hot gas from waste-heat boiler is corroding exchanger tubes. REFERENCE: Sec. 6-2(d).		
1. Combustion gases are carrying sulfur.	*g*. Analyze gas or corroded tubes.	*g*. Retube the exchanger using Monel metal tubes. Switch to a sulfur-free fuel. Substitute steam or a high-temperature water in place of waste-heat gas.
COMPLAINT: Coils submerged in boiler fail to deliver enough water at peak demands. REFERENCE: Sec. 6-2(e).		
1. There is not enough storage in boiler coil. Hot water is soon exhausted.	*h*. Take temperature of leaving hot water at start of run and note length of time before it cools.	*h*. Install a storage tank.
2. The temperature of water in boiler is too low.	*i*. Note temperature of boiler water and how increase of temperature is possible.	*i*. Step up temperature of boiler as necessary or permissible.

6-3. ELECTRIC WATER HEATERS

(a) *Discussion.* Since electric water heaters require no chimney connection or piping for fuel and are relatively easy to install, they have been widely accepted by the domestic market.

Troubles with electric water heaters duplicate those encountered with other types of heaters. They develop scale; they corrode; relief valves leak; the output of hot water is, or becomes, inadequate; hot water is a long time in reaching the faucets; water is rusty.

When a trouble call comes in, the first step of the troubleshooter is to ascertain the time the unit has been in service. If more than 5 years and if still within the guarantee period, it will probably be most economical for the owner to replace the defective unit with a new one.

If less than 5 years of service has been completed, the nature of the trouble will suggest what is to be done and how far the corrective procedure can go and still be the best economy. One aspect of the complaint should always be kept in mind: owners rarely take the trouble to drain sediment and scale from the heater drain valve every 60 days as they should. Thus the discharge of rusty water can indicate simply a dirty heater that can be cleaned readily by the serviceman. Liming of a safety relief valve, likewise, can be a result of the no-drainage program.

The Troubleshooter's Field Guide (Table 6-3) summarizes what

Table 6-3. Troubleshooter's Field Guide for Electric Domestic Water Heaters

Possible cause	How to check for cause	Recommended ways to correct
COMPLAINT: 1. Loud knocking follows the opening of a faucet.		
COMPLAINT: 2. Water is rusty.		
COMPLAINT: 3. Water is too hard.		
COMPLAINT: 4. Hot water is slow to appear at faucet.		
COMPLAINT: 5. Dishwasher gets cold water.		
COMPLAINT: 6. Relief valve spills.		
COMPLAINT: 7. Water carries a sediment.		
REFERENCE: All these complaints are examined and corrections suggested in Table 6-1 for fired domestic water heaters.		
COMPLAINT: Water is dripping from the heater somewhere under the insulation.		
REFERENCE: Discussion, Sec. 6-2(a) and 6-3(d).		
1. Heater tank has corroded through.	*a*. Note age of tank and if tank contains a zinc or magnesium sacrificial rod.	*a*. Replace heater in total.
COMPLAINT: Water does not heat.		
REFERENCE: Sec. 6-2(a) and 6-3(a).		
1. One or both of the electrodes has burned out.	*b*. Check electric service energy to electrodes for interruption of service. If service is energized, pull electrodes and replace defective units.	*b*. Restore service as necessary. Replace defective electrodes as necessary.
2. Fuse has burned out or breaker has de-energized the heater circuit.		

FIG. 6–9. Horizontal storage type of electric water heater. (*Patterson-Kelley Co., Inc.*)

can go wrong with the electric heater and suggests the methods of correction.

See Fig. 6-9 for a horizontal electric storage heater.

6-4. SOLAR WATER HEATERS

(a) ***Discussion.*** Originally, solar water heating was assumed to be restricted to the predominantly sunny areas of the country, such as Florida, California, and Arizona. In more recent time, solar installations are being made for summer vacation homes wherever the summer is reasonably accompanied by clear weather.

Solar systems, to be satisfactory, must include a storage tank large enough to provide hot water for at least one overcast day and one night in a row. The size of the storage tank can be estimated by this simple formula:

$$\text{Tank size, gal} = \frac{\text{av daily use} \times 0.50}{0.75}$$

where av daily use = gal per person × no. of persons

0.50 = assumed storage minimum of 50 percent of daily use

0.75 = percentage of tank water that can be drawn off hot before it begins to grow too cool for intended use

Thus if five people use 30 gal a day each, storage must be at 50 percent of this or 75 gal. Since only 75 percent of those 75 gal can be counted on to stay hot, 75 gal ÷ 0.75 = 100 gal, the proper size storage tank for this house. Larger tanks can be used, of course. All tanks should be well insulated to prevent a too-fast cooling of the water.

The other units in the system are a coil of copper or iron pipe, usually painted black to increase the absorption of sun heat, and a "hot box" to contain the coil.

Placement of the hot box should be as necessary to run the tubes of the coil from east to west and to avoid the impact of house or tree shadows on the box at all hours of the day. This means a location on the south side of the building.

Figure 6-10 illustrates a solar arrangement with the coil mounted on a porch roof and the storage tank in the attic, as close to the roof peak as possible. This is to give the circulation from tank to coil and tank again the maximum of circulating head.

Other relative positions are possible, both the tank and coil being ground-set when necessary, but in this arrangement, the natural cir-

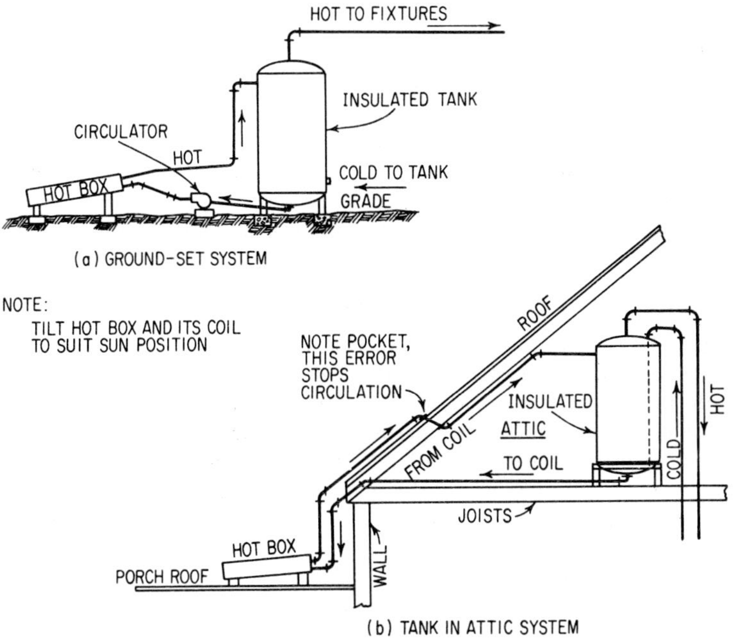

FIG. 6–10. Solar water heating.

culating head will be insufficient to overcome pipe friction, and a circulating pump must be included.

As might be expected, complaints usually take the form of insufficient hot water. Here are the reasons for such conditions and what may be done about them.

Causes

1. Circulation is sluggish as a result of too much pipe friction. This excessive friction may follow the installation of piping that is too small between the tank and coil. Or the friction can be contributed by an excessive footage of horizontal pipe runs.
2. There is shade trouble. Relocating of the coil to keep out of the shade might be the only solution.
3. Coil, tank, or both have been undersized.
4. Glass cover of hot box is not air tight, allowing a loss of heat by air movement through the box.
5. Inside walls of box are not painted white to reflect heat against the coil.
6. Circulating riser from coil outlet to tank does not rise everywhere and dips down in one or more places, creating an air pocket that blocks the circulation. See Fig. 6-10.

Recommended Actions

1. Inspect the entire piping run to the tank, using a level if expedient, and note the presence of any air pockets or overlong horizontal pipe runs. Correct as necessary.
2. Check the coil box for shade and paint. Relocate to keep out of shade and paint the inside of the box as noted and the coil black.
3. Check the size of the storage tank and the capacity of the coil and hot box. Refer to the latest edition of the ASHRAE Guide for information and guidance on the sizing of the coil and box.

6-5. HOT WATER DISTRIBUTION SYSTEMS

(a) ***Discussion.*** Troubles sometimes charged against the hot water distribution system actually may be originating elsewhere, with the symptoms apparently in the distribution. For example, consider the situation of an apartment house in which there is a definite peak of hot water demand. If this occurs simultaneously with a condition of low pressure on the main supply, many of the users will find themselves short of water. The causes of such complaints follow.

Causes

1. City water supply systems tend to outgrow their capacities as the city grows. This shows up in distribution street mains that are

too small to carry sufficient water at a sufficient pressure to reach the upper floors of the apartment house.

2. Much less frequently, the building distribution piping is undersized; consequently, those faucets at the end of the line find themselves without water or at least without sufficient water. The undersizing in these cases is usually the result of calculating pipe sizes on a certain expected pressure at all times and of this pressure failing to arrive.

Recommended Actions

1. If the city cannot increase the pressure, the alternative is for the apartment house to install a hot water circulating pump to boost the pressure and delivery as needed. This contingency is so common that architects and engineers are accustomed to designing large buildings with water circulating pumps from the beginning.

2. Provide a hot water peak tank, mount it on the roof, and thus feed the distribution system from both ends at peak periods.

3. Occasionally the troubleshooter will come across a recirculating system that depends for its recirculation on gravity, that is, the difference in density between hot water leaving the tank and the cooler water returning. Circulation in this sort of system can be boosted by insulating the hot lines while leaving the return piping bare. For gravity systems already insulated in this manner, the only sure way to improve circulation is to install a circulating pump.

(b) *Complaint.* Although the storage tank is full of hot water, lavatory faucets discharge only cold. The system is a gravity return. Sometimes hot water appears in the showers.

Cause

Cold water is getting into the hot water piping as the result of some cross connection. Because of its greater density, the cold water takes over control of the flow, either diluting the hot water or blocking it off entirely.

Recommended Actions

1. Examine the shower mixing valves. A valve as illustrated in Fig. 6-11 is likely to develop leaky tendencies as it grows old in service and to permit the cold water supply to push the hot water right back out of the valve. The only cure for this situation is to replace the valve internals, if these are available, or else replace the entire mixing valve with a new one.

2. "Walk the hot water line" from tank to faucets, meaning follow the hot piping with the bare hand to locate a point of cold water junction. This may be difficult if the hot water piping leads through attic and crawl spaces or is buried underground. If the

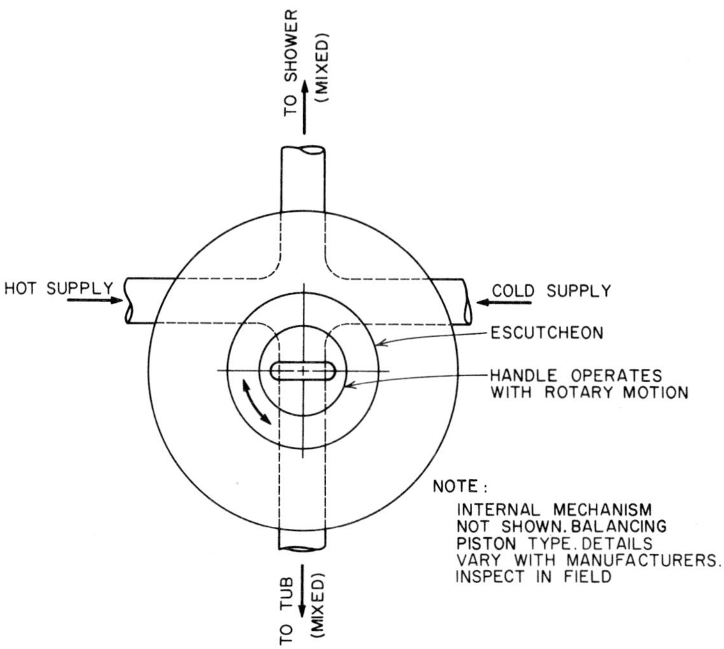

FIG. 6–11. Shower mixing valve.

evidence, without digging up a buried pipe, indicates reasonably that the cold dilution is being accomplished underground, consider replacing the buried section with a new length to be installed above ground, the buried section being blanked off. This may be a cheaper cure than excavation.

6-6. QUESTIONS AND ANSWERS

QUESTION 1. Are the automatic water hammer arresters satisfactory for a domestic hot water system? Are they worth the cost?
ANSWER 1. Yes, to both questions. However, when installing the air-cushion type of arrester, be sure that it is mounted in an accessible location, since this style may need an air recharge from time to time.

QUESTION 2. Are plastic dip tubes for domestic water heaters acceptable under the various codes?
ANSWER 2. Some, but not all. Check the codes applicable at the site of the water heater installation. In any event, make sure the

heater will not generate hot water in excess of the dip-tube temperature rating.

QUESTION 3. How do you clean a hot water generator?

ANSWER 3. For a vertical storage tank type of heater, a regular schedule of draining until the water appears clear at the drain valve usually will be enough if carried out every 60 to 90 days, depending on the character of the water. A sequence for scale removal is (1) drain tank; (2) connect hose to drain valve and wash out tank with water under pressure; and (3) if practical, vary temperature of wash-out water from hot to cold, with the temperature change being made abruptly, in order to crack loose the scale.

For heaters in which the water being heated passes through a coil which is heated by direct combustion, flush coil with a 50-percent solution of muriatic acid. Alternatively, several proprietary cleaners are available which may be used instead of the muriatic acid. An acid-circulating pump is recommended for service with the proprietary fluids; this is a small unit weighing less than 4 lb, and in no dimension greater than 5 in.

QUESTION 4. Is a drip pan considered essential under a storage type of water heater?

ANSWER 4. Depends on tank location. If leakage from the tank should occur and damage floors or property, the provision of the drip pan with a suitable drain pipe is absolutely necessary as simple insurance. This is particularly true for attic locations where leakage has been known to seep through the floor and damage the room below and its furnishings. A drip pan with a 1-in. raised lip is inexpensive and can avoid later complaints.

DIVISION 7

Warm Air Heating Systems

7-1. GRAVITY SYSTEMS

(a) ***Discussion.*** Since a gravity air movement works on the principle that warm air rises and relatively cool air sinks, gravity furnace installations are found only where the furnace can be located in a basement or where the spaces to be heated are above the furnace.

This form of heating system is now considered obsolete, and new installations are rare, but since many older buildings are still heating on the gravity cycle, the troubleshooter will meet a gravity problem sooner or later.

For furnace and combustion troubles, refer to Division 2 and the appropriate section. Distribution troubles are examined in the following subsections.

(b) ***Complaint.*** One or more rooms fail to heat in severe cold, although the dampers in the supply ducts are wide open.

Causes

1. Registers in these chilly rooms are at the ends of long horizontal supply ducts. Friction in these long horizontal ducts is too great for an adequate flow of warm air to overcome.

2. Return registers are too high in the wall or are wrongly placed, so that air stagnates on the floor or seeks another route of return, perhaps through an adjacent hallway.

3. If installation has an outside air makeup duct, the damper in this duct is wide open, thus reducing the effectiveness of the furnace.

4. Air temperature leaving the furnace is too low. Generally, the desired temperature at the room registers is 175 F.

Recommended Actions

1. Check air temperature at the register outlet. If much below

175 F, increase combustion rate if furnace and chimney have the increased capacity. Remember that fires can result from overheated chimneys and smoke pipes, and check temperatures carefully.

2. Examine horizontal ducts for possibilities of increasing the upward pitch. Consider insulating the horizontal ducts to retain heat and installing larger horizontal ducts to reduce friction.

3. Balance the entire system and adjust dampers to prevent oversupply of warm air to favored ducts.

Caution: One leader from the furnace must be nondampered and open at all times to protect the furnace from overheating.

(c) ***Complaint.*** The system is dirty and distributes dust throughout the building.

Causes

1. There is no effective filter on the air entering or leaving the furnace.

2. Air passages around the furnace radiator are not being cleaned periodically. These accumulations of dust are the source and reason for the explosions that occasionally eject smoky air from gravity registers.

Recommended Action

Clean the furnace air passages thoroughly at least once every 60 days. Air filters are not practical for gravity systems because of the frictional resistance they introduce into the air circuit.

(d) ***Complaint.*** Floor furnace located in central hallway supplies heat only to the immediately adjacent rooms. Remainder of building stays cold.

Cause

Warm air from a floor furnace cannot be expected to circulate effectively through more than one doorway.

Recommended Actions

1. By setting an electric fan in the doorway between a warm room and a cold one, the warm air can be induced to enter the cooler room in limited volumes.

2. Install some form of auxiliary heater to serve the cold areas of the building. This can be in the form of an electric unit, a gas-fired room heater, or, as decided in some instances, a second floor furnace.

7-2. FORCED CIRCULATION SYSTEMS

(a) ***Discussion.*** Circulating warm air through ducts by means of a fan produces a great improvement over the atmospheric control obtained with gravity heating. However, forced circulation introduces the additional sources of trouble that come with the new components of the system, such as the fan or blower, the filters, the diffusers, grilles and registers, and the automatic control arrangements.

Troubles originating in the furnace are discussed in Division 2; ailments elsewhere, with their causes and recommended treatments, are examined in the following paragraphs and sections. Combination heating and cooling system troubles are cataloged in Division 9, which covers air conditioning, primarily because new forms of machinery are involved in air conditioning and because ducts designed strictly for heating are too small for service on an air-conditioning cycle. This is because a pound of air on cooling duty with a temperature differential of 20 deg is handling only 4.8 Btu, whereas that same pound of air working on a difference of 80 deg between 150 F at the register outlet and a room temperature of 70 F, delivers 19.2 Btu. In short, we cannot do so much with a pound of air on cooling as we can on heating.

(b) ***Complaint.*** Ceilings and walls are smudged around the diffusers and registers.

Causes

1. Warm air is carrying dust that settles out when the air makes contact with a relatively cool surface. For example, the ceiling around a diffuser, and above a room at 72 F, probably has a surface temperature of about 82 to 85 F, and the air emerging from the diffuser is about 150 F.
2. The air filter is dust-logged and requires cleaning or replacing.
3. The interiors of the ducts have not been cleaned for perhaps years and have served as accumulators for dust, dirt, and lint. These reservoirs of dirt provide the materials that ride on the air stream and stick to the ceilings and walls.
4. Room dust is picked up by secondary air currents and comes to rest on the ceilings and walls.

Recommended Actions

1. Set up a wiping schedule under which the surface areas around diffusers and registers are wiped clean with a damp cloth at least once a month. Certain proprietary cleaning solutions may be used on the cloth with good results.
2. Vacuum the diffuser and register outlets monthly.
3. Clean or replace the filters.

(c) ***Complaint.*** System is noisy.

Causes

1. Velocity of air is too high through ducts, through return grilles, or through both.
2. Blower noise is being transmitted through the ducts.

Recommended Actions

1. By use of an anemometer or velometer, determine volumetric

output of each noisy register and relate this to the size of the supply duct and to the clear area of each register. If velocity is higher than is recommended for the particular environment involved, as presented in the ASHRAE Guide, evaluate suitability of reducing the velocity.

2. For return grilles, maximum face velocity for silent operation is not over 500 ft per minute. Verify performance against this standard.

3. For the transmission of machinery noise, check mounting of machinery and note whether vibration is causing noise and whether this vibration is getting into the building structure as well as through the ducts. In any event, vibration can be stopped, so far as transmission is concerned, by setting the machinery on vibration insulators.

4. Install sound absorbers in the supply ducts if these are large enough. Otherwise such absorbers may require new and larger ducts to accommodate them.

(d) ***Complaint.*** System delivers odors to various spaces in the building.

Causes

1. Odors are coming from dust, dirt, and lint accumulations in the ducts.

2. A small animal such as a rat or mouse has crawled into a duct and died.

3. The odors are originating outside of the building, seeping in, and being circulated by the fan or blower.

Recommended Actions

1. Clean the ducts and outlets. Schedule this cleaning to be repeated every 60 days.

2. If odor is so obnoxious as to justify the expense, install an activated-charcoal air purifier as an integral part of the system.

(e) ***Complaint.*** Furnace operates most of the time, but various rooms remain chilly.

Causes

1. Warm air outlets are in wrong locations. See Fig. 7-1 for errors of location.

2. Heater is undersized, system is unbalanced, or both.

3. Rooms are drafty as a result of leaky windows and doors.

4. Volume of air provided in the design for the cold rooms is insufficient to meet the heat demands.

5. High ceilings are bleeding the heat upward and out of the comfort zone.

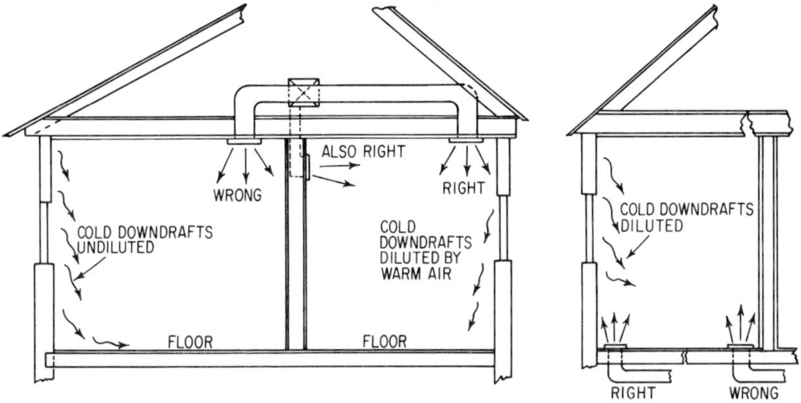

Fig. 7–1. Register placements.

Recommended Actions

1. If wrongly located outlets cannot be relocated, provide auxiliary heating or install new outlets.

2. If heater rating appears too small on inspection, calculate the building heat demands and compare with the rating. Do not attempt to increase the heater firing without first carefully checking the capacity of the chimney and the heater and their abilities to operate under increased firing rates.

3. Install auxiliary peak heating where this seems to be the most economical improvement, such as in the case of excessively high ceilings.

4. Check and correct the balance of the system, taking readings at every outlet.

5. For drafts and leakage, recommend that owner install weather stripping and storm doors and windows, as indicated.

(f) ***Complaint.*** Fan operation is accompanied by a rattle.

Causes

1. Fan is loose on shaft.

2. A fan or motor bearing is cracked or broken.

Recommended Actions

1. Verify position of fan on the shaft. Tighten key or setscrew as necessary.

2. Replace any broken bearings.

3. Test fan for static balance. If out of balance, clean blades and examine for a bent blade. Correct as necessary.

(g) ***Complaint.*** Rooms lose their heat and cool off when room doors are closed.

Causes

1. Many residential systems are designed without individual room returns and are served by one or two common return grilles in hallways or other open areas. When room doors are closed, the rooms are segregated from the return grilles, and the back pressure which develops in the closed room prevents the effective entry of the warm air supply.

2. Rooms do not have individual thermostats but depend on a common or zone thermostat, from which the particular room is isolated by the closing of the door.

Recommended Actions

1. Provide air circulation louvers in the lower panel of the room doors.

2. Plane off to provide clearance of at least 1 in. beneath the door.

3. Open a window by 1 in. This is wasteful but permits warm air to enter the room from its register, so long as the zone or master thermostat is calling for heat.

(h) ***Complaint.*** Wintertime air within the building is too dry to be comfortable.

Cause

Winter air is naturally very dry, and the amount of in-leakage is too much for the normal activities within the structure to maintain a comfortable level of humidity. This condition usually signifies a loosely built structure without an everywhere tight vapor barrier.

Recommended Action

Vaporizers of proprietary design are available on the market. Some are simply water slingers for installation in the ducts, and others are electric. The installation of an automatically controlled humidifier is the most effective means for relieving this condition.

(i) ***Complaint.*** Humidity conditions are so bad that condensation forms on the windows, doors, and other cool-to-cold surfaces.

Cause

Excessive humidity is common in structures that are tightly built with effective vapor barriers. All the water vapor generated by laundries, kitchens, and shower baths is unable to escape and accumulates within the building.

Recommended Actions

1. Operate the ventilating fans in the laundry, kitchen, and all

bathrooms while these facilities are in use. If this does not suffice, try operating these ventilating fans continually.

2. Provide a ventilating fan large enough to move air from inside to outside at a rate that will completely evacuate the inside atmosphere in from 2 to 3 hours. Control this fan automatically by humidistat. When fan operates, make sure that some outside opening can be uncovered long enough to admit a sufficient volume of dry air. Care must be exercised to avoid pulling down the inside temperature to uncomfortable levels.

(j) ***Complaint.*** Smoke is emerging from the warm air supply registers. Heater is oil-fired or gas-fired.

Causes

1. The heat exchanger is leaking at a welded joint or has corroded through in some area and is admitting the gases and smoke of combustion from the furnace into the warm air distribution ducts.

2. Dust and dirt in the filter have become overheated and are burning.

3. Combustible material left in the ducts from construction has ignited.

Recommended Actions

1. Stop the heater oil burner immediately. This condition, if not promptly corrected, can result in a call to the fire department with a need to extinguish fires in either the ducts or the rooms, or it can smoke the building interior so badly that a complete redecorating job is required.

2. Remove the air filter and examine for clogging, oil charring, or any other evidence of deterioration. Clean thoroughly and replace, or install a new filter.

3. If source of smoke is found to exist in ducts, use a vacuum cleaner and endeavor to withdraw the smoky material. Lack of success with this expedient means that one or more sections of the duct system must be opened up and cleaned.

4. All other checks proving negative, dismantle furnace and verify integrity or lack of integrity in the furnace heat exchanger.

5. If system has a fresh air intake connection, examine this as a possible entry for outside smoke.

(k) ***Complaint.*** A group of unit heaters, perfectly satisfactory when installed in an industrial plant several years ago, no longer emits sufficient heat, although the fans appear to be running at their designed speed. The heaters are served with forced circulation hot water.

Causes

1. The evidence of decreasing capacity indicates that the heaters

are in dire need of a cleaning. The exchange surfaces of unit heaters should be vacuumed at frequent intervals, say a minimum of twice a heating season. Before actuating in the fall, another wipe-off cleaning is recommended to remove summer accumulations of dust, dirt, and lint.

2. Heaters have become air-bound.

Recommended Actions

1. Initiate a regular cleaning program for the heaters.

2. Check hot water connections for air pockets. If an automatic air-vent valve is supposed to be operating, see that it is. Hand feel the supply piping for cold areas and correct where found.

3. For unit heaters on a steam system, in addition to cleaning, check the condensate for excessive back pressure over the heater trap. Track down cause of back pressure and correct.

(l) ***Complaint.*** Referring to Fig. 7-2, rooms on the first floor turn drafty and cool almost immediately after the thermostat stops the heater in the attic.

Causes

1. When fan stops running, circulation tends to reverse itself, the warm air in the supply ducts rising while the cooler air column in the

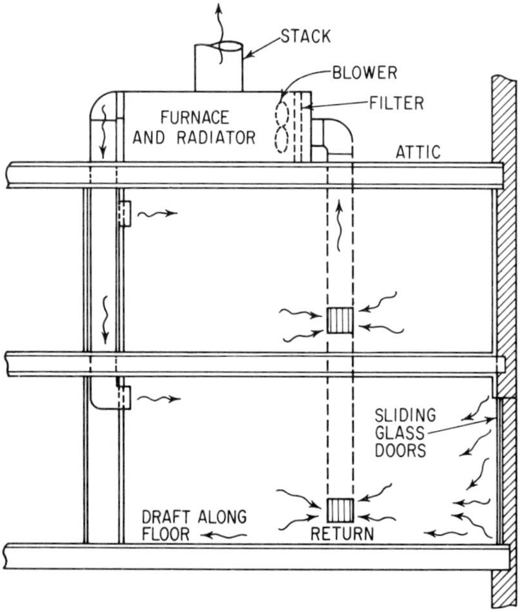

Fig. 7–2. Warm air heater in attic.

return, no longer under fan effect, sinks back into the lower rooms.

2. There is obviously considerable infiltration of cold outside air at the floor level. This expedites the upward evacuation of the warm air through the supply ducts.

Recommended Actions

1. Locate the sources of cold air infiltration and check with weather stripping or storm doors and storm sash. Sliding glass doors are particularly likely to leak cold air.

2. Replace the blower motor with a two-speed unit, the lower speed being selected to maintain the correct directional circulation at all times. A speed that simply prevents reverse circulation will do.

Table 7-1. Troubleshooter's Field Guide for Warm Air Heating Systems

Possible cause	How to check for cause	Recommended ways to correct
COMPLAINT: On a gravity system, one or more designated rooms fail to heat.		
REFERENCE: Sec. 7-1(a).		
1. Supply registers are at the ends of long horizontal leaders or ducts.	*a.* Inspect supply pipes. Note whether pipe has a pitch upward from the furnace.	*a.* Consider larger duct. If not insulated, apply insulation. Raise temperature at registers to 175 F, if furnace and chimney can accommodate it.
2. Return registers in chilly rooms are high above the floor.	*b.* Inspect.	*b.* Provide active return register or install auxiliary heating.
3. Outside air-intake damper is open too wide.	*c.* Check damper position, particularly if under manual control.	*c.* Close or throttle the damper.
4. System is unbalanced. Other rooms are getting more than their share of heat.	*d.* Balance system with velometer or anemometer to check output from each register.	*d.* Set all dampers as needed to bring system into balance.

Table 7-1. Troubleshooter's Field Guide for Warm Air Heating Systems (*Continued*)

Possible causes	How to check for cause	Recommended ways to correct
COMPLAINT: Dust and dirt from registers are distributed through the building.		
REFERENCE: Sec. 7-1(c).		
1. Ducts have accumulated dirt and dust over years.	*e.* Open and inspect two or more of the leaders from the furnace.	*e.* Clean out all the ducts. Use a vacuum.
2. Air passages around furnace radiator are fouled with dust and dirt.	*f.* Open air plenum and inspect.	*f.* Wipe out and clean by vacuum.
COMPLAINT: Floor furnace heats immediate area only.		
REFERENCE: Sec. 7-1(d).		
1. Warm air from a floor furnace does not pass satisfactorily through more than one doorway.	*g.* Inspect arrangement of furnace relative to other rooms.	*g.* Install auxiliary heat in unheated areas. Direct warm air to cool areas by means of fan.
COMPLAINT: With forced circulation system, walls and ceilings are smudged around diffusers and registers.		
REFERENCE: Sec. 7-2(a) and (b).		
1. Ducts, grilles, and registers have not been cleaned for months and are filled with dirt, dust, and lint.	*h.* Inspect interiors of ducts. Note condition of registers and diffusers.	*h.* Open ducts as necessary and clean by vacuum. Do the same for outlets.
2. Air filter is dirt-logged and not removing matter from air passing through.	*i.* Pull out filters and inspect.	*i.* Clean filters or replace with new units.
3. Secondary circulation of room air is conveying room dust to walls and ceilings.	*j.* With smoke-maker, note room air currents and whether they lead to smudges. Run finger under the	*j.* Suggest regular cleaning schedule for the room.

Table 7-1. Troubleshooter's Field Guide for Warm Air Heating Systems (*Continued*)

Possible causes	How to check for cause	Recommended ways to correct
	rungs of a chair or table. If finger is smudged, call for a thorough cleaning of the room.	
COMPLAINT: The system is noisy. REFERENCE: Sec. 7-2(c).		
1. Velocity of the air through ducts is too high.	*k*. Listen for rushing sound through ducts. Check volume of discharge from each outlet. Compare with allowed volume for the maximum velocity.	*k*. Consider reducing velocity coupled with increase of the air temperature to counteract the lower volume.
2. Vibration of blower or other machinery is being transmitted through ducts.	*l*. Verify source of noise. Feel for blower vibration.	*l*. Set vibrating machinery on vibration absorbers. Install sound absorbers in ducts.
COMPLAINT: System delivers odors to heated spaces. REFERENCE: Sec. 7-2(d).		
1. The accumulation of dirt, dust, and lint accumulated in ducts can be source of odors.	*m*. Smell ducts. Open and inspect. Ask when ducts were last cleaned.	*m*. Clean out ducts by vacuum.
2. Small animal, such as a rat or mouse, has died in a duct.	*n*. Inspect ducts' interiors.	*n*. Clean the ducts. Use a hoe with extension handle to rake out heavy material.
3. Odors originate outside of building.	*o*. Trace the odor outdoors.	*o*. If very obnoxious, install activated-carbon purifier on warm air system.

Table 7-1. Troubleshooter's Field Guide for Warm Air Heating Systems (*Continued*)

Possible causes	How to check for cause	Recommended ways to correct
COMPLAINT: Furnace operates most of the time bı certain rooms stay chilly. REFERENCE: Sec. 7-2(e).		
1. Warm air outlets are in wrong locations. See Fig. 7-1.	*p*. A warm discharge should sweep outer walls and windows.	*p*. Install auxiliary peak heating equipment. Install a properly placed outlet.
2. Heater is undersized, system is unbalanced, or both.	*q*. Compare heater rating with your computed building load. If large enough, check for unbalancing by use of velometer or anemometer.	*q*. For undersized unit, provide an auxiliary peak heater. If out of balance, proceed to balance.
3. Structure is drafty.	*r*. With smoke-maker, locate source of drafts. Note absence of weather stripping, storm doors, etc.	*r*. Recommend weather stripping and storm sash for leakage.
4. High ceilings take heat from comfort levels.	*s*. Take ceiling temperature and temperature at 5 ft above floor. The difference should not exceed 6 to 8 deg.	*s*. Prescribe a form of auxiliary heater to make up balance at comfort level.
COMPLAINT: Fan operation is accompanied by a rattle. REFERENCE: Sec. 7-2(f).		
1. Fan is loose on shaft or has broken bearing.	*t*. Inspect and determine which cause is present.	*t*. Tighten up key or screw. Install new bearing.
2. Fan is out of balance.	*u*. Check static balance.	*u*. Clean and straighten blades. Put into balance.
COMPLAINT: Rooms cool off when doors are closed. REFERENCE: Sec. 7-2(g).		
1. Rooms are segre-	*v*. Inspect relative posi-	*v*. Provide louvers in

Table 7-1. Troubleshooter's Field Guide for Warm Air Heating Systems (*Continued*)

Possible causes	How to check for cause	Recommended ways to correct
gated from return grilles in the halls or elsewhere when doors are closed.	tions of return grilles and rooms.	bottom panels of doors. Trim off bottoms of doors by 1 in. as an alternative. Open window as temporary move to establish circulation of warm air into room.
2. Control thermostat is so located that it is not affected by conditions in rooms behind closed doors.	*w*. Check relative positions.	*w*. Act as suggested above.

COMPLAINT: Air in building is much too dry. Humidity is generally lower than 40 percent.
REFERENCE: Sec. 7-2(h).

1. Dry winter air is leaking into building. Sources of indoor water vapor are unable to satisfy the desired conditions.	*x*. Note relative humidity with humidistat.	*x*. Increase the relative humidity by installation of a suitable water injection device. Keep device under automatic humidistat control.

COMPLAINT: Air in building is usually so humid that condensation forms on windows and other cool-to-cold surfaces.
REFERENCE: Sec. 7-2(i).

1. Building is tightly built and is contained within an effective vapor barrier.	*y*. Note existing relative humidity by humidistat.	*y*. Install small ventilating fan in ceiling and attic. Run ventilating fans in kitchen, baths, and laundry when these facilities are in service. Install a proprietary dehumidifier only if all other acts are unsuccessful.

Table 7-1. Troubleshooter's Field Guide for Warm Air Heating Systems (*Continued*)

Possible causes	How to check for cause	Recommended ways to correct
COMPLAINT: Heat output from a group of hot water unit heaters, originally satisfactory, has fallen off to an inadequate level.		
REFERENCE: Sec. 7-2(k).		
1. Heat exchange surfaces are dirty.	*z*. Inspect. Run finger over surface.	*z*. Clean by vacuum and wiping with a cloth.
2. Heaters have become air-bound.	*aa*. Feel hot water connections for cold spots. If auto vent valve is in place, remove, inspect, and make sure valve is functioning properly.	*aa*. If pocket comes from a pipe sag, eliminate the sag. Replace a faulty vent valve.
3. Steam unit heaters are operating against excessive back pressure over thermostatic trap.	*bb*. Open trap with care and check presence of excess pressure. Be cautious of spouting and hazards of scalding.	*bb*. Identify source of back pressure. Isolate with check valves if practical. Consider installing a vacuum return pump.
COMPLAINT: Infiltration and reverse flow of supply air chills rooms.		
REFERENCE: Sec. 7-2(k) and Fig. 7-2.		
1. Stack effect of supply duct is setting up reverse flow.	*cc*. Check air flow through registers by anemometer. If instrument runs backward during fan's off periods, reverse flow is occurring.	*cc*. Install two-speed motor on fan to maintain right flow during off periods of heater.
2. Excessive infiltration at floor level is expediting the reverse circulation in warm air supply ducts.	*dd*. Check air movement at floor level by use of smoke-maker. Take temperature of draft to identify source as outdoors.	*dd*. Weather-strip or install storm windows and doors. Recommend draw curtains over sliding glass doors.

Table 7-1. Troubleshooter's Field Guide for Warm Air Heating Systems (*Continued*)

Possible causes	How to check for cause	Recommended ways to correct
COMPLAINT: Smoke is being emitted from the warm air registers. Heater is fired by oil or by gas. REFERENCE: Sec. 7-2(j).		
1. Heat exchanger is leaking or corroded through. Smoke is escaping from furnace into duct system.	*ee.* Furnace must be shut down, casing removed, and exchanger inspected.	*ee.* Weld up or replace with new part or new furnace.
2. Dust and dirt in filter are burning.	*ff.* Remove filter and examine.	*ff.* Clean filter or replace with new.
3. Combustible material in ducts has ignited.	*gg.* Stop furnace but keep fan running. If register discharge does not clear up, a fire in the duct is a possibility. Trace duct and feel for excessive hot spot.	*gg.* As precaution, call fire department, if fire is sure. For stoppage of smoke, use vacuum to withdraw material from duct. Open the hot section of duct. Clean out.
4. If system has fresh-air-intake, examine for smoke and fumes entering system from outside.	*hh.* Note condition of air entering intake.	*hh.* Provide fresh air filter. Stop system until outside conditions clear up.

NOTE: Smoke coming from warm air registers instead of clean air can be a danger signal. Stop the burners immediately and keep stopped until the source of the smoke is identified.

7-3. QUESTIONS AND ANSWERS

QUESTION 1. Can a gravity warm air system be converted into a forced circulation system without replacing the ducts and heater?

ANSWER 1. Yes. A blower may be introduced into a common return plenum or, alternatively, mounted on the warm air outlet plenum. This writer prefers the return side mounting, since the cooler air presents less volume for the fan to handle. As noted previously, one of the warm air leaders must be left undampered so that air circulation over the furnace radiator is sure to continue and protect the radiator from overheating. The same conditions that would cause the radiator to glow red hot have a strong tendency to ignite chimney fires. In all cases, the new air filter for such a conversion is mounted in the return.

QUESTION 2. Can a forced air system originally designed for heating only be used for summer air conditioning as well?

ANSWER 2. It can be used, but not well. The reason is that two or three times as much air must be circulated on cooling as on heating. The heating ducts cannot handle the greater cooling volumes except under conditions of greater fan horsepower and greater velocity of air in the ducts. Obviously, the greater horsepower for the blower motor costs more to buy and more to operate. Also, the high air velocity could become a noise nuisance. See the ASHRAE "Guide and Data Book" for velocity limits, from the noise viewpoint, in various environments.

In addition, insulation for warm air ducts does not normally include a vapor barrier, whereas the chilled air ducts do. This means that the entire duct system must be sealed in with a new vapor envelope for the cooling cycle; otherwise, the insulation will become waterlogged and useless. Fortunately, the vapor barrier is always placed on the warm side of the insulation, and the heating insulation therefore need not be replaced, just vapor-sealed.

QUESTION 3. A new technique is being proposed and used for discovering a leaky heat exchanger in a gas-fired, warm air, heating system. It is based on injecting a salt spray into the combustion air that is being supplied to the gas burners. Then, if there is a leak in the exchanger and the salt spray is being carried on the warm air to the duct system and the rooms, testing the warm air with a halide torch will cause the torch flame to turn yellow. This is because sodium causes a halide flame to react to its presence by a change in color.

This technique is said to be extremely sensitive, and even a small leak in the exchanger will produce a reaction. Is this technique dependable?

ANSWER 3. Reports from the field say that this test can be even too sensitive. The flame will change color from a leak of only one or two ounces of combustion products a year. One contractor states that such infinitesimal leaks are frequent even with new units right off the assembly line. Obviously, a troubleshooter who replaces such an insignificant leak with a new furnace can find himself in a most embarrassing situation should the new furnace match the reaction of the old one. And this can happen.

Perhaps the old smoke bomb test, or a visual inspection, would produce more certain evidence of a significant leak.

The salt spray technique has been endorsed by the National Warm Air Heating and Air Conditioning Association.

DIVISION 8

Fuel- and Ash-handling Systems

8-1. COAL DELIVERY, STORAGE, AND COMBUSTION FEED

(a) *Large Plants.* In these plants, which consume coal in quantities that are best expressed in tons per hour, the handling of the coal from delivery to the site, through storage, and finally to the stokers or burners is usually quite complex. Each step in the procedure has its own particular troubles with its own peculiar cures.

To illustrate, delivery to the plant may be accomplished by trucks that dump the coal into a hopper, or coal may arrive in a hopper car that drops its load into a track hopper, or perhaps the whole car is turned upside down and shaken empty by a car dumper, or the coal to riverside plants may arrive in a barge which is emptied by a crane and bucket.

From the point of delivery, it may go to an outdoor storage pile that is shaped by a bulldozer or drag scraper, or it may be picked up by a series of conveyers or a skip hoist and stored in an overhead bunker or silo. Movement from the overhead storage to the stoker or pulverizer mill is generally by gravity down a feed pipe or chute.

Troubleshooting from delivery to burning is the responsibility of the plant maintenance or operating people, except in those instances that may involve questions of defects or malperformance contrary to the guarantees. Then the manufacturer's representatives are called in, as experts in that area of trouble.

Since these people all know their equipment by reason of constant attention to its care, a discussion of troubles, symptoms, and corrections would be telling them what they already know. Therefore troubleshooting techniques will be limited to small-plant applications.

(b) *Small Plants.* In this category, we place the low-pressure boilers for heating systems and the low- and high-pressure boilers in those boiler rooms that are shut down overnight or have only casual attention during the day, such as components of some process or other industrial plant activity. For these installations, fuel- and ash-handling troubles frequently require outside correction, the regular force usually being unqualified in any but the simplest of maintenance procedures. For that reason, an examination of potential troubles will be made in subsequent sections.

(c) ***Complaint.*** Coal dust from outside storage has become a neighborhood nuisance.

Recommended Actions

This is the most obvious trouble with a coal storage system from the neighbors' viewpoint, since winds tend to distribute the dust indiscriminately. Here are some ways to combat this situation:

1. Spray the surface of the coal pile with a mineral oil or other combustible and viscous fluid.
2. Surround the coal pile by a solid fence or by a line of compact trees, such as cedars. These will not stop the blowing absolutely, but they will reduce the weight carried off.
3. Not recommended is wetting down the coal, since the water sometimes reacts with the sulfur in the coal, producing a sulfuric acid that generates heat and may ultimately end up as the start of a coal-pile fire. This prohibition does not apply to underwater storage, a condition in which fire is impossible.

(d) ***Complaint.*** Fire has started in the coal pile, bunker, or silo.

Recommended Action

There is only one sure way to eliminate a fire in coal: dig it out. A blanket of foam over the surface of the coal may or may not smother the fire, but chances are it will not. There is too much oxygen in the coal, and too much air between the particles of coal. The fire usually goes on undisturbed. This is true also of CO_2-blanketing. Theoretically, CO_2, with a density $1\frac{1}{2}$ times that of air, should seal over the surface of the coal and, when the available oxygen has been consumed, starve the fire of fresh oxygen and thus extinguish it. In the experience of this writer, starvation of oxygen simply has never occurred, and the fire continued to smolder until dug out. Similarly, both foam and CO_2 have been ineffective against fire in a tank of soy beans.

(e) ***Complaint.*** Coal slags over furnace walls.

Recommended Action

The first action is to begin feeding coal with a higher ash melting point than is characteristic of the slagging coal. Melting points can be obtained from the coal supplier, and in all cases, coal should be bought only on specification and not on the premise that the price is a bargain.

This slag can seriously affect the boiler output, especially if it is slagging on waterwalls. Removal of the slag is a manual job, using steam lances.

(f) ***Complaint.*** Downfeed pipes from bunker to hopper erode.

Recommended Action

Protection for the metal will be necessary wherever the coal flow changes direction in an elbow. Check the market for protecting coatings and special lining metals.

(g) ***Complaint.*** Segregation in the hopper is causing an unbalanced fire in the furnace.

Recommended Action

The market offers a nonsegregating coal spout that should be considered in this situation. Manual distribution of the coal across the hopper is possible but less desirable than the automatic equipment.

(h) ***Complaint.*** Stoker grates burn out too fast.

Recommended Actions

1. Check the temperature of the underfire air. This is a common trouble observed in preheated air installations. Lowering the windbox temperature by as little as 25 deg will sometimes eliminate the burnouts.

2. Take the furnace temperature and compare this with the maximum allowable temperature for the stoker metals.

3. Carry a little more ash on the grates. This will protect the metal to some extent.

8-2. FUEL-OIL PROBLEMS

(a) ***General Notes.*** Fuel oil most frequently reaches the storage tank by way of tank truck for the small- and medium-size plants, but it can arrive in a barge, in a tank car, or through a pipeline. However it arrives, storage tanks, valves, piping, and pumps have something to do with it before it reaches the burners. Trouble aspects of the delivery equipment are the problems of the specialists supervising such equipment; what happens after the oil is accepted by the user is examined in the subsequent sections.

(b) ***Complaint.*** The oil becomes contaminated with water, and ignition is uneven and sometimes lost altogether.

Causes and Recommended Actions

1. Check vent as a point of entry for rain. See that the gooseneck is not only turned down, but is not subject to a splash-up from the ground or other adjacent surface. Do the same for the fill pipe.

2. Take a sample of the oil as delivered and check it for water.

3. Because of fluctuating atmospheric humidity and temperatures, especially during summertime, moisture from the air over the surface of the oil in the tank may condense and gradually build up a pond inside the tank. When this pond level submerges the suction pipe of the feed pump, water is lifted to the burners as an oil emulsion, and erratic combustion follows.

The recommended treatment for this situation is to pump the water from the tank. Then fill the tank with oil and keep the oil level high thereafter.

(c) ***Complaint.*** In cold weather, oil in storage tank becomes so viscous that burner feed is uncertain.

Recommended Actions

1. "Heating pads" may be applied to the tank, using steam, hot water, or a heat exchange oil. Figure 8-1 shows this kind of warming device on a small tank. Figure 8-2 illustrates the pads unassembled. Treatment for a large tank appears in Fig. 8-3.

2. Another form of pad is electric, resistance wires being incorporated in a suitable material.

3. A third possibility is the use of a heating cable wrapped around the tank or around the piping that conveys the oil to the burners.

The use of a torch on either the piping or the tank should be avoided. The concentrated high temperatures so rapidly developed with a torch have been known to produce an explosion. Also,

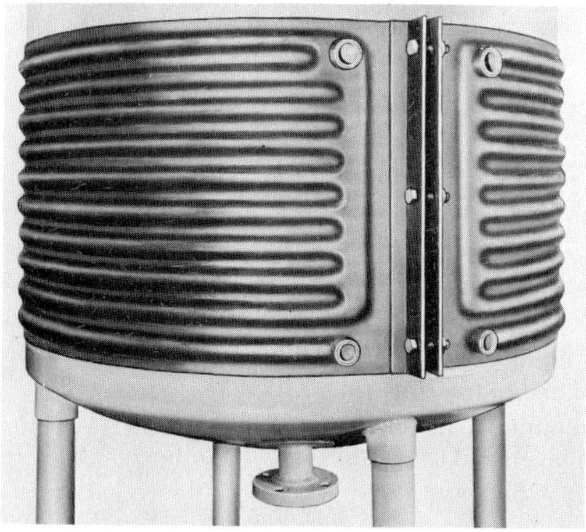

FIG. 8–1. "Heating pad" in place around exterior of small tank. (*Dean Products, Inc.*)

Fig. 8–2. Sections of heating pad before placement around tank. (*Dean Products, Inc.*)

these hot spots promote the deposition of soot and scale on the inside surface of the heated area. Permanent heating coils for the tank are recommended in climates where chilling of the oil becomes a handling problem.

(d) ***Complaint.*** Oil piping between the storage tank and the fuel-oil burner pump has become choked with sludge, a condition which threatens to reduce the capacity of the burner.

Cause

This is a most unusual condition, but it does happen. The sludge is the result of contaminated fuel oil in the storage tank.

Recommended Actions

1. Clean the tank. Lift open end of suction pipe to an elevation at least 3 in. above the bottom of the tank, thus keeping the outflow above the settling area.

2. Replace the entire run of piping. This is cheaper than trying to clean the piping with solvents and produces surer results.

3. Replace the oil filter with a new one. This will be foul, too.

(e) ***Complaint.*** Oil burner pump requires frequent priming and is difficult to operate at full capacity.

FIG. 8–3. Arrangement of heating pads around exterior of large oil tank. (*Dean Products, Inc.*)

Causes

1. Most frequent cause is a combination of lift and pipe friction that interferes with free flow of oil into pump.
2. Fuel-oil pipe is choked with sludge.

Recommended Actions

1. See Sec. 2-7(d) for details of treatments.
2. Consider installation of a two-stage burner pump if pump giving trouble is a single-stage unit.
3. See preceding Sec. 8-2(d) for handling of clogged oil line.

8-3. FUEL GAS, NATURAL

(a) ***General Notes.*** Trouble with a gas-fired system is generally the result of someone's carelessness, the incorrect adjustment of burners and pressures, or interruptions in the supply for which the supplier, not the system owner, is responsible. To counteract supply interruptions or loss of pressure due to an overloaded supply network, some utilities offer the user an "interruptable rate" which is substantially lower than the regular rates but relieves the gas utility company of responsibility for the gas supply during certain designated periods of heavy consumption in the territory.

When gas pressure is found by the troubleshooter to be too low, a pressure check on the main supply is in order. A meter check of the consumption over a period of time, say 15 minutes, will also be useful in determining the source of trouble, since this may be compared with the normal consumption of the burners.

(b) ***Complaint.*** The pilot light does not stay lighted.

Causes

1. Set up the pilot adjustment to increase the size of the pilot flame. The gas feed to the pilot may be too restricted.

2. Interruptions in the main gas supply can extinguish the pilot.

3. Downdrafts plus the lack of a draft diverter, or an incorrectly placed diverter, can cause this trouble.

Recommended Actions

1. Check the size of the pilot flame. Increase it if necessary.

2. Check for downdrafts. Make sure diverter is functioning correctly.

(c) ***Complaint.*** The burning gas causes an odor throughout the building.

Cause

This is not the fault of the gas. See Sec. 2-8 for a discussion of gas burner troubles and their correction. Probably the system has a burned-out or leaky heat exchanger, or the vent pipe is too low, allowing combustion gases to drift into the building.

8-4. SAFETY IN HANDLING GASES

(a) ***Precautions Applicable to All Gases.*** The troubleshooter or serviceman, in arriving at the scene of the service call, should immediately take the following steps for safety:

1. Shut off all the gas, closing the valve between the gas source and the point of use. Unless the trouble is in the pilot and requires simply an adjustment of the pilot flame, the pilot should be shut off also.

2. Notify *all* gas users on that particular service that the gas is being interrupted. Make sure all burners are off.

3. If a leak is known to be present or is suspected, use a hand-held flashlight. If electric lights are burning, do not touch the switch, but let them continue to burn. If no electric lights are burning, do not attempt to turn them on; the consequence of ignoring this caution can be a spark at the switch, well-calculated to blow the serviceman and the area into oblivion.

4. Never work alone in areas, or on the type of jobs, to which good practice assigns two men. The saving in manpower may just happen to require payment in the form of the single worker's health or life.

5. See that *all* downstream valves are closed before restoring ser-

vice at the main supply valve. This is to avoid spillage from an open line or burner prior to ignition.

6. Do not permit smoking, the striking of matches, or the ignition of a cigarette lighter in the vicinity of the gas piping, even if the piping is known to be empty of gas.

(b) ***Precautions Applicable to Liquefied Petroleum Gases.*** In addition to all the precautions recommended in Sec. 8-4(a), observe the following for liquefied petroleum:

Since these gases are heavier than air and sink to the lowest level available, a leak can be present without being readily realized. These gases normally have no odor; therefore, by law they are odorized and when present emit the smell of rotting cabbage. A prudent action therefore is to use the nose, as close to ground level as possible, in the pursuit of possible leaks. Where liquefied petroleum gas-fired water heaters are in an attic space, the troubleshooter is advised to sniff between the floor joists before taking any other step toward identifying and correcting the trouble. This writer had one case of this nature in which delayed ignition finally caught up with such a leak and burned out the attic. The gas had accumulated between the joists.

8-5. GAS FUEL, LIQUEFIED PETROLEUM GAS

(a) ***General Notes.*** Liquefied petroleum gas is delivered from a truck to permanent on-site tanks or reaches the point of use in charged cylinders that replace emptied cylinders.

Under the governing codes of the "National Board of Fire Underwriters Standard Liquefied Petroleum Gases" and the API-ASME Code, and finally the rules of the Interstate Commerce Commission, handling and using liquefied petroleum gas is a safe operation.

The major trouble, or rather complaint, is that gas is lost by opening of the tank safety valve. When this happens, the probabilities are that the tank is in a location that is subject to sun heat or other ambient high temperatures. Valves can be defective, but each tank must be retested before being refilled, and the likelihood of a defective valve slipping past is small.

A hazard of this gas is that it is heavier than air and may travel considerable distances from a leak before making contact with an open flame or spark and then ignite with more or less explosive violence.

Possible sources of complaint follow.

(b) ***Complaint.*** The gas will not vaporize in cold weather.

Cause

At atmospheric pressure, the vaporizing temperature of butane is

15 F. This means that at lower temperatures, common in many parts of the country, the liquid remains liquid and no gas emerges for service to the burner.

Recommended Actions

1. Replace the butane with propane. Propane vaporizes all the way down to −51 F.

2. Provide heating means to the tank, such as a steam or hot water line around the tank which is made operative at temperatures below freezing.

If the tanks are located at substantial distances from the burners, tracing the gas piping with the steam or hot water warming piping may be necessary in addition to the tank heating.

(c) ***Complaint.*** Gas does not burn well.

Cause and Recommended Action

Check type and orifice dimensions of the burner. If burner was originally designed for a natural or manufactured gas, it will not perform well, if at all, on the LPG.

8-6. GAS FUEL, MANUFACTURED OR CITY GAS

(a) ***General Notes.*** In some areas of the country, this form of fuel gas is also called "water gas." The name comes from a step in the manufacturing process in which steam is blown into and through a bed of hot coke or selected coal; the steam dissociates and reacts with the carbon to produce an interim gas called "blue gas." The blue gas is relatively low in thermal value; therefore it is enriched with a petroleum gas formed by spraying oil over hot checkerbrick in the middle section of the gas machine called the carburetor. The blue gas is formed in the first section, termed the generator, and the mixture of blue and oil gas is fixed in the final section named the superheater.

The average thermal value of this commercial gas is about 550 Btu per cu ft, or approximately half the value of natural gas and perhaps one-fourth that of a liquefied petroleum gas.

Industrial gases, meaning gases produced as by-products of industrial activities, may be used as fuel for other processes in the plant of origin. The troubleshooter rarely can expect problems that involve blast furnace gas or producer gas. These are handled by the regular operators. The heating value of the former is very low, about 100 Btu per cu ft, but the volume is tremendous, about 127,000 cu ft per ton of iron.

Producer gas, created by the reactions of steam, air, coal, coke, or peat, shows a calorific value of 150 Btu per cu ft on the average. This gas is close in character to the blue gas but is leaner in value.

(b) ***Complaint.*** Gas pressure is frequently too low for satisfactory combustion.

Causes

1. Gas from the mains of a city distribution system can be affected by sudden excessive draws by users of the service. This condition is likely during extra-cold days of winter when all the heating loads reach a simultaneous peak.

2. The system may be overloaded, and the capacity of the mains too low to handle the demand.

3. Accumulations of tar and water in low spots of the supply main can restrict the flow of gas through the main.

Recommended Action

Request an investigation from the gas supplier.

(c) ***Complaint.*** Consumption of gas appears to be excessive.

Causes

1. Original estimate of consumption overlooked some load factor and was therefore too low.
2. Calorific value of gas is lower than expected.
3. Fuel-air mixture is not in optimum adjustment.

Recommended Actions

1. Recheck the calculated load.
2. Verify calorific value of gas through supplier.
3. Check combustion efficiency by means of stack temperature and CO_2 readings. Readjust mixture as necessary.
4. Check for leaks between the meter and the burners. With all use of gas shut down, watch and time the meter for at least ½ hour. Amount of leakage can be told from meter registration during nonuse period.

8-7. ASH HANDLING

(a) ***General Notes.*** Ash from the burning of coal is encountered in two forms: bottom ash and fly ash. The latter is the more likely to develop trouble as a neighborhood nuisance; the former is the more common.

Bottom ash, meaning ash that is taken from the bottom of a stoker-fired or hand-fired furnace, is currently removed by means of manual action; by dumping from hopper gates into cars or trucks; by a combination of ash pit and skip hoist, or an ash pit and a hydraulic or pneumatic conveyor system.

Some of the large power boilers operate with what is known as a "wet bottom furnace," the ash being removed directly from the furnace in the form of a slurry. The slurry is pumped to a fill area, as in Fig. 8-4, where the water drains out and the ash may be allowed to remain or is removed by crane bucket or other suitable means.

FIG. 8–4. Hydraulic ash disposal system spilling into a fill area. (*Allen-Sherman-Hoff Co.*)

Alternatively, the hydraulic systems may end in an overhead bin from which the water is drained, the dewatered ash then being gravity fed to a collecting truck or car through dump gates in the bottom of the bin.

(**b**) ***Complaint.*** The abrasiveness of the ash is rapidly destructive of the liners in the disposal piping. Replacement is a continuing and substantial expense.

Recommended Action

Establish experimental lengths in the piping, and in these lengths, install liners or piping of various materials and time the life of each material. The degree of abrasiveness of ash varies with the characteristics of the coal, the effectiveness of the clinker grinder, the temperature of the ash, and the velocity of flow. For these reasons, an abrasion-resistant material satisfactory in one plant may be completely unsuited for service in another.

Proprietary materials having a hardness approximately equal to that of the diamond are available under the trade names of Carborundum, a silicon carbide made in the electric furnace, and Abresist, a volcanic basalt produced in Germany.

Frequently the greatest erosion occurs in elbows and other curvatures of the line. Protecting these surfaces with a highly resistant material will solve the problem satisfactorily in some instances; in others, replacing or relining the entire system may be necessary. Counteracting this abrasiveness of the ash is the one great problem of handling systems for bottom ash.

(c) ***Complaint.*** Fly ash from a pulverized coal-firing system is falling over the neighborhood, creating criticism of the plant operation and bringing threats of legal action.

Cause

The coal feed is fine to begin combustion and produces a very lightweight ash that is readily airborne and carried from the chimney.

Recommended Actions

Corrective methods involve problems of design beyond the scope of the general troubleshooter. This is a pressing subject, intertwined with the movement to reduce air pollution, and projected actions may require the approval of local or federal antipollution agencies.

Present techniques available for fly-ash reduction include electrostatic precipitators, combination baffle and settling boxes, cyclones, special material filters for temperatures up to 600 F, and wet scrubbers or washers.

Although this writer has had excellent results with sprays in flue gases, using stainless steel pipe, the possibilities of the contaminated water simply transferring pollution from the atmosphere to streams, lakes, and ground waters indicate that washing must be considered with the highest caution.

Dry removal also can be quite satisfactory, the form chosen being the result of careful engineering studies. Recovered fly ash is being used in some areas as a building material; this possibility produces a credit against the cost of the separating installation. For additional comment on fly-ash troubles, see Sec. 2-10(e).

Table 8-1. Troubleshooter's Field Guide for Fuel- and Ash-Handling Systems

Possible cause	How to check for cause	Recommended ways to correct
COMPLAINT: Dust from an outside storage field for coal blows over the area and constitutes a nuisance.		
REFERENCE: Sec. 8-1(c).		
1. High winds and absence of windbreaks around the coal pile.	*a.* Note wind velocity and direction and whether condition is chronic or sporadic.	*a.* Spray coal with a mineral oil or other combustible and viscous fluid. Erect windbreaks or barrier fences around pile.

Table 8-1. Troubleshooter's Field Guide for Fuel- and Ash-handling Systems (*Continued*)

Possible causes	How to check for cause	Recommended ways to correct
COMPLAINT: Fire occurs from time to time in the coal bunker or silo. REFERENCE: Sec. 8-1(d).		
1. Coal is permitted to lie undisturbed for more than 90 days.	*b*. Balance the storage weight against rate of use. Examine silo or bunker for dead corners in which coal may lie undisturbed indefinitely.	*b*. Expedite movement of the coal through bunker or silo. Modify system to keep transit time inside 90 days.
2. Coal is of a high sulfur analysis.	*c*. Obtain analysis of coal. Note whether sulfur content is excessive, average, or low.	*c*. With a high-sulfur coal, provide thermometer wells to monitor temperature of coal.
3. Bunker or silo structure is neither air- nor watertight and coal becomes damp, thereby generating heat.	*d*. Stop leaks through which water is entering the bunker or silo.	*d*. Use patching and plugging materials suitable for the structure.
COMPLAINT: Ash in furnace slags over the furnace waterwalls. REFERENCE: Sec. 8-1(e).		
1. Melting temperature of ash is too low for the furnace temperatures reached.	*e*. Obtain ash-melting temperature from a coal supplier. Take a pyrometer reading of the furnace temperature.	*e*. Switch to a coal of a more suitable analysis.
COMPLAINT: Coal downspouts from bunker to stoker hopper corrode and erode. REFERENCE: Sec. 8-1(f).		
1. Inside surfaces of spouts are not protected against the abrasion by a moving stream of coal.	*f*. Inspect inside of spouts. Obtain analysis of metal in contact with flowing coal.	*f*. Line the spouts with abrasion-resisting material if spouts are large enough. For small spouts, replace

Table 8-1. Troubleshooter's Field Guide for Fuel- and Ash-handling Systems (*Continued*)

Possible causes	How to check for cause	Recommended ways to correct
		with new spouts designed to resist abrasion.
2. Coal presently being run to hoppers may have a high abrasive index.	*g.* Have coal supplier furnish information on hardness of coal. If coal is harder than spout lining, the lining will go.	*g.* Consider the possibility of a different analysis coal being used.
3. Coal is damp and also high in sulfur, and spout is being attacked by sulfuric acid.	*h.* Take a scraping of the spout lining and analyze for acid deterioration. Analyze coal for water content by weight.	*h.* Line the spout with acid-resisting material, dry the coal, or buy another analysis.
COMPLAINT: Coal is segrating in stoker hopper, producing an uneven fire. REFERENCE: Sec. 8-1(g).		
1. Coal-spout design and arrangement permits segregation.	*i.* Examine design of spout. See whether design is directed at avoiding segregation.	*i.* Replace the spout with the nonsegregating kind.
2. Wet coal is a contributing factor.		*j.* Dry the coal before feeding down to the hopper.
COMPLAINT: Stoker grates burn out too fast. REFERENCE: Sec. 8-1(h).		
1. Grates are overheated from preheated air.	*k.* Take furnace temperature. Compare with allowable temperatures specified by stoker manufacturer.	*k.* Reduce the preheated air temperature.
2. Coal is a low-ash type.	*l.* Obtain ash content of coal. If 4 % or less, preheated air can be destructive.	*l.* Burn coal with more ash.

Table 8-1. Troubleshooter's Field Guide for Fuel- and Ash-handling Systems (*Continued*)

Possible causes	How to check for cause	Recommended ways to correct
COMPLAINT: Fuel oil is being supplied to boiler burners from a storage tank. The oil reaching the burners is contaminated with water. How can this condition be corrected? REFERENCE: Sec. 8-2(b).		
1. Rainwater is entering the tank by way of the vent.	*m*. Examine the vent. Note whether gooseneck is turned down to prevent rain entry.	*m*. Adjust the vent pipe, also the fill pipe, to keep out rain and splash water.
2. Oil as delivered to tank is contaminated with water.	*n*. Take a sample of oil. Let stand for 24 hours. Note whether water settles in bottom of container.	*n*. Pump out water from bottom of storage tank. Check new oil for water content before filling tank.
3. Water has settled in tank as result of condensation from humid atmospheric air.	*o*. Note level of oil in tank and air space above surface of oil.	*o*. Keep tank filled so that humid air cannot enter in significant volume.
COMPLAINT: Fuel oil in storage tank becomes so viscous in cold weather that pumping is difficult. REFERENCE: Sec. 8-2(c).		
1. Most fuel oils do become viscous in cold weather.	*p*. Take temperature of oil in storage tank. From supplier, obtain viscosity at this temperature.	*p*. Provide heating for the storage tank. Change to an oil with better temperature characteristics.
COMPLAINT: Oil piping between storage tank and burner becomes choked with sludge. REFERENCE: Sec. 8-2(d).		
1. Oil tank contained foreign material before oil was put in and sludge resulted.	*q*. Open the oil feed line to the burner. Inspect inside surfaces for gummy or hard coatings.	*q*. Replace the entire line from tank to burner. Replace filter.
COMPLAINT: Oil feed line to burners is such as to require frequent priming of the feed pump. REFERENCE: Sec. 8-2(d).		
1. Oil line is choked with sludge.	*r*. See above.	*r*. Correct as above.

Table 8-1. Troubleshooter's Field Guide for Fuel- and Ash-handling Systems (*Continued*)

Possible causes	How to check for cause	Recommended ways to correct
2. Feed of oil is erratic as result of high lift and high line friction.	*s.* Check lift. If over 14 ft, it is a potential cause of trouble. If less than 14 ft, inspect inside of oil line for sludge or other reason for high friction.	*s.* Install a two-stage pump or set a booster pump close to tank. Replace feed line if interior is in poor condition.
COMPLAINT: With a gas-fired boiler, pilot light does not stay lighted. REFERENCE: Sec. 8-3(b).		
1. Too little gas is being supplied to pilot.	*t.* Inspect size of flame.	*t.* Increase gas to pilot.
2. Interruptions are occurring in the main gas supply.	*u.* Observe the performance of main burner flame. Watch for fluctuations.	*u.* If interruption is verified, notify gas supplier.
3. Downdrafts are extinguishing the pilot.	*v.* Check draft diverter for performance and placement. Test with smoke-maker, noting whether smoke is drawn in or blown out of diverter.	*v.* Correct size, placement, and type of draft diverter as necessary.
COMPLAINT: Gas combustion produces an odor throughout the building. REFERENCE: Sec. 8-3(c). This is an equipment malfunction. Refer to Sec. 2-8(h) for causes and corrections.		
COMPLAINT: Liquefied petroleum gas will not vaporize in cold weather. REFERENCE: Sec. 8-4(b).		
1. Ambient temperature is lower than rated vaporizing temperature of gas.	*w.* Check gas rating with air temperature.	*w.* Switch to a gas with a lower rating, as from butane to propane. Alternatively, heat gas storage area and pipeline as necessary.

Table 8-1. Troubleshooter's Field Guide for Fuel- and Ash-handling Systems (*Continued*)

Possible causes	How to check for cause	Recommended ways to correct
COMPLAINT: Consumption of gas is excessive. REFERENCE: Sec. 8-6(c).		
1. Original estimate was too low.	*x*. Calculate the load.	*x*. Show error if any.
2. Original estimates were made of a gas with higher Btu.	*y*. Get Btu value of gas from supplier. Compare with original.	*y*. Reestimate the consumption on new Btu, if different from original.
3. Fuel-air mixture is out of efficient adjustment.	*z*. Take a reading of CO_2 and flue gas temperature.	*z*. If former too low and latter too high, adjust burner.
4. There are gas leaks between the meter and burners.	*aa*. Shut down burners and pilot. Note meter registration, if any, for at least ½ hour. Registration indicates leakage.	*aa*. Find leaks and correct them.
COMPLAINT: Ash from a boiler furnace is so abrasive as to destroy its disposal piping every few months of operation. REFERENCE: Sec. 8-7(b).		
1. Materials of piping or pipe lining are unsuited to this ash.	*bb*. Observe points of destruction, whether in piping run or curves and elbows. Get laboratory check of hardness of the ash. Calculate velocity of flow in line.	*bb*. Replace with a more resistant material. Slow down velocity. Turn elbows into long radius curves if practical. Try a coal having different ash characteristics.
COMPLAINT: Fly ash falling over neighborhood is bringing the liability of lawsuits. REFERENCE: Sec. 8-7(c). Various corrective measures are discussed in this section, and also in Sec. 2-10(e). This problem is not within the scope of the field troubleshooter, being a matter of basic design arrangements and characteristics.		

8-8. QUESTIONS AND ANSWERS

QUESTION 1. In some water-tube boilers fired with oil, pits have developed in the exterior surfaces of the fireside tubes. Reports have been made by investigators, charging the pits to contaminants in the fuel oil. Is this correct? What can be done about it?

ANSWER 1. Researchers have blamed the presence of vanadium in the fuel oil as a prime cause of the trouble. High velocity of the combustion gases sweeping the tubes has also been accused of contributing to the severity of the pitting. The use of oils testing low in vanadium has been recommended, plus a reduction in the load. Boilers frequently are operated at high overloads, as much as 400 percent of rating, when the rating is given in boiler horsepower. Large utility and industrial boilers generally are sold on a pounds per hour output which is not exceeded, or very little exceeded, in regular operation.

QUESTION 2. Can soot in boiler passages be removed successfully by means of chemicals injected into the furnace?

ANSWER 2. Yes. However, such chemicals should not be used until after such use has been cleared through the local fire department. If a very hot fire is created to burn off the soot, there is a possibility of flying sparks spreading over the neighborhood, and some localities do not permit it at all or else limit the chemical per application, or by name. Handling of the material is reasonably safe prior to use. One brand approved by the New York City Fire Department has a flash point of 148 F.

QUESTION 3. Does bagasse require a special type of furnace?

ANSWER 3. Yes. A Dutch oven type is necessary, and the bagasse must be dried before use. Cleaning also is recommended to remove earth and dust.

DIVISION 9

Comfort Air-conditioning Systems

9-1. DOMESTIC INSTALLATIONS

(a) *General Notes.* The forms of domestic air conditioning are window units; through-the-wall units; package units; and split systems with the air fan and evaporator coils inside the house and the compressor and condenser outside, usually but not necessarily on the ground. Also, domestic installations may obtain their cooling from a heat pump.

At one time there was a common tendency to oversell the capacity of air-conditioning installations, expressed in motor horsepower rather than Btu; presently, units are being rated by the Air-Conditioning and Refrigeration Institute, Washington, D.C., and the capacity in tons of refrigeration is clearly stated.

Since each of these named forms depends on a motor-driven refrigerant compressor to develop the necessary cooling, the troubles encountered by the troubleshooter on one system may well be met on another. These troubles will be discussed.

(b) *Complaint.* The compressor runs all the time but does not produce adequate cooling.

Causes

1. If this is a machine built earlier than 1960, there is a possibility of its being undersized for the load.
2. The condenser coils and the air filter are fouled with dust and lint.
3. Because of resistance or need for lubrication, the fan is running slower than its design calls for, and the right volume of air is not being circulated.
4. It may be low in refrigerant.

Recommended Actions

1. Note the capacity of the compressor in Btu per hour as it appears

on the data plate. Compare this capacity with the cooling load, calculating this load if it is not known. If compressor is too small, consider replacing the unit with a larger one or reducing the load by improving insulated areas, providing sun shading, arranging a smaller electric light load, and ventilating the kitchen. Determine whether additional heat-generating apparatus of any kind has been added since the cooling unit was installed.

2. With a stiff brush and vacuum cleaner, go over the filters and coils. Remove all dust and lint.

3. Using a tachometer, check the speed of the fan and compare this reading with the rated speed of the fan. Note temperature of fan motor, whether normal or high.

4. Make note of both the suction pressure and the head pressure. If the former is high and the latter is low, the unit probably is low in refrigerant. Add refrigerant.

5. In some instances, an undersized compressor has been speeded up by means of a faster motor, a change in gearing, or a change in pulleys for belt-driven machines. This method of obtaining increased refrigeration *is not recommended.* The life of the compressor is shortened thereby and, in any event, efficiency decreases.

(c) ***Complaint.*** The compressor short-cycles.

Causes

1. Most likely, evaporator coils need defrosting.
2. High head pressure is causing cutout to operate.
3. Strainer in expansion valve needs cleaning.
4. Thermostatic bulb on coil has become loose or has slipped out of position.
5. Check and adjust pressure cutout.

Recommended Actions

1. Inspect evaporator coils. Defrost if necessary. If there is an automatic defroster, check for a failure to operate. Repair automatic as necessary.

2. High head pressure is the result of several conditions, individually or together. First check should be of the condenser, noting temperature of air or water into the condenser and the temperatures of these media on leaving. Since the condensing temperature of the refrigerant is always from 10 to 20 degrees higher than the condensing medium, a high condensing medium means high refrigerant temperature and a consequently high pressure, as determined by this temperature. Clean condenser coils.

3. Inspect condition of expansion valve strainer. Clean or replace as necessary.

4. Readjust thermostat and pressure cutout as necessary.

5. If condensing water is cooled in a tower and temperature of cooled water is higher than was specified, check performance of tower. Inspect for presence of tree limbs, wind-blown papers, leaves, grass, and other foreign material blocking in part the free passage of air through the tower.

(d) ***Complaint.*** Compressor is noisy.

Causes

1. Knocking sounds result from liquid refrigerant getting into the compressor through the suction. They may also come from a burned-out, or otherwise defective, crankshaft bearing.

2. Slapping and rattling sounds originate in worn pistons, in out-of-round cylinders, loose wrist pins, broken piston rings, and defective valves.

3. Squealing noises are likely to be found coming from the shaft seal.

4. Vibrations show up when the compressor is loose on its foundation or other support and when flywheels and pulleys, if the drive is so arranged, are loose on their shafts. The loudness of these vibration noises is usually enough to produce a quick shutdown by the owner, who may believe the machine is on the verge of violent disintegration.

Recommended Actions

1. To identify interior noises in the compressor with exactness, use the stethoscope on the machine, listening to crankcase, cylinder block, and valve section in order. Probability is that the compressor must be taken to shop for repairs.

2. For external noises and vibrations, tighten on the holding down bolts. Check the tightness of keyways and shaft screws.

3. Note level of oil in crankcase. If excessively low, bring up to proper level, start machine, and note whether internal sounds are eliminated. A continuation of the sounds, coupled with a previously low oil level, is fair indication that some damage has been done to the bearings and that opening the compressor is required.

(e) ***Complaint.*** Low oil conditions in the crankcase are resulting from failure of oil to return to the compressor.

Causes

1. Compressor is considerably higher in position than the evaporator coil. Oil settles out in the bottom of the coil.

2. Leaky valves, worn piston rings, or an out-of-round cylinder, either singly or severally, are permitting the compressor to pump out oil at an excessive rate.

3. Expansion coil is being overflooded with refrigerant, and liquid refrigerant is dropping oil abnormally into the bottom of the coil.

Recommended Actions

1. Compare elevations of compressor and evaporator coils. If the

former is more than 6 ft higher than the latter, the relationship is a potential cause of trouble. The oil and refrigerant are miscible, with the former tending to separate when the mixture is permitted to stand. Consult with the compressor manufacturer on the possibilities of changing the oil being used for one having better refrigerant-oil characteristics.

2. The stethoscope can be used to discover the existence of leaky valves, worn rings, and possibly an out-of-round cylinder.

3. For an overflooded expansion coil, adjust the expansion valve to maintain the proper rate.

(f) ***Complaint.*** Compressor refuses to start.

Causes

1. Electrical. A blown fuse. A burned-out or defective switch. A burned-out capacitor, possibly resulting from lightning strikes in the area.

2. Compressor crankshaft is frozen in the bearings from overheating and lack of oil lubrication.

3. For belt-driven compressors, belt may have become too tight, thus overloading the motor.

4. Pressure control may have taken charge, either from high head pressure or low suction resulting from a leakage of refrigerant.

5. Low line voltage.

Recommended Actions

1. Check the electrical system, including the capacitor. Renew defective items as found.

2. If motor and compressor refuse to turn over by hand, check shaft seal for binding. If seal is satisfactory, compressor must be opened up for inspection. Disconnecting motor from compressor allows each to be tested by turnover, thus showing whether the trouble is in a particular unit or is common to both.

3. Check tightness of belt for belt-driven machines.

4. Note the head and suction pressures. If conditions are abnormal, the pressure control may be in charge, and the correction will take the form of establishing normal pressures by charging system with additional refrigerant, by purging air pockets, etc.

5. Check line voltage by voltmeter. If more than 10 percent below rated voltage, a burnout of the motor is possible. Do not attempt to restart until normal voltage is restored.

(g) ***Complaint.*** Crankcase sweats and develops frost.

Causes

1. Liquid refrigerant is getting back to compressor.
2. The system is overcharged with refrigerant.
3. Piston rings are worn and permitting refrigerant to get past.

Recommended Actions

1. Examine trap on suction line ahead of compressor. Install larger trap if line is flooding with liquid refrigerant.

2. Test functioning of expansion valve. If passing too much liquid refrigerant, check position and operation of the thermostatic bulb that controls the opening of the expansion valve. A loose bulb can result in a flooding of the coil, with air layers intervening where the coil should be tight against the bulb.

3. Listen with the stethoscope for a blow-by of refrigerant past the piston rings in any of the cylinders.

(h) ***Complaint.*** Compressor is pulling an abnormally low suction pressure resulting in the pressure control cutting out compressor operation.

Causes

1. System is low in refrigerant.

2. Expansion valve is fouled with dirt or other foreign matter that restricts the orifice. This condition is most likely to occur when low temperatures in the expansion coil are desired and the expansion valve orifice is of relatively small area.

3. Compressor is overcharged with oil, and too much oil is being carried through the system.

4. Water in the refrigerant is likely to cause a freeze-up at the orifice of the expansion valve and either restrict or stop refrigerant-feed through the valve.

Recommended Actions

1. Check the functioning of the expansion valve and its controlling thermostatic bulb. Clean the valve orifice and its strainer. If ice is the trouble, it will be seen on this inspection.

2. Verify the adequacy of the charge of refrigerant, pumping down if necessary.

3. Note oil level in crankcase. If too high, or shows water present, bleed off to a normal level after the water is removed.

4 If refrigerant is under the amount required, build up and inspect system for leaks. The halide torch may be used if the area and atmosphere are safe for an open flame, otherwise check with a soap solution.

(i) ***Complaint.*** The building heat-pump installation is not delivering sufficient cooling.

Causes

1. All the troubles that affect a refrigeration compressor are applicable to the compressor in a heat-pump system. Refer to Sec. 9-1(a) through (h).

2. In some cases the heat sink exchanger is required to operate at a temperature higher than intended. For example, a coil designed to

operate at a maximum ambient temperature of 100 F is so installed as to receive the direct rays of sunlight. Because of the heat from the sun, the effective operating temperature becomes perhaps 115 F, thereby increasing the volume of the gas and seriously reducing the capacity of the compressor.

Recommended Actions

1. Consider the recommended actions in Sec. 9-1(a) through (h).
2. Provide shading against the sun heat or increase the total area of the heat exchange surface. Do not speed up the compressor since this shortens the life of the machine.

9-2. COMMERCIAL INSTALLATIONS, MECHANICAL

(a) ***General Notes.*** A commercial design of frequent occurrence is composed of an indoor air-handling casing containing the expansion coils, blower, filter, and zone control dampers, while the compressor, condenser, and water tower, if any, are mounted outside, frequently on the roof.

In another favored arrangement, the expansion coil is buried in a water chiller where the water is reduced to a range of 40 to 48 F, on the average, and then circulated by pump to air-cooling coils. One or more blowers are contained in the cabinet with the coils, and these blowers force the cooled air through ducts to the areas of discharge. The compressor and condenser circuit, together with a water tower or other device for disposing of the waste heat, are located as noted before in a convenient spot such as the roof.

In another system, known as the absorption system, heat is used to generate chilled water. The key units in this system are shown in Fig. 9-1(a) and (b). The absorption system is best suited to applications where waste heat or low-pressure steam is available for boiling out the water from the lithium bromide solution in the generator. In these circumstances, the absorption system tends to be least expensive and therefore is likely to be encountered where the exhaust from a diesel engine or gas turbine is available.

High-pressure steam, from 100 psig up, is used occasionally for creating a partial vacuum over water in a tank similar to the evaporator tank in the absorption system. Evaporation from the surface of the water is thus promoted, and chilling of the remaining water is of course the natural consequence. Figures 9-2 and 9-3 illustrate steam jet arrangements. The circulating pump delivers the chilled water to the air-cooling coils; the condensate pump returns the steam condensate to the boiler for a repeat of the cycle.

The vacuum maintained above the surface of the water in the tank

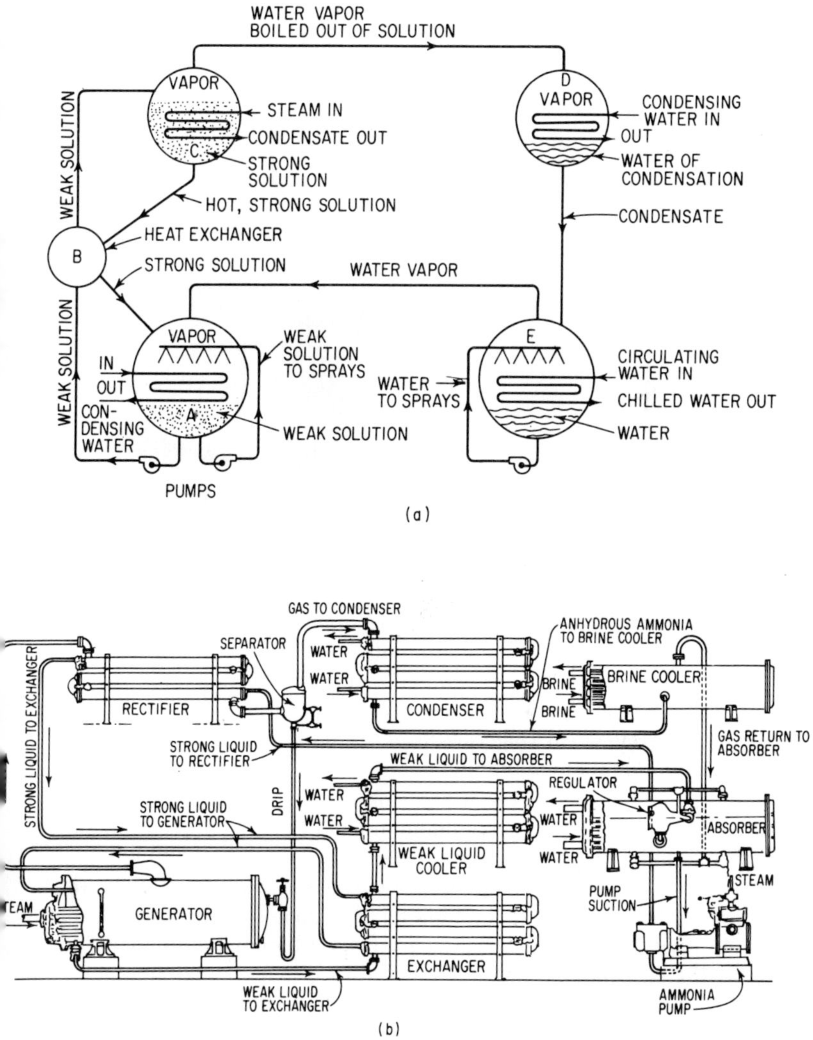

FIG. 9–1. Absorption cooling cycle. (a) Absorption cycle. A, absorber where lithium bromide absorbs water; B, heat exchanger; C, generator where heat is applied to boil off water; D, condenser for water vapor; E, evaporator where evaporation chills water. Partial vacuum in evaporator is produced by vapor absorption in absorber. Pumps improve efficiency of cycle. (b) Diagram of the Vogt absorption refrigerating machine. (*Henry Vogt Machine Co., Inc.*)

FIG. 9–2. Steam jet water chiller with barometric condenser producing 200 tons of refrigeration. (*Ingersoll-Rand Company.*)

ordinarily is designed to produce a chilled water temperature of 40 F or thereabouts. The practical minimum is 35 F, corresponding to a vacuum of 29.9 in. of mercury, approximately.

The vacuum necessary for any given water temperature can be readily calculated from the steam tables. It is done by subtracting the absolute pressure in inches of mercury from a barometer of 29.922, this being the equivalent of atmospheric pressure at sea level.

The vertical arrangement in Fig. 9-2 is needed when the steam condenser is of the barometric design. Since the tail pipe must extend at least 34 ft above the level of the water in the hot well, space requirements of the design generally limit its use to industrial installations.

Although it may be less expensive than a surface-type condenser, it presents a possible drawback from its operating characteristic of mixing the condensing water with the steam. This mixture may not be acceptable for direct boiler use, since the condensing water could contain calcium, magnesium, mud, and organic matter.

Notes on Centrifugal Compressors. Figure 9-4 shows a two-stage centrifugal design used with a water chiller. Available in an almost unlimited range of sizes, this particular unit is rated at 500 tons of refrigeration.

Centrifugal machines are built in both open and hermetic forms. A point of potential trouble in the former is in its shaft seals. Loss of refrigerant can cut out the machine automatically.

Safety features which the troubleshooter should understand include a cutout switch for low pressure in the lubricating oil system; a cutout switch for excessive condenser pressure; a cutout switch for a condition of low water or brine in the liquid chiller; and automatic stoppage if the temperature of the refrigerant drops to an abnormal value.

Some machines will be fitted with an overload, or load-limiter device, that takes charge during start-up and at pull-down and prevents overloading of the motor.

Notes on Package Coolers. Figure 9-5(a) and (b) illustrates a package unit, in which everything from compressor to the conditioned air blower is packed or "packaged" within a common case. Since the condenser is set within the casing, some form of cooling water apparatus is needed, usually a cooling tower set on the ground or on the roof, as convenient. A pump is necessary to circulate the water from the condenser to the tower and back to the condenser.

When an air-cooled condenser is provided, the condenser moves out of the casing and goes outdoors, thus creating what the trade calls a split system.

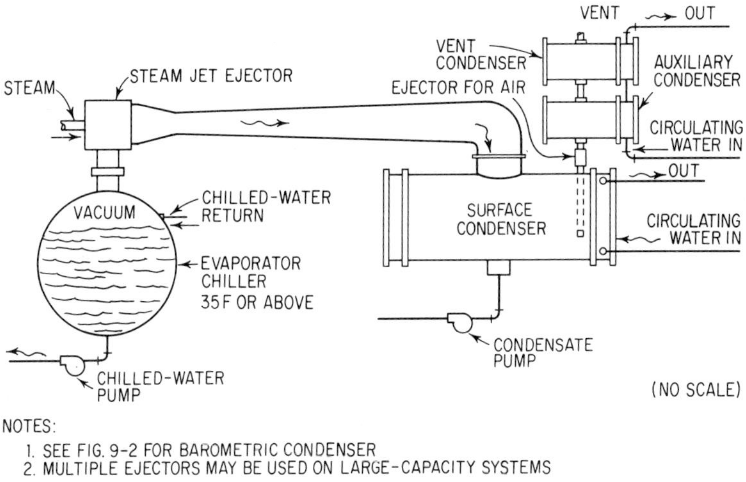

FIG. 9–3. Surface condenser and steam jets (no scale).

FIG. 9-4. Two-stage centrifugal compressor operating with a water chiller. (*Airtemp Division, Chrysler Corporation.*)

Package units can be used with or without distributing ducts. Without ducts, the discharge is from the upper regions of the casing, arranged to blow horizontally a distance of from 30 to 40 ft. Duct systems involve discharge registers placed in suitable areas for the most effective cooling. The advantage of a duct system is that the ducts can double as heating components in winter, thus becoming part of an all-year system.

Notes on Rotary Compressors. Figure 9-6 is a sketch illustrating the principle of the rotary method of compression. Whereas the piston in a reciprocating compressor operates on a back-and-forth cycle, the gas in a rotary machine is compressed by being squeezed between a rotating shaft and the housing or cylinder that encloses it. The rotating shaft, to serve its purpose, is mounted off-center with regard to the axis of the housing. Rotary compressors are found most frequently in the smaller appliances, such as deep freezes, conventional refrigerators, and the like.

Notes on Air-handling Units. As noted previously, the air-handling components of the system include the filter, the expansion coil or chilled water coil, and the blower or blowers that force the air across the coil and out to the space being conditioned. These are all con-

FIG. 9–5. (a) Cabinet for package-type air conditioner. (b) Package air conditioner with panel removed. (*York Corporation.*)

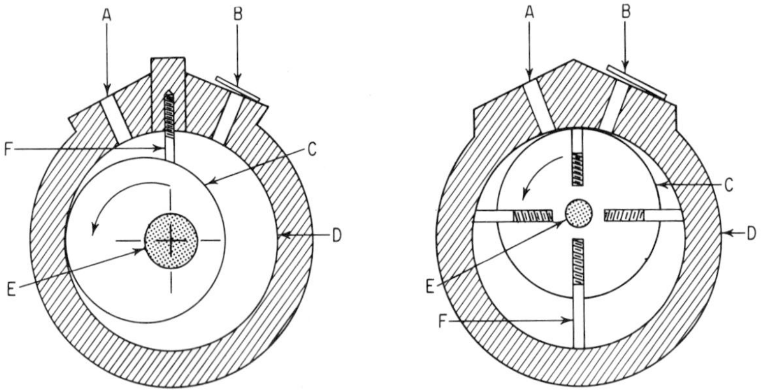

FIG. 9–6. Two forms of rotary compressor (no scale). A, inlet port; B, discharge valve; C, rotor; D, cylinder or casing; E, rotor shaft; F, sealing blades, spring-loaded.

tained within the casing or cabinet of the package design, but in split-system installations, they are cased by themselves.

In the package illustration in Fig. 9-5, the blowers cannot be seen, but the discharge openings in the top of the cabinet and a section of the

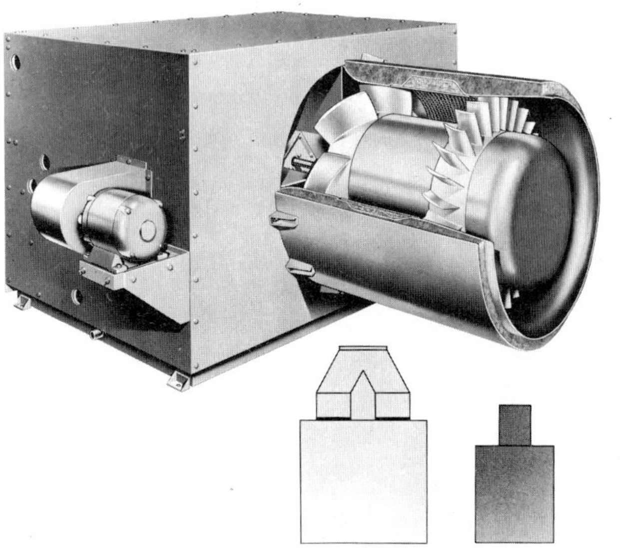

FIG. 9–7. Recent unusual design and arrangement of air-handler blower. (*The Trane Company.*)

return side filter are shown. The compressor does not appear either, the space assigned to it being empty.

Table 9-1. Common Refrigerant Control Devices

Form of device	Normal condition evaporator coil	Compressor operation controlled by
Automatic expansion valve	Dry. Should contain refrigerant gas.	Thermostat and pressure gage.
Thermostatic type of expansion valve	Mostly dry. Should contain refrigerant gas.	Thermostatic bulb attached to outlet end of coil. Temperature of coil outlet determines opening of expansion valve. Pressure gage controls pump operation, but temperature cutoff may be provided also.
Capillary tube *	Flooded. A dry coil shows capillary is overloaded.	Thermostat in conditioned space.
Low-side float valve	Flooded. Refrigerant goes directly from compressor to float valve in evaporator which admits liquid to evaporator as needed.	Thermostat and pressure gage.
High-side float valve	Flooded. Refrigerant receiver, under pressure from compressor, contains float valve which passes liquid through tubing to evaporator as needed.	Thermostat.

* Originally used mostly on small domestic appliances, the simplicity and economy of the capillary tube as a refrigerating control device has promoted its adoption by much larger systems. Installations of 10 tons or more have performed successfully.

Figure 9-7 shows the most recent design of air handler, in which the blower is designed to a type and size that reduces the overall space needed by the handler.

Troubleshooting the air handler will be discussed as a part of the overall system performance, since any malfunction of the handler immediately affects the quality, quantity, or both, of the delivered conditioned air.

File the eight items of site data in Table 9-2 for present guidance and future reference. They are the supports of the investigation.

Table 9-2. The Eight Pillars of Refrigeration Troubleshooting

On arriving at the site of an air-conditioning complaint, the first action of the troubleshooter (unless the cause of the trouble is already known) is to record significant conditions as he finds them. This program immediately indicates the areas in which performance is ailing or even disrupted, and the data so collected often point with reason to the underlying cause.

Here are the data to be recorded:

1. *Note the temperature of each conditioned space.*
   For zoned installations, record temperature of each zone.
2. *Note the status of each control thermostat.*
   List as satisfied or not satisfied and calling for cooling.
3. *Note temperature of air from discharge of air handler.*
   Also note temperature of return air entering the handler.
4. *Note refrigerant suction pressure and temperature.*
   Compare this with setting of the low-pressure cutout.
5. *Note liquid pressure and temperature leaving the condenser.*
   Compare with control settings of the cutout switch.
6. *Note condenser conditions.*
   (a) For air-cooled, note air-in and air-out temperatures.
   (b) For water-cooled, note temperatures of water in and out.
7. *If compressor seems to be short-cycling, note length of run in minutes.*
8. *Note the presence of any unusual sounds.*
   If possible, designate apparent source, such as compressor, blower, or expansion valve.

(b) ***Complaint.*** The conditioned area is too warm.

Causes

1. If compressor is running all the time, refer to Sec. 9-1(b).
2. Compressor is short-cycling; see Sec. 9-1(c).
3. High head-pressure cutout switch is functioning because of ice in expansion valve or other obstruction to the normal movement of refrigerant.
4. Condenser conditions are unsatisfactory. Temperature of air or water is too high as a result of dirty coils and passages.

5. Air in the system.

Recommended Actions

1. Consider recommendations given in Sec. 9-1(b).
2. Check entire refrigerant system for evidence of obstruction somewhere along the pipeline.
3. Make sure expansion valve is not frozen.
4. Clean the heat-exchange coils of the air condenser.
5. Compare rpm of condenser fan with the rated speed.
6. For water-cooled condenser, examine water circuit for algae. Clean as necessary.
7. Inspect tower for obstructing leaves, branches, and other foreign items.
8. Confirm satisfactory operation of the circulating pump.

(c) ***Complaint.*** Compressor is noisy. See Sec. 9-1(d).

(d) ***Complaint.*** Compressor will not start. See Sec. 9-1(f).

(e) ***Complaint.*** With an open-type, motor-driven compressor, the motor runs too hot.

Causes

1. Belt connecting motor to compressor is too tight.
2. Motor is undersized. See Sec. 9-1(b).
3. Bearings need lubrication.
4. Line voltage may be a little below that for which the motor was designed. Note whether lights dim momentarily when motor starts.

Recommended Actions

1. Slack off on driving belt. Adjust pulleys.
2. Replace the motor and compressor with units of adequate size.
3. Oil all bearings.
4. Check line voltage. If more than 5 percent below the motor design, investigate ways to reduce the load. For example, increase the size of the service wiring.

(f) ***Complaint.*** Compressor motor burns out with abnormal frequency.

Causes

1. It is operating with voltage too low. See above.
2. If all three phases of the incoming electric service are not protected by grounded lightning arresters, surges on the line from near-hits may burn out the capacitor.
3. After first burnout, system was not adequately flushed out and cleaned. This left contaminating acids and sludge in the system, which may have caused subsequent failures.

Recommended Actions

1. Correct low-voltage conditions.
2. Inspect service for proper lightning arresters. Install if all three phases are not protected.

3. Install a dryer in the suction line to the compressor if none is present. If present in the line immediately after a burnout, remove and replace with a new, noncontaminated unit.

(g) ***Complaint.*** Conditioned space is warmer than desired and a hissing sound is audible in the area of the expansion valve.

Causes

1. Hissing results from some partial obstruction in the valves or strainer. Check for the presence of ice.
2. The system is low on refrigerant.
3. Expansion valve is too small, or the orifice is out of adjustment.

Recommended Actions

1. Confirm free condition of refrigerant piping.
2. If low in refrigerant, add as necessary. Look for a point of refrigerant leakage, using the halide torch.
3. Verify size and model of expansion valve. Replace or correct setting as necessary.

(h) ***Complaint.*** Air-conditioning system that performed well for several years no longer seems able to meet the comfort requirements. This system uses an evaporative condenser.

Causes

1. Condenser spray nozzles are clogged.
2. Pump discharge pressure has fallen off.
3. Condenser fan is running below design speed.
4. Exchange surface of coil is dirty.

Recommended Actions

1. Increase water pressure enough to blow out the spray nozzles, or remove the nozzles and clean by hand. Check for advanced corrosion that might require replacement.
2. Check pump operation for speed, discharge head, and suction lift or pressure as appropriate. Determine how much pump efficiency has fallen off. If more than 15 percent, replacement of the pump may be involved.
3. Clean the fan with special attention to accumulations on the blades. These may be enough to seriously reduce the speed of the fan and therefore its capacity.
4. Clean the heat exchange surface. Also kill and flush out all algae in the condenser basin.

(i) ***Complaint.*** Conditioned space is generally too cold. Adjustments to the controlling thermostats produce conditions that are too warm.

Causes

1. If there is more than one zone, this complaint indicates a lack of proper balancing, possibly due to wide variations in the conditioned-air requirements of adjacent zones.

2. One of the control thermostats is defective in its operation.
3. The zone dampers are incorrectly set.

Recommended Actions

1. Check the correctness of the response of the control thermostats.
2. Check the positions and movements of the zone dampers.
3. Balance the air movements through the ducts to the various conditioned spaces.
4. Check rpm of blower in air handler.

(j) ***Complaint.*** System uses an excessive amount of makeup water as reflected by recent increases in city water bills.

Causes

1. With a cooling-tower system of condenser cooling, a leak in the basin or elsewhere in the system is requiring excessive city makeup.
2. With a chilled-water system functioning with air-cooling coils, the leak may be in the coils.

Recommended Actions

1. Check all water-using components of the cooling system for leakage.
2. If no leak is found, request a check on the accuracy of the water meter.

(k) ***Complaint.*** Cost of system operation appears to be excessive. Power bills are much higher than was predicted by the contractor who furnished and installed the entire system.

Causes

1. Capacity of the equipment may be inadequate for the existing climatic conditions, resulting in uninterrupted operation. Concurrently, the cooling effects may be reported as unsatisfactory.
2. If originally satisfactory power bills have become excessive, there may be a clogged and dirty filter or an ailing condition of the air-cooled condenser or of the water-cooled condenser.
3. Investigate possibilities of putting system on a more favorable power rate. Discuss this with power company.

(l) ***Complaint.*** System is excessively noisy. Sounds are projected from the machinery room, through the ducts, and into the conditioned spaces. See Sec. 7-2(c) for a similar complaint in heating.

Causes

1. For compressor origin, refer to Sec. 9-1(d).
2. Air is moving at excessive velocity through ducts.
3. Air velocity emerging through registers or diffusers is exceeding the limits required by noise control.
4. On return side of system, face velocity of air at inlet grilles is too high, indicating inlets have been undersized.

Recommended Actions

1. Correct compressor operation as indicated in Sec. 9-1(d).
2. Check air velocity through ducts, diffusers, and return grilles.

Compare with allowable speeds from viewpoint of sound control. Note the maximum sound level acceptable in each of the air-conditioned spaces.

3. If velocity is higher than the design calls for, determine speed of blower and adjust accordingly.

4. Consider lining the duct with sound-absorbing material for the first 6 or 8 ft of the duct as it leaves the air-handler connection. Material must be at least ½ in. thick, and a greater thickness is preferred. Alternatively, replace a short section of the supply duct with a proprietary type of sound trap.

5. In some instances, replacement of the existing diffuser or grille with one operating at a lower sound level is practical and desirable.

6. If duct walls are vibrating, stiffen with strap or angle iron.

(m) ***Complaint.*** Walls and ceiling around various outlets, diffusers, and registers are developing black smudges. See also Sec. 7-2(b).

Causes

1. Air leaving the outlet is carrying dirt or dust.

2. Room air, already burdened with dust, is induced by low-pressure areas on the ceiling adjacent to the diffusers to settle out in those areas. This air is nonfiltered air.

Recommended Actions

1. Clean or replace all the air filters.

2. Clean the ducts and use a vacuum cleaner and a brush on the outlets themselves.

3. Wipe off all smudges from wall and ceiling surfaces, and instruct system owner and operator to do the same, preferably every 2 weeks and never less frequently than once a month.

(n) ***Complaint.*** Conditioned air emerging from diffuser and register outlets carries a noticeable and unpleasant odor. See also Sec. 7-2(d).

Causes

1. If odor develops soon after new system is put into operation, probabilities are that some foreign material has been left in the ducts by the erecting workmen. Common findings include sandwiches, fruit, paper coffee cups, cake, pie, dirty waste, rags, gloves, caps, or anything that might have been set temporarily in the duct and then forgotten. Other items include the bodies of dead mice, rats, birds, and occasionally a snake, all of which had the misfortune to choose a hiding place that was shortly afterward closed up and thus became a smelly coffin.

2. The air filter is dirty.

3. Refrigerant is leaking from the air-cooling coil.

4. Odors may be originating outside the air-conditioning system, being brought into the cycle with the fresh air makeup.

Recommended Actions

1. Clean all air filters.

2. Open ducts at appropriate points and clean with brush and vacuum cleaner as appropriate. Remove all foreign materials and objects.

3. Check the refrigerant circuits for leakage. Use the halide torch or other leak detector as available.

4. If odor is from outside source, one possibility is to bring in outside air through an activated carbon filter. Alternatively, investigate the possibility of relocating the intake to an area where the contamination is either absent or acceptably less concentrated.

9-3. COMMERCIAL INSTALLATIONS, ABSORPTION

(a) ***The Components of an Absorption System.*** The troubleshooter will meet absorption installations in which each of the basic components of generator, condenser, evaporator, and absorber (Fig. 9-1) is separately housed. He may meet them all fitted neatly into a single horizontal tank, or the "machine," as it is frequently called in the trade, will be designed for, and placed in, a vertical position. The refrigerant is usually lithium bromide or ammonia.

Here are comments on the various units of an absorption system, not all of which are encountered everywhere. The troubleshooter should obtain, if possible, a diagram of the installation he is about to service.

1. *Evaporator or Cooler:* This cools the medium to be cooled by the evaporation of the refrigerant. In Fig. 9-1, note that the water being chilled is circulated through a coil submerged in the water that is being continually evaporated in this unit. Evaporation is the result of a high level of vacuum over the surface of the water, this vacuum being maintained by the absorption of the water vapor in the adjacent absorber. This part of the cycle is comparable to the steam jet system, the steam jet creating a vacuum over the water in place of the absorber.

2. *Condenser:* This unit liquefies the water vapor that is being driven off the refrigerant-water solution in the generator. Cooling water for the condenser coil may be obtained from a cooling tower, an evaporative condenser, or even a cold lake or river, if one is handy.

3. *Absorber:* Here the water vapor from the evaporator comes into contact with the refrigerant and is absorbed thereby. The mixture is termed the "weak solution," naturally, since the "strong solution" from the generator is diluted by the water.

Because heat is produced in the absorber by the reaction, cooling coils are provided in the absorber, the cold water for these coils being obtained from any convenient source, including cooling towers and evaporative condensers. Figure 9-1 shows the strong solution being introduced into the absorber from a nozzle; other designs spread the strong solution over the cooling coils by means of spray nozzles, and the water vapor from the evaporator is discharged into contact with the spray.

4. *Generator:* This is the part of the system in which the heat of operation is applied. Figure 9-1 shows the heat being applied through steam coils; however, in some designs, the generator may be direct-fired, using oil or gas as convenient or economical.

In the generator, the water vapor that was originally created by being driven out of the solution in the generator, then was condensed in the condenser in order to be revaporized in the evaporator, was absorbed by the strong solution in the absorber, and now is back in the generator as part of the solution, again feels the heat of distillation, and once more, in the form of vapor, sets out on a repetition of the cycle.

In case this all seems a little complicated and repetitious, the key to the entire cycle is the necessity of removing heat from the water in the evaporator by creating a vacuum over it and then disposing of the heat of vaporization by transferring it to various cooling and condensing coils. Figure 9-1 shows where these coils are.

5. *The Heat Exchanger:* This unit is interposed between the absorber and the generator. The strong solution, returning to the absorber from the generator, transfers much of its heat to the weak solution, which is traveling in the opposite direction to be boiled, ultimately, in the generator. In this way, heat is conserved, increasing the economy of the generator operation.

6. *Purger:* Since the high vacuum necessary for effective operation would be affected adversely by noncondensable gases or air, a purge unit may be provided in the form of a vacuum pump or eductor. The purger is not shown in Fig. 9-1.

7. *Controls and Protective Devices:* These include a low temperature cutout, which stops the machine if the evaporator temperature is too low; liquid-level float switches that shut down operations if the liquid level in the evaporator becomes abnormal, either too high or too low. In addition, ammonia-water machines may be subject to a high-pressure switch, and direct-fired generators may be protected by a high limit temperature switch.

(**b**) ***Preliminary Findings.*** The troubleshooter should note, as his first action in arriving at the trouble site, the temperature of the chilled water entering the evaporator and the temperature of the chilled water leaving the evaporator. He later can compare this with the design temperatures, noting the departures from the design.

Other data of importance include the condensing-water temperatures at the various components, both in and out, and the temperature and pressure of the steam to the generator coils.

(**c**) ***Complaint.*** Performance of the absorption system is erratic.

At times it is excellent; at other times, it seems to be entirely inadequate.

Causes

1. The supply of low-pressure steam or of the exhaust gases from diesel engines or gas turbines is erratic.

2. If bypass dampers are used with a waste-heat boiler or other form of heat exchanger, the control mechanism of the dampers may be in need of adjustment.

3. One of the circulating pumps handling a solution is performing below its rating.

Recommended Actions

1. Compare the temperatures of the steam and of the exhaust gases with the temperatures called for by their design. These may fluctuate with the load on the engines or on the waste boiler.

2. Compare the chilled-water temperatures as observed with the temperatures called for by the design.

3. Make what adjustments and changes are necessary to establish a satisfactory performance, in accordance with the design values. Clean heat exchangers and waste boiler tubes of accumulated soot. Note whether cleaning of the tubes produces a drop in exhaust-gas temperatures leaving the exchangers or boilers.

4. Verify settings of bypass dampers and other controls.

5. Check appearance of sprays for possible partial blockage of the nozzles.

6. Check balance of refrigerant with water. Too much water can affect performance adversely. Note whether volume of solution is enough to meet the load requirements.

(d) ***Complaint.*** Condensing-water temperature is so high as to reduce the cooling capacity of the machine.

Causes and Recommended Actions

1. If it is the evaporative type of condenser for cooling the water to the machine condenser, refer to Sec. 9-2(h).

2. If a cooling tower is used, note temperatures of water on entering tower and on leaving. Verify the speed and capacity of the pump.

3. If it is a natural-draft tower and trouble seems to develop on extra hot days, consider replacement with a forced- or induced-draft tower. Look for leaves, branches, and other obstructions to a free passage of the air through the tower.

9-4. COMMERCIAL INSTALLATIONS, AMMONIA ABSORPTION

(a) ***General Notes.*** The major difference between an ammonia absorption system and the lithium bromide cycle is that with the lithium bromide, the water is the refrigerant, and the lithium bromide, the absorbing material. In an ammonia system, the ammonia is the refrigerant, and water the absorber.

Referring to Fig. 9-1, instead of water vapor leaving the generator and going to the condenser, ammonia vapor is driven off the strong liquid and follows the circuit to the condenser.

Another difference is that with water as the refrigerant, evaporation is induced by the establishment of a high vacuum in the evaporator. In the ammonia machines, the cycle operates under a positive pressure comparable to the pressures of the mechanical compressor systems. This being true, ammonia can be used to produce subzero temperature conditions, an impossibility with water, which of course faces a 32-F limit in the evaporator. An ammonia system arrangement is shown in Fig. 9-8.

Basic components are the same as in the lithium bromide cycle; however, several units peculiar to ammonia installations should be noted. These are:

1. *Brine Cooler:* This is used when the cooling medium for circulation is a brine, usually calcium chloride but on occasion sodium chloride or even a glycol solution. If chilled water is being circulated, the brine cooler is omitted.

2. *Analyzer:* Ammonia vapor leaving the generator on its way to the condenser, passes through a heat exchanger which permits it to transfer some of its heat to a counterflowing weak solution. The weak solution is on its way to the generator, and in some designs, the vapor bubbles through the liquid weak solution, thus giving actual contact in the exchange. The purpose of the analyzer is economic, the cost of the analyzer being made up for by the increased efficiency of condenser performance.

3. *Rectifier:* This is another heat exchanger immediately preceding the condenser, usually of shell-and-tube construction. The rectifier is expected to pick up any economies that might have been missed by the analyzer.

4. *Brine Pump:* Where chilled water is the cooling medium, the brine pump is not needed and is replaced by the chilled water circulating pump.

Terminology

Anhydrous ammonia is a "dry" ammonia and contains no more than ½ of 1 percent moisture.

Aqua ammonia refers to a "wet" solution of ammonia. Percentage of water in the ammonia varies and is determined by the design needs of the system.

Ammonia regulator is a valve that automatically controls the flow of weak liquid from the generator to the absorber.

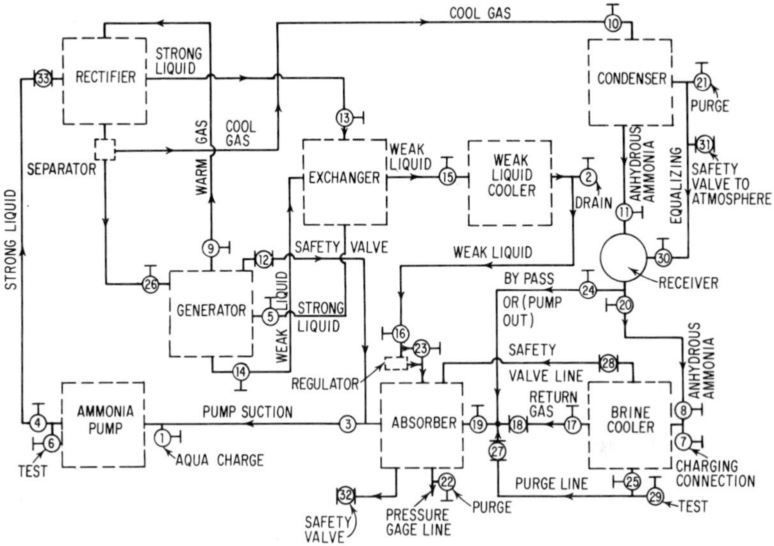

VALVES

1. Aqua charge
2. Weak liquid drain
3. Pump suction
4. Pump discharge
5. Strong liquid to generator
6. Sample valve
7. Charging valve
8. Expansion valve
9. Gas from generator
10. Gas to condenser
11. Liquid to receiver
12. Safety valve from generator
13. Strong liquid to exchanger
14. Weak liquid from generator
15. Weak liquid to weak liquid cooler
16. Weak liquid to regulator
17. Return gas on brine cooler
18. Check valve return gas line
19. Return gas to absorber
20. Liquid stop valve
21. Purge from condenser
22. Purge from absorber (in pressure gage line)
23. Weak liquid by pass around regulator
24. By pass of condenser pump-out
25. Purge valve from brine cooler
26. Drain from separator
27. Check valve in purge line
28. Safety valve from brine cooler
29. Drain or test valve from brine cooler
30. Equalizing line condenser and receiver
31. Safety valve on receiver
32. Safety valve on absorber
33. Check valve in pump discharge line

FIG. 9–8. Piping diagram of the absorption system showing the flow of liquids and gases. In this diagram, any reference to valves which are to be opened and closed means a single valve (if there is only one) and all the valves (if there are more than one) on any particular line. Apparatus-like condensers, weak liquid coolers, and absorbers, consisting of a number of stands, are equipped with individual valves for each stand. If there is no king valve in the single line, leading to such apparatus then, all the said valves must be closed where the instruction states that any valve is to be closed. No attempt has been made to show any valves in their true position on the machine. Identify all valves by numbers only. (*Henry Vogt Machine Co., Inc.*)

(b) ***Complaint.*** The performance of the ammonia system is failing to meet its design capacity.

Causes

Any one or a combination of the following conditions can produce a lack of capacity:

1. Insufficient steam pressure on the generator.
2. An insufficient charge of ammonia.
3. Air or noncondensable gases in the system.
4. Leaks in the rectifier, or any of the other heat exchangers. These are sometimes audible, producing a hissing sound.
5. Blow-by from leaky rings in the ammonia pump.
6. Condensing water is either too warm or being circulated in insufficient volume. The symptom is excessive pressure in the condenser or in the absorber.

Recommended Actions

1. For finding ammonia leaks in air, check with a lighted, sulfur-coated candle. If ammonia is present, a white vapor will be created.
2. For ammonia leaks in liquid, test the liquid with white litmus paper. Ammonia will cause it to change color.
3. To detect the presence of ammonia in brine, mix a small amount of caustic soda with a cup of brine. Then attach a piece of white litmus paper to one side of a piece of glass and lay the glass over the cup of brine-caustic mixture with the litmus paper down and exposed to any vapors that might rise from the brine in the cup. The litmus will turn red within a few minutes if any ammonia vapor rises from the mixture.

(c) ***Caution in Handling Ammonia Checks and Corrections.*** Ammonia even in small concentrations is highly irritating to the eyes and mucous passages, and poisonous besides. Therefore, the troubleshooter, before undertaking corrective actions at the site of a complaint, should make sure that one or the other of these protective devices is at hand:

1. A standard ammonia helmet. In some ammonia plants, one or more of these helmets is kept available at all times. If in doubt about such availability, the troubleshooter should take one with him from his own shop.
2. A cloth large enough to cover nose and mouth and well-saturated in vinegar. A large sponge may be used in place of the cloth if desired.

In pumping down a system, in order to avoid overfilling of the receiving cylinder, limit the inputs of the anhydrous ammonia to the amounts given in Table 9-3.

Another precaution, during the procedure of pumping down, is to maintain a good flow of cold water to all condensers in order to hold down the pressure and to completely liquefy the gas. Ammonia pres-

Table 9-3

Dimensions of receiving cylinder		Maximum safe input, lb *
Length, ft and in.	Diameter, in.	
2′-4″	10	20
3′-6″	10	40
3′-10″	10	50
7′-0″	10	100
7′-0″	12	150

* Determine the tare weight of the empty cylinder in each case before filling with ammonia.

sure passes the 100 psia level at a temperature of only 57 F and exceeds 200 psia at 97 F. An overfilled cylinder in a warm ambient can be as dangerous as a stick of dynamite with the fuse lighted.

9-5. STEAM JET SYSTEMS

(a) ***General Notes.*** As pointed out previously, a steam jet produces chilled water by establishing a high vacuum above the surface of the water in a tank termed an evaporator. The high vacuum causes rapid evaporation from the surface of the water in the tank, the heat of evaporation being taken from the body of the water.

Since each pound of water evaporated at an absolute pressure of 0.24 in. of mercury abstracts 1071 Btu from the body of the water, the water left in the tank is soon chilled down to 40 F.

The key to the development of the necessary vacuum is the steam jet, and we must have a steam pressure of from 70 to over 100 psia. The volumes of vapor are relatively enormous. For example, a pound of water vapor at 40 F occupies a space of 2,444 cu ft. Low steam pressures are unable to handle such large volumes of vapor; consequently, one of the problems of a steam jet installation is that of steam pressure.

Failure of a steam jet system to meet its rated performance can usually be tracked down to one of these causes:

1. Steam pressure is too low. In some industrial installations, a drop in steam pressure is periodic, occurring only when process demands for steam coincide with a strong need for comfort cooling. In these instances, the fault may be mysterious at first if the pressure happens to be high and adequate at the time the troubleshooter arrives at the job. If the boiler is fitted with a steam chart, the dips on the chart marking the drop in pressure may solve the problem.

2. The condensers are being overloaded as the result of the condensing water being too warm or too skimpy. This condition can be illuminated by checking the condenser supply water temperature and

doing the same with the condenser discharge. The net effect of course is to reduce the vacuum on the evaporator and thus reduce the refrigerating capacity of the machine.

3. Loss of evaporating capacity is resulting from erosions of the high-pressure steam nozzles. If this is occurring frequently, consideration should be given to using nozzles of a special alloy. Stainless steel is a good possibility under these conditions.

9-6. EVAPORATIVE COOLING

(a) ***General Notes.*** Evaporative cooling in the everyday sense refers to a system of materials and devices designed to promote the evaporation of water under atmospheric conditions. Since there is no vacuum involved, the cooling effect of the evaporation is limited by the existing wet-bulb temperature. The wet bulb often varies from hour to hour as well as from day to day; consequently, we should not be surprised if an arrangement that is satisfactory today fails to deliver enough cooling tomorrow. This is an important fact for the troubleshooter to understand when he answers a call from the owner of an evaporative cooling installation.

While an efficiently designed system comes pretty close to the wet bulb, we can expect it to deliver air from 2 to 3 deg warmer than the wet bulb. Thus if we have an existing wet bulb of 60 F, the ultimate temperature of the air being delivered to the conditioned spaces will be about 62 or 63 F. With an outside dry bulb of say 90 F, this 28-deg lowering of temperature is quite worthwhile. However, the pleasantness of that cool air is somewhat diminished by its burden of water vapor; it has a relative humidity of more than 90 percent.

Evaporative cooling is what takes place in a cooling tower and thus becomes an important auxiliary to many mechanical systems of air conditioning. When it does the whole job, without the aid of compressors, absorbers, or condensers, and the water it cools removes the sensible heat of the air being delivered to the conditioned spaces, we have the kind of arrangement common in the dry areas of the country.

One form of installation is the air washer. Another involves the passage of the air through a matting material which is kept wet constantly by water trickling through it and over its surfaces.

Because of the basic simplicity of the system, evaporative cooling troubles, if and when they occur, are likely to be obvious. However the following have been reported, which might not be readily discovered:

1. Providing an exhaust system to move out room air so that the conditioned air can enter may have been overlooked, and in consequence, the fan is blowing nowhere, and the conditioned air never reaches its

destination. A recommended exhaust arrangement is to direct the departing vitiated air through the attic, thereby keeping attic temperatures within reason.

Incidentally, a rule-of-thumb method for determining the amount of cooled air necessary is the formula:

$$\frac{\text{Dry-bulb temp} - \text{wet-bulb temp}}{10} = \text{minutes between air changes}$$

For example, $(95 - 60)/10 = 3\frac{1}{2}$ minutes or 17 changes per hour. Use of this quick calculation will help the troubleshooter in checking the performance of the fan.

2. Some of the fan ratings, as given in the manufacturers' catalogs, may be unrealized in actual performance because of some peculiarity of the environment or atmospheric conditions. In these instances, a fan of larger capacity must be installed if satisfaction is to be achieved. Use the anemometer to determine the fan output.

(b) ***Two-stage Evaporative Cooling.*** This arrangement, intended to provide lower air temperatures than are possible with the single unit evaporative cooler, typically provides a heat-exchange coil in the air stream followed in series by the evaporative unit. Cool water circulated through the coil is obtained by means of a cooling tower, perhaps a well, a convenient lake, or a river. To be most useful, the precooling coil should lower the air temperature entering the coil to at least 15 deg above the outdoor wet bulb.

Two-stage arrangements of course increase the variety of equipment that may give rise to trouble. These include the cooling tower, the cool water circulating pumps, the additional piping runs, float control valves, and other controls. Refer to these items under their own names in previous sections of this handbook for guidance in identifying and correcting their operating troubles.

(c) ***Applications.*** Evaporative cooling service calls are not frequent to the average troubleshooter, except in the Southwest and other dry areas of the United States. The troubleshooter should know that evaporative cooling systems, despite their limited applications, have commercial and industrial applications where a measure, if not an absolute condition, of environmental comfort is needed.

Applications therefore include tobacco factories, steel mills, greenhouses, agricultural and citrus storage plants, and the spot cooling of machinery.

9-7. WATER PROBLEMS IN REFRIGERATION SYSTEMS

(a) ***General Notes.*** Water is used in comfort cooling systems as a transfer medium for heat. In a chilled water system, it is being constantly recirculated and is not normally exposed to the possible effects and contaminations of the atmosphere. When used in cooling towers or evaporative condensers, a percentage of the water is lost to windage and evaporation, causing a continuing concentration of impurities and undesirable contaminants despite the automatic makeup of the losses from a clean supply.

Complaints that are generated directly or indirectly from the condition of the water almost exclusively concern the system subject to contamination.

(b) ***Complaint.*** Water is hard and is promoting deposits of calcium and magnesium in the system, thus affecting heat exchange and the functioning of controls.

Cause

Trouble from hardness is not common in refrigeration systems, since depositions of the calcium and magnesium salts generally follow heating of the water rather than cooling. However, it does develop as evaporation proceeds in cooling towers and evaporative condenser and leaves the mineral salts behind.

Recommended Actions

1. Manually remove the accumulated scale found in the water basin of the tower or condenser. Examine float-valve internals and the openings of pipe and fittings for hard layers of "lime." In some instances, "liming" may be so thick and hard as to require a complete replacement of the plugged or partially plugged item.

2. If not already established, set up a regular schedule of water blowdown. This is intended to flush out the hard scale before it has time to build up to trouble-causing concentrations in the basins, piping, and valves. Valves are particularly susceptible to operation-crippling accumulations.

3. Initiate a program of water treatment. This will require the services of a water-conditioning expert, but where makeup water is hard and automatic softening equipment is not available or economically preferable, the expert is well-justified.

4. Arrange to switch the source of makeup water to a soft supply. Many industrial plants, commercial establishments, and hotels operate water-softening equipment for their domestic and process water needs. The troubleshooter should check the capacity of the existing softening plant as a possible new source of makeup for his tower or condenser.

(c) ***Complaint.*** The water is corrosive.

Cause

Atmospheric gases of a corrosive nature, when in contact with water surfaces, become absorbed by the water. The concentration increases as the water evaporates; consequently, the basin reservoir develops metal-attacking characteristics, which are carried through the system by the pumps.

Recommended Actions

1. Invite the services of a water expert to prescribe methods and concentrations of a corrosion inhibitor, with introduction of the inhibitor into the system to be automatic.

2. Do not scrape clean the corroded surfaces until some form of corrosion-inhibiting treatment is ready for application. Otherwise sound metal under the corrosive material is subjected to immediate attack, with no protection whatever.

(d) ***Complaint.*** Algae and slime accumulations are a constant source of trouble in the tower or condenser water system. They clog nozzles.

Cause

Organisms of one kind or another are nurtured by sun and warmth of the water. Algae are almost exclusively plant growths, while slime usually contains a substantial proportion of microscopic animal cells.

Recommended Actions

1. Manual cleaning removes the worst of these contaminants but leaves enough seed items to reproduce an undesirable condition in a short time.

2. A water-treatment expert will prescribe a program of chemical cleaning capable of reaching through the piping circuits to water jackets, spray heads, and other manually inaccessible areas. Common chemicals are chlorine and copper sulfate, the proper dosage being established correctly only after tests to determine the needed concentration. The sign of successful treatment is the change in color of the algae and slime from green or blue-green to dead or rusty brown.

9-8. SPECIAL COMPLAINTS IN AIR-CONDITIONING SYSTEMS

(a) ***Complaint.*** A battery of air-cooled condensers is excessively noisy and generating complaints from neighboring buildings.

Causes

1. Selection of fan equipment did not adhere to the noise standards established by the Air Moving and Conditioning Association, Inc. This may be a reason the fans are noisier than desired.

2. Condenser and fans are mounted in the open without any sort of sound barrier or sound absorber between their location and the neighboring buildings.

3. Condenser foundations and bolts holding it down are undersized, loose, or both, and the equipment is vibrating excessively.

4. Air screens need stiffening because of their large diameters and light metal construction.

Recommended Actions

1. Check the condition of all foundations, holding-down bolts, and unstiffened surfaces. Correct as necessary.

2. Check fan speeds against fan performance curves and consider possibilities of decreasing fan speeds.

3. Consider the erection of one or more sound barriers as sketched in Fig. 9-9.

(b) ***Complaint.*** "Hide-away" air-conditioning unit, concealed in closed space over a closet, is causing condensate to appear in the closet.

Cause

Humid air is reaching the evaporator coils and producing water that does not drain off.

Recommended Actions

1. Examine the condensate pan under the unit and determine why the condensate does not drain by way of the copper drain tube. Look for a pan that is tilted away from the drain hole and correct its pitch.

2. If possible, provide natural ventilation through the unit enclosure. This will permit the reevaporation of the condensate.

(c) ***Complaint.*** Water droplets emerging from the ceiling registers are staining the ceiling around the registers. Alternatively, several high wall registers are showing similar stains.

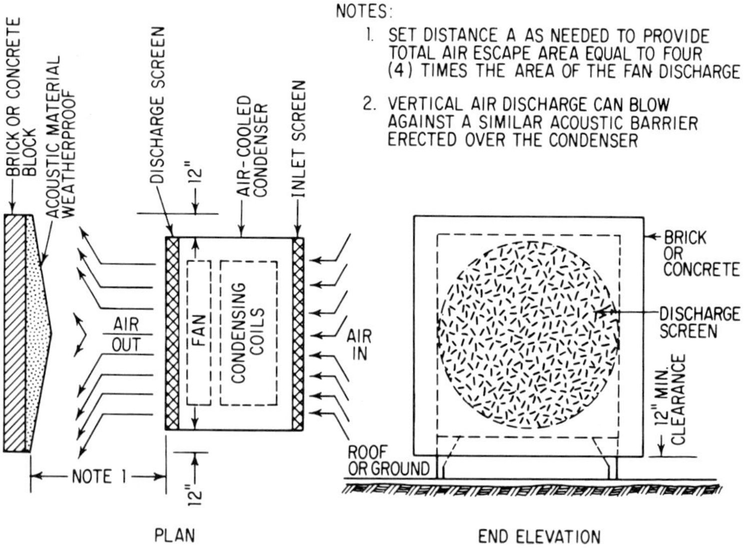

FIG. 9–9. Noise baffle, air-cooled condenser (no scale).

Causes

1. The temperature in the evaporator coils is too low. Excessive condensation is taking place on these coils and is being carried through the ducts.

2. As the result of outdoor, untreated air entering the conditioned spaces, the wet-bulb temperature in these spaces is higher than the temperature of the conditioned air emerging from the registers.

3. The water is condensation formed on noninsulated areas of the duct system and thence draining down the branches to emerge from the registers.

Recommended Actions

1. Step up the evaporator coil temperature to a wet-bulb depression of not over 30 deg from the outside dry bulb.

2. Locate the points of entry of the outside, untreated air and restrict them.

3. Inspect the entire duct system. If noninsulated areas are found, direct their covering with adequate materials including a vapor barrier.

4. Check the speed and delivery of the fan or fans in the air handler. Excessive velocity of the chilled air leaving the coils can easily carry moisture with it. Look for evidence of the fan speed having been increased or being of larger capacity than called for by the original design.

(d) ***Complaint.*** Vapors from a cooling tower are contributing to the formation of a fog, or industrial smog, in the vicinity of the tower. How can this condition be relieved?

Cause

For the vapors of a cooling tower to become visible, a certain combination of climatic conditions must occur. This may arise only once or twice a year and then perhaps for only a few hours. This combination consists of an atmosphere in which the relative humidity stands at 100 percent, while simultaneously there is a cool breeze or other drop in the dry-bulb temperature. The resulting fog can be quite annoying and even hazardous if it drifts over a heavily traveled highway. If the tower is atop a building, the fog may lie against the windows and obscure both the light and the view.

Recommended Action

The only effective treatment, discovered by a large refinery, is to mount several gas burners around the perimeter of the tower, so that when lighted, the products of combustion rise and mix with the water vapors, superheating these vapors and thus preventing fog formation.

9-9. HEAT PUMP INSTALLATIONS

(a) ***General Notes.*** The principle of the heat pump is simply that, on its heating cycle, it uses the heat of vaporization of the refrigerant to deliver heating comfort to a heat-needing space. On the cooling cycle, the heat pump functions as any other cooling system employing a mechanical compressor.

Figure 9-10 shows schematically how heat is taken from the outside air (or water) by permitting liquid refrigerant to vaporize in an outdoor air coil or in a coil submerged in water.

More heat is added to the vapor by the work of the compressor, and this total load of Btu is condensed in an inside coil. Thus the heat of vaporization and whatever has been added by the compressor are transferred to the inside air which is being blown across the condenser coil and is warmed accordingly.

The heat pump does its best work when the outside temperature range comes between 40 and 65 F. Below 40 F, the amount of heat in the outside air steadily decreases, and from about 35 F on down, some form of supplementary heat is needed to satisfy the heat demands. Usually this is in the form of resistance heaters that are actuated by a temperature control in steps of 4 kw each.

Several terms peculiar to the heat pump should be familiar to the troubleshooter. These include:

Balance Point: This is the minimum temperature at which the heat pump can meet the heating demands unaided. A drop of 1 deg from this point actuates the first step of the supplementary resistance heaters.

Coefficient of Performance (COP): This is the ratio of the energy purchased for operating the compressor, plus the energy used for supplementary heating in activating the electric strip heaters, and the total heat delivered to the condenser. For example:

$$\text{COP} = \frac{\text{80,000 Btu per hour to condenser}}{\text{8000 Btu to compressor} + \text{36,000 supplementary heat}}$$

$$\text{COP} = \text{1.81, approximately}$$

Heat Source: The outside air becomes the heat source when the evaporator is located outdoors and the heat of vaporization is taken from that air. Other heat sources are water, obtained from a city water system, from a nearby lake or river, or occasionally from the ground, evaporator coils being buried in the earth. Ground coils tend to be troublesome as a result of corrosion and also of ground moisture freezing around the coils and thus effectively insulating them against an adequate heat absorption.

Heat Sink: This term designates the place of heat disposal when the heat pump is operating on the cooling cycle. A common arrangement is to use the outdoor evaporator of the heating cycle as the heat sink for the heat removed from the indoor spaces on the cooling cycle. This is accomplished by reversing the refrigerant flow through the facility

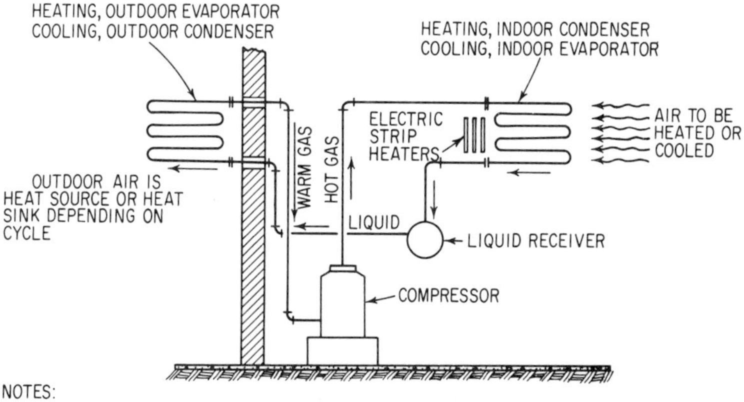

NOTES:

1. ELECTRIC STRIP HEATERS NOT ENERGIZED AT TEMPERATURES ABOVE 35 F
2. FOUR-WAY REVERSING VALVE, EXPANSION, AND OTHER VALVES NOT SHOWN

FIG. 9–10. Heat pump flow diagram, heating cycle (no scale).

of a four-way valve. Thus, in effect, the heating-cycle evaporator becomes the cooling-cycle condenser.

The troubleshooter generally should answer a service call aware of the following aspects of heat pump service:

1. Approximately 75 percent of the troubles develop in the motor. This may be the result of protracted short-cycling or, in hermetically sealed jobs, either the lubricating oil or the refrigerant or both might have deteriorated the motor insulation. Continued overheating of the motor indicates that an abnormal internal condition exists.

2. The life of the compressor cannot be expected to match the operating life of a similar compressor on strictly cooling service, so far as elapsed time is concerned. The heat pump compressor operates all year; the cooling unit only about 5 months. The ratio of annual operating hours is about 8,760 to 3,650, with heat requirements more than double the cooling requirements.

3. Next to motor troubles, erratic operation of the automatic controls is most frequent. In some climates where heat is needed in the morning and overnight and cooling is needed in the afternoons, failures of the controls to make the switchover day after day without a miss show up with some frequency. Another trouble area is the evaporator which becomes frosted and loses capacity following an interruption of the defrosting control.

4. Heat pumps come in the form of package units or split arrangements. Either form is subject to the generalities noted in the previous three paragraphs.

5. Trouble involving the shutdown of the heat pump is more likely to occur on the heating cycle than on the cooling. For this reason, and in view of the discomforts and hardships that can follow a protracted shutdown, service work should not be undertaken for jobs that are more than 50 miles distant from the repair shop.

(b) ***Running Down Complaints.*** All the troubles that can develop with a heat pump installation have their counterparts in conventional systems. Therefore in tracking down and correcting malperformance of a compressor, fan, evaporator, or condenser, the troubleshooter is referred to Secs. 9-1, "Domestic Installations"; 9-2, "Commercial Installations, Mechanical"; and 9-7, "Water Problems in Refrigeration Systems"; and the two Field Guides for mechanical systems, Tables 9-4 and 9-5.

9-10. ICE STORAGE SYSTEMS FOR AIR CONDITIONING

(a) ***General Notes.*** Not in general use, these ice-bank installations are designed for the primary purpose of holding electric service demand charges to a relatively low level. Since the freezing of the ice is spread over many hours, applications of the system are pretty well confined to activities that experience short high peaks. Illustrative of potential users are churches, funeral homes, and similar buildings of relatively brief but concentrated operation.

In addition, an ice bank may be considered for peak booster service; for example, if comfort must be maintained continuously and the general level of refrigeration demand is exceeded for an hour or two each day, the excess could be satisfied simply by cutting in an ice bank as an auxiliary source of added comfort. The ice from the ice bank is replaced by the base load refrigeration system.

The system design employs a water tank in which are submerged a number of refrigerated plates. To prepare the tank for its hours of duty, ice is built up on the plates to a predetermined thickness. In operation, water in the tank is kept cold by melting of the ice on the plates, and this water is pumped through fan coil units, thereby cooling the air blowing through the coils. The temperature of the water circulated is usually about 45 F, being diluted by warmer water through a mixing valve if necessary.

Refrigeration of the hollow plates is accomplished by mechanical refrigeration and direct expansion valves feeding each row of plates. Figure 9-11 shows the arrangement of a tank and its plates.

(b) ***Complaints and Their Handling.*** As with any other mechanical refrigeration system, complaints will be found to originate in the compressors, the condenser, the expansion valves, the controls, and the fans and blowers.

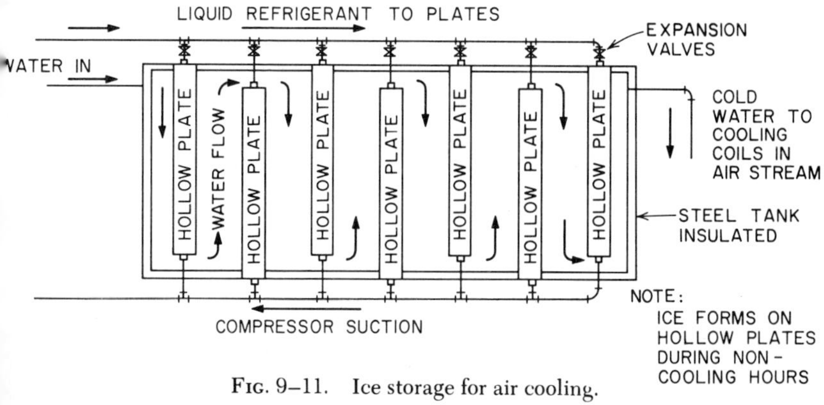

FIG. 9–11. Ice storage for air cooling.

In addition, the ice-bank installation may introduce conditions of mixing-valve malfunctions, producing a too cold or too warm environment.

Sweating of the tank may appear also if the insulation has not been presealed in a vapor barrier.

For the correction of the mechanical refrigeration troubles, refer to those sections of this division appropriate for each component of the refrigeration system.

For faulty water temperatures, adjust the control equipment as may be necessary. Examine the water mixing valves and replace these items or reset them, if practical, to the correct outlet temperatures. For commercial absorption systems, see Field Guide, Table 9-6.

Table 9-4. Troubleshooter's Field Guide for Comfort Air-conditioning Systems (Domestic)

Possible cause	How to check for cause	Recommended ways to correct
COMPLAINT: Compressor operates continually, but cooling is inadequate. This is a domestic installation.		
REFERENCE: Sec. 9-1(b).		
1. Machine is too small.	*a.* Compare capacity with connected load.	*a.* Replace with machine of right size.
2. Condenser coils and air filter are dirty.	*b.* Inspect.	*b.* Clean. Replace air filter with new one.
3. Circuit is low in refrigerant.	*c.* Suction pressure is lower than normal.	*c.* Recharge.

Table 9-4. Troubleshooter's Field Guide for Comfort Air-conditioning Systems (Domestic) (*Continued*)

Possible causes	How to check for cause	Recommended ways to correct
4. Fan is running too slowly.	*d.* Measure fan discharge in cfm.	*d.* Lubricate bearings.
5. Low voltage to motor.	*e.* Check voltage.	*e.* Correct voltage.
COMPLAINT: Compressor short-cycles. REFERENCE: Sec. 9-1(c).		
1. Frost on evaporator coils is excessive.	*f.* Inspect coils.	*f.* Defrost. Adjust thermostat.
2. High head pressure stops machine.	*g.* Note head pressure. See Table 9-2.	*g.* For details, see Sec. 9-1(c).
3. Strainer in expansion valve is dirty.	*h.* Inspect.	*h.* Clean or replace.
4. Condenser water is too warm.	*i.* Take temperature of water in and water out.	*i.* Clean cooling water tower of all leaves, branches, etc.
COMPLAINT: Compressor is noisy. REFERENCE: Sec. 9-1(d).		
1. Piston is knocking.	*j.* Listen with stethoscope.	*j.* Take compressor to shop. Repair.
2. Shaft seal is squealing.	*k.* Inspect for worn or misaligned seal.	*k.* Replace or repair as appropriate.
3. Conditioner is making a rattling sound from loose flywheel, or the compressor is loose on its foundation.	*l.* Stop machine. Test each fitted component for tightness.	*l.* Tighten as necessary. Make new keys.
COMPLAINT: Lubricating oil is failing to return to compressor in volume. REFERENCE: Sec. 9-1(e).		
1. Compressor occupies an elevated position in relation to evaporator coil.	*m.* Measure difference in elevation of coil and the compressor crankcase.	*m.* See Sec. 9-1(e) for corrective suggestions.

Table 9-4. Troubleshooter's Field Guide for Comfort Air-conditioning Systems (Domestic) (*Continued*)

Possible causes	How to check cause	Recommended ways to correct
2. Wrong kind of lubricating oil is being used.	*n*. Compare oil characteristics with oil called for by compressor manufacturer.	*n*. Replace oil with a suitable kind.
3. Coil is being flooded with refrigerant and is dropping oil.	*o*. Inspect coil outlet and suction line for excessive frosting.	*o*. Adjust expansion valve for proper flow.
COMPLAINT: Compressor will not start.		
REFERENCE: Sec. 9-1(f).		
1. Electrical trouble.	*p*. Inspect for blown fuse, burned-out capacitor, short circuit, ground, etc.	*p*. Correct as found necessary with new fuse, new switch, etc.
2. Crankshaft is frozen in burned-out bearings.	*q*. Disconnect compressor from motor. See which refuses to turn over.	*q*. Take to shop and repair. For hermetically sealed unit, replace in total.
3. Abnormal conditions of suction or head pressure have made compressor stop as a safety measure.	*r*. Inspect settings of high- and low-pressure cutouts. High head pressure after system is shut down indicates presence of air or noncondensable gases.	*r*. Relieve condition that gives the cutout the charge of operation. Purging may do it.
COMPLAINT: Crankcase is heavily covered with frost.		
REFERENCE: Sec. 9-1(g).		
1. Expansion valve is passing too much refrigerant.	*s*. Inspect expansion valve. See if suction pressure is high.	*s*. Adjust valve if necessary or replace if defective.
2. System has been overcharged.	*t*. Note head pressure. Will be higher than normal.	*t*. Bleed off enough refrigerant to establish proper operation.

Table 9-4. Troubleshooter's Field Guide for Comfort Air-conditioning Systems (Domestic) (*Continued*)

Possible causes	How to check for cause	Recommended ways to correct
COMPLAINT: Low suction pressure is causing pressure switch to cut out machine.		
REFERENCE: Sec. 9-1(h).		
1. System is low in refrigerant.	*u*. Suction line may be warm back to coil.	*u*. Add refrigerant to establish a normal charge.
2. Expansion valve is dirty, or the orifice may be choked with ice.	*v*. Inspect valve.	*v*. Clean, deice, or replace if defective.
3. There is too much oil in system.	*w*. Inspect oil level in crankcase.	*w*. Bleed off enough oil, and possibly water, to establish correct oil level.
COMPLAINT: Heat pump is not delivering sufficient cooling.		
REFERENCE: Sec. 9-1(i).		
1. Compressor trouble.	*x*. Refer to Sec. 9-1(a) through (h).	*x*. See section as noted.
2. Heat sink exchanger may be forced to operate at temperatures higher than originally designed for.	*y*. Examine exchanger for heat impact, as from sunlight.	*y*. Relieve exchanger of the excess heat.

Table 9-5. Troubleshooter's Field Guide for Comfort Air-conditioning Systems (Mechanical-Commercial)

Possible cause	How to check for cause	Recommended ways to correct
COMPLAINT: Conditioned area is too warm.		
REFERENCE: Sec. 9-2(b).		
1. If compressor is running continually, refer to Sec. 9-1(b).		

Table 9-5. Troubleshooter's Field Guide for Comfort Air-conditioning Systems (Mechanical-Commercial) (*Continued*)

Possible cause	How to check for cause	Recommended ways to correct
2. If compressor is short-cycling, refer to Sec. 9-1(c).		
3. High head pressure switch is taking control. See Sec. 9-2(o).	*a*. Compare the existing head pressure with design pressure. See Sec. 9-2(o).	*a*. Locate cause of excessive pressure and correct. See Sec. 9-2(o).
4. Condensing water temperatures are not satisfactory.	*b*. Compare temperatures as found with design temperatures.	*b*. Examine condensing water cooling system for dirt, foreign material, algae, and clear it up.
5. Air or noncondensable gas is in system.	*c*. Combination of high pressure and a temperature lower than is consistent with pressure indicates foreign gas is present.	*c*. Purge system. Recharge.

COMPLAINT: Compressor operates with excessive noise.
REFERENCE: For procedures and recommended actions, see Sec. 9-1(d).

COMPLAINT: Compressor will not start.
REFERENCE: For causes and recommended corrective measures, see Sec. 9-1(f).

COMPLAINT: With an open type of compressor, motor runs hot.
REFERENCE: Sec. 9-2(e).

Possible cause	How to check for cause	Recommended ways to correct
1. Line voltage may be low.	*d*. Check with voltmeter. Note whether lights dim when motor is started.	*d*. Correct the voltage.
2. Motor is undersized.	*e*. Note relationship of motor horsepower to refrigeration capacity.	*e*. See Sec. 9-1(b).

Table 9-5. Troubleshooter's Field Guide for Comfort Air-conditioning Systems (Mechanical-Commercial) (*Contined*)

Possible cause	How to check for cause	Recommended ways to correct
3. Driving belt, motor to compressor, is too tight.	*f*. Inspect. Note whether belt is taut or sags.	*f*. Loosen if too tight by adjusting pulley.
4. Bearings are dry of lubrication.	*g*. Disconnect compressor. Try both compressor and motor separately for hand turning.	*g*. Lubricate as necessary. Hand-try again after lubrication.
COMPLAINT: Compressor motor has burned out several times. REFERENCE: Sec. 9-2(f).		
1. Voltage fluctuations leave motor at times with under-voltage service.	*h*. Check service with voltmeter.	*h*. Request investigation of low voltage by power company.
2. Lightning surges on service wiring.	*i*. Inspect service for lightning arresters on all three wires.	*i*. Install arresters if not present.
3. After each burnout, system was not purged of contaminated refrigerant.	*j*. Examine drier for fouling. Analyze sample of refrigerant.	*j*. Purge system thoroughly. Install new drier. Recharge.
COMPLAINT: Cooling is not satisfactory. Hissing sound is audible in area of expansion valve. REFERENCE: Sec. 9-2(g).		
1. Hissing indicates partial blockage in expansion valve.	*k*. Shut down system. Inspect for ice or other fouling in valve and strainer.	*k*. Clean valve and strainer or replace with new units.
2. Expansion valve is too small, or orifice is defective.	*l*. As described above.	*l*. Replace with new valve of proper size.
COMPLAINT: Air-conditioning system has been gradually losing capacity. Condenser is of the evaporative type. REFERENCE: Sec. 9-2(h).		
1. Spray nozzles are clogged.	*m*. Shut down condenser. Start water	*m*. Clean or replace plugged nozzles.

Table 9-5. Troubleshooter's Field Guide for Comfort Air-conditioning Sytems (Mechanical-Commercial) (*Continued*)

Possible cause	How to check for cause	Recommended ways to correct
	gradually. See which nozzles do not spray as the pressure increases.	
2. Exchange coil is dirty.	*n*. Inspect coil; feel for slime, mud, etc.	*n*. Clean thoroughly. Use dilute acid if necessary.
3. Pump head or capacity has deteriorated.	*o*. Measure the pump speed and discharge pressure. Compare with requirements.	*o*. Identify trouble as wear, foreign obstruction, change in suction, or discharge condition.
4. Condenser fan is under speed.	*p*. With shaft tachometer, determine fan speed in rpm. Refer to fan curve for extent of capacity fall-off. Disconnect from motor and measure motor rpm.	*p*. Clean all fan blades. Lubricate. If motor is running slow, replace.
COMPLAINT: Conditioned spaces are either too cold or too hot. It seems thermostat adjustments are not accurate. REFERENCE: Sec. 9-2(i).		
1. Zone balancing is faulty.	*q*. With a velometer, determine air volume delivered to each conditioned space in cfm.	*q*. Adjust zone dampers to achieve balance.
2. One of the control thermostats is defective and is upsetting balance.	*r*. Check each thermostat separately for correctness of response.	*r*. Replace defective instrument.
COMPLAINT: Cost of system operation is much greater than predicted at time of installation. REFERENCE: Sec. 9-2(k).		
1. Equipment is undersized for the load.	*s*. Calculate the load. Compare with installed capacity.	*s*. See Sec. 9-1(b). Replace with larger unit.

Table 9-5. Troubleshooter's Field Guide for Comfort Air-conditioning Systems (Mechanical-Commercial) (*Continued*)

Possible cause	How to check for cause	Recommended ways to correct
2. Power rate is not favorable.	*t*. Examine power company rate schedule for a better rate.	*t*. Call power company to see if a better rate is available.
3. Extra load being imposed by dirty filter, poor performance of condenser, blowers, etc.	*u*. Examine each component for detracting conditions.	*u*. Correct conditions as found by cleaning, adjusting, etc.
COMPLAINT: System is excessively noisy. REFERENCE: Sec. 9-2(l). See Sec. 7-2(c) for similar complaint on heating cycle. See also Sec. 9-1(d) for compressor noise.		
1. Air ducts and registers are too small.	*v*. Measure output in cfm.	*v*. Reduce volume, or lower temperature.
2. Shaft seal is squealing.	*w*. Inspect for worn or misaligned seal.	*w*. Replace or repair as appropriate.
3. Low level of crankcase oil is causing seal to run dry.	*x*. Check level of oil in crankcase.	*x*. Add oil.
COMPLAINT: Discharge of air from various registers and diffusers is laying black smudges on adjacent ceilings and walls. REFERENCE: Sec. 9-2(m).		
1. Dirty air is being blown from outlets.	*y*. Inspect air filters. Test by wiping with clean, white cloth.	*y*. Clean or replace filters as needed. Vacuum ducts.
2. Room air is dirty and is being induced to rise by jets from the outlets.	*z*. With smoke test, determine directional movement of room air, as from floor to ceiling or horizontal.	*z*. Set up program of regular cleaning for soiled surfaces.
COMPLAINT: Unpleasant odors are being emitted by air-conditioning system. REFERENCE: Sec. 9-2(n).		
1. The air filters are dirty.	*aa*. Inspect filters. Test by wiping with clean, white cloth.	*aa*. Clean or replace filters. Clean ducts.

Table 9-5. Troubleshooter's Field Guide for Comfort Air-conditioning Systems (Mechanical-Commercial) (*Continued*)

Possible cause	How to check for cause	Recommended way to correct
2. Foreign organic material has been closed up in a duct and is decomposing.	*bb*. Locate area of maximum odor, remove register or grille face, and probe, using a hooked wire.	*bb*. Remove odorous matter if possible. Install activated carbon odor absorber.
3. Odor is originating outside the building.	*cc*. Sniff outside building to discover direction from which odor is coming.	*cc*. Relocate fresh air intake. Treat fresh air with activated carbon.

Table 9-6. Troubleshooter's Field Guide for Comfort Air-conditioning Systems (Commercial-Absorption)

Possible cause	How to check for cause	Recommended ways to correct
COMPLAINT: Rating of an absorption system of refrigeration is not being met in operation.		
REFERENCE: Sec. 9-3(a)–d and 9-4(a) and (b).		
1. Heat supply to generator is not sufficient.	*a*. Take temperature of refrigerant vapor leaving generator, whether water or ammonia.	*a*. Increase steam pressure or, if fired, increase burning rate or get a "hotter" fuel.
2. Heat exchangers between generator and condenser, such as analyzers and rectifiers, are dirty and sluggish in exchange.	*b*. Check temperatures in and out of each exchanger.	*b*. Clean the exchangers.
3. If heat is from exhaust gas, evaluate positions of the gas dampers.	*c*. Compare volume of gas from stack with the volume passing through waste heat boiler.	*c*. Adjust dampers as necessary.
4. Condenser is not supplying enough cooled water.	*d*. Record temperatures and volumes of water entering and	*d*. Clean tower and pipelines. Bring back tower performance to

Table 9-6. Troubleshooter's Field Guide for Comfort Air-conditioning Systems (Commercial-Absorption) (*Continued*)

Possible cause	How to check for cause	Recommended ways to correct
	leaving the cooling tower.	design values.
5. Circulating water pumps are not meeting required volume and head.	*e.* Check pump discharge pressure and speed. Note suction lift and evaluate its effect on pump performance. Compare with pump curves.	*e.* Repair pumps as necessary.
6. The relationship of the refrigerant absorbent is out of balance.	*f.* Determine weight of ammonia or lithium bromide presently in system and compare with design weights.	*f.* Exercise caution in pumping down system; weigh with care and add absorbent or refrigerant as necessary to reestablish balance.

DIVISION 10

Ice Making and Cold Storage

10-1. REFRIGERANTS AND THEIR PROPERTIES

(a) ***Refrigerant Properties.*** Certain materials are used as refrigerants because they possess low boiling points, the boiling point being the temperature at which the material ceases to be a liquid and turns into a vapor. Heat known as the heat of vaporization, the heat of evaporation, or the latent heat, is absorbed in the change, and this removal of heat from the liquid produces in it a drop in temperature.

These materials that change their state from liquid to gas are termed "primary refrigerants." Other materials that do not change their state but have their temperatures changed by the action of the primary refrigerants and are used to carry heat from a point of origin to a point of disposal are called "secondary refrigerants." These secondary refrigerants are commonly water or a chemical brine, such as sodium chloride or calcium chloride.

Brines are frequently employed in ice-making plants and in cold-storage operations; water in air-conditioning arrangements is founded on the circulation of chilled water through the living spaces to be cooled. Water limitations of course are set by the freezing temperature of 32 F. The brines can be kept fluid to subzero temperatures, as noted in Tables 10-3 and 10-4.

Secondary refrigerant temperatures have no relationship to pressure; increasing their temperatures brings no reaction on the pressure gage. With the primary refrigerants, temperature and pressure are interlocked; we can defer the vaporization of a liquid refrigerant simply by increasing the pressure under which it is contained. To illustrate, ammonia, which at atmospheric pressure boils or vaporizes at -28 F, will remain a liquid up to 100 F if we just hold it under an absolute pressure of 212 lb per sq in. (Since the difference between absolute and gage pressures is 14.7 pounds, the ammonia *gage* at 100 F should show 197.3 lb.)

This relationship between boiling points and pressure for the primary refrigerants is shown in Table 10-1.

(b) ***Propylene Glycol.*** This is one of several glycols frequently used in providing antifreeze protection. Propylene glycol is a stable mixture, noncorrosive, and may be substituted for any of the other refrigerants without the need to make any significant changes in the system. Where corrosion has become a problem, the replacement of the existing primary refrigerant with propylene glycol is likely to cure the trouble.

While Table 10-1 shows the minimum boiling point of this refrigerant to be −50 F, this temperature may be lowered substantially by increasing

Table 10-1. Temperature-Pressure of Common Refrigerants

	Pressure, psia *				
Boiling point, F	Freon 12	Carbon dioxide	Ammonia	Sulfur dioxide	Propylene glycol
−109		14.7 *			
−50		118.3			14.7 *
−28		185.0	14.7 *		27.0
−21	14.7 *	212.0	17.8		33.5
−10	19.2	257.3	23.7		40.0
0	23.9	305.5	30.4		47.0
10	29.3	360.2	38.5		58.0
13	31.2	378.1	41.3	14.7 *	61.0
20	35.7	421.8	48.2	17.2	68.5
30	43.2	490.8	59.7	21.7	81.0
40	51.7	567.8	73.3	27.1	96.0
50	61.4	653.6	89.2	33.4	112.0
60	72.4	748.6	107.6	40.9	130.0
70	84.8	853.4	128.8	49.6	150.0
80	98.8	968.7	153.0	59.7	174.0
87.8 †		1070.0			
90	114.3		180.6	71.2	196.0
100	131.6		211.9	84.5	240.0

* The pressure of 14.7, representing atmospheric pressure, where it appears under a given refrigerant, indicates the lowest boiling point for that particular refrigerant.

† The critical temperature for carbon dioxide (CO_2). At higher temperatures, carbon dioxide cannot be maintained in a liquid state regardless of the pressure under which it is held.

the percentage of glycol in the solution to something above 60 percent by volume.

(c) ***The Freons.*** The most recent member of this family to appear in the market place is Freon 502. According to the manufacturer, this is a fluorocarbon refrigerant that combines the most desirable properties of Freon 12 and Freon 22 in one refrigerant. As a result of these improved characteristics, compressors are claimed to be more reliable in their performance and may be smaller in their size. Characteristics of the Freon family are shown in Table 10-1.

The Freons are generally nontoxic and nonflammable, but they will dissociate at temperatures of 900 F plus and produce a gas that is irritating to mucous membranes. This condition might develop if the freon refrigeration system is surrounded by fire, but otherwise it should be a rarity.

(d) ***Carbon Dioxide*** **(CO_2).** This refrigerant is odorless, colorless, nonpoisonous, and nonflammable. Pressure is the main problem with handling CO_2, as Table 10-1 shows. Applications also may be limited by the need to keep liquid temperatures below 88 F. For example, marine installations find sufficiently cool condensing water to be critical or even nonexistent in some tropical areas.

The procedures for turning CO_2 into dry ice are the following:

1. CO_2 gas is compressed to 1,000 psia by a compressor and, under this pressure, is passed through a condenser where the heat of vaporization is removed and the gas is converted to a liquid.

2. The pressure on the liquid is now dropped, and this drop of pressure causes vaporization to take place, the heat of this vaporization being harvested from the body of the liquid, thus chilling the liquid. The vapor generated in this step is withdrawn from over the liquid by a compressor which discharges through another condenser in which the heat of vaporization is transferred to cooling water.

3. The chilled liquid CO_2 is piped to a press chamber, entering the chamber through a nozzle. Since pressure in the press chamber is at atmospheric or even a little less, the liquid emerging from the nozzle immediately begins to evaporate, and the removal of heat from the emerging liquid is so swift that much of the liquid freezes into a fluffy snow as it leaves the nozzle and accumulates in the press chamber. Again, the CO_2 vapor in the press chamber is removed by a compressor and returned to the original source of the gas in preparation for a continuation of the cycle.

Troubles with dry ice systems may be expected to approximate those generic to the equipment used, that is compressors, condensers, pumps, and controls. In answering trouble calls, field guides and discussions that apply to these familiar components should be consulted for identifying symptoms and for deciding on the form of correction. Particular attention should be given to the temperature of the condensing water, since a failure to abstract heat from the gas results in nonliquidation of the gas and brings a shutdown of the procedure.

(e) ***Sulfur Dioxide*** (***SO_2***). Not generally favored for large refrigerating systems, the property of this refrigerant most important to the troubleshooter is its toxicity. A concentration of as little as ½ of 1 percent can be deadly to its inhaler after as short a time as 5 minutes. As a hazard, the Underwriters set it in Group 1. For information concerning its other properties, see Table 10-1.

Sulfur dioxide may be used in household refrigerators of obsolete design. Modern equipment does not use it, and its replacement by all new appliances is recommended wherever it is encountered.

10-2. SAFETY IN HANDLING REFRIGERANTS

(a) ***The Freons and What Can Happen***

1. Freon getting into the eyes may cause serious injury by freezing the moisture in the eyes.

2. Liquid Freon making contact with the skin may produce a frostbite.

Recommended Safety Precautions

1. Whenever handling a Freon refrigerant, as when pumping down a system or charging one, wear goggles to protect the eyes and gloves to protect the hands.

2. For first aid after eye contact with Freon, wash out the eyes with a weak solution of boric acid.

3. Obtain a medical examination of the eyes as quickly after contact as possible.

(b) ***Ammonia and What Can Happen***

1. In the presence of from ½ to 1 percent ammonia by volume, unprotected men and animals may be killed in a short time.

2. Exposure to the fumes for a period of from 30 to 60 minutes is dangerous if the concentration ranges from 0.25 to 0.45 percent by volume.

Recommended Actions

1. Do not attempt to enter or work in an atmosphere containing enough ammonia fumes to be irritating. A gas mask, if properly chosen, may suffice to permit the closing of critical valves but for no longer. If a gas mask is at hand, make sure it is suitable for ammonia.

2. Ventilate the area with fresh air. Stay out of it until the concentration drops to nonirritating levels.

3. If obliged to enter a space containing ammonia fumes, keep the

head as low as possible. The fumes are lighter than air; consequently, the maximum safety is near the floor.

4. Among the items of the troubleshooter's equipment, include a bottle of 2 percent boric acid solution for washing the eyes, a quart bottle of saturated aqueous solution of picric acid for application to ammonia burns on the skin, and a 1-oz bottle of liquid petrolatum with a medicine dropper for palliating the eyes after washing with the boric acid. A sponge saturated with vinegar may be useful briefly in a contaminated atmosphere if held over the nose and mouth.

(c) ***Sulfur Dioxide and What Can Happen.*** Although sulfur dioxide can be deadly if inhaled, the relatively few installations encountered by the troubleshooter reduce his exposures to negligible numbers. Sulfur dioxide and moisture form sulfuric acid which attacks both the metals of the refrigeration system that contains it and the respiratory system that breathes it.

The immediate treatment for a person exposed to SO_2 is to get him into an atmosphere of uncontaminated fresh air. Then obtain medical attention.

Recommended Action

The presence of sulfur dioxide is instantly recognized by its foul odor. Do not attempt to work in such an atmosphere and stay clear until positive ventilation has removed all danger.

(d) ***Safe Handling of Refrigerant Cylinders.*** General safety rules should be observed in storing, charging from, and draining to the pressure cylinders for any refrigerant. These rules are:

1. When opening or closing cylinder valves, use only the wrenches or other tools provided by the cylinder manufacturer.

2. If heating a cylinder to expedite the discharge, do not let the temperature rise above 150 F. Fusible plugs, when mounted in cylinders, generally are designed to melt at any temperature between 150 and 175 F and actually may soften at several degrees lower.

3. Store all cylinders in the upright position. This keeps any foreign material within the cylinder out of the normal discharge path.

4. Do not drop, jar violently, or strike a cylinder. A slight shock on a cylinder already charged with its maximum pressure can result in an explosion and possibly loss of life.

5. Ammonia cylinders must never be left attached to a system after charging or draining has been completed.

6. Do not handle or work with cylinders not properly fitted with a relief valve. This is of the built-in type with four discharge ports.

7. Store cylinders so that they *are not* subjected to the direct rays of the sun nor to accumulations of ice and snow in winter.

8. Storage areas should be safely distant from flammables such as gasoline, oil, and other chemicals.

9. Do not attempt to lift and transport cylinders by electromagnets. The danger of a drop followed by an explosion is always present.

10. Avoid overfilling of cylinders. For the maximum safe loading of *ammonia* cylinders, see Table 9-3 and Sec. 9-4(c). For sizes and weights of safe *Freon* cylinders, see Table 10-2. Carbon dioxide cylinders of the high-pressure type contain 50 or 75 lb of the refrigerant, depending on the size of the cylinder. Low-pressure storage is accomplished by refrigerating coils in the container, which maintains the liquid at 0 F and 305 psia. This type of container is well insulated and protected against a sudden rise in pressure by means of a safety relief valve designed to open at 345 psia. Low-pressure storage structures can be designed for very large capacity, and units holding over 100 tons are not unusual.

The high-pressure cylinders, which the serviceman is much more likely to meet, are of steel and tested up to 3,000 psig. Depending on the ambient temperature, the pressure of these cylinders while in storage will range from 1,000 psig up.

Table 10-2. Safe Loading for Freon Cylinders *

Cylinder dimensions, in.		Max. load, lb
Diameter	Length	
4 3/16	25 1/4	10
6 3/4	19	25
8 1/4	26 3/4	50
10 1/4	46 3/4	145

* All pressure cylinders involving shipment over state lines must meet the requirements of the Interstate Commerce Commission in design and construction.

10-3. ICE MAKING AND HANDLING

(a) ***General Notes.*** The designer of an ice-making plant usually observes certain guidelines, such as a maximum velocity of brine circulation, standard sizes of block ice cans, and a high, low, or medium pressure of air for agitating the water in the cans. If the troubleshooter is aware of these guides in answering a call for service, his task is likely to be made easier. These can be, in many instances, the values against which he measures the degree of malperformance of the installation. Here are some of them:

1. Brine is circulated through the freezing tank and around the ice cans at a steady rate of less than 35 ft per minute, the purpose being to make as uniform as possible the rate of freezing in all cans. Higher speed is a waste of horsepower on the propeller, and from 15 to 25 ft per minute is generally considered most suitable.

2. The standard size in the United States of a block ice can is 11 in. deep and 22 in. wide; it produces a 300-lb block of ice. In evaluating ice tank performance, a first step is to ascertain whether the cans are standard and to make the proper allowances if they are not.

3. If the agitating air nozzle is attached in any way to the can, the air pressure is likely to be from 15 to 25 psig. Low-pressure air at 2 psig is introduced into the cans, usually from a hose nozzle suspended from above the can.

4. Tonnage of the refrigerating compressor required to produce 1 ton of ice is based on the temperature of the water filling the cans and the temperature of the brine, both in degrees F. For example, with water at 70 F and brine at 20 F, approximately 1.6 tons of installed compressor is needed. This ratio can be improved by forecooling the water and the refrigerant, prior to their entry into the tank. The effects of such forecooling have been reported under some conditions to reach 30 percent.

5. The temperature of the brine necessary to freeze a given number of *standard* cans in 24 hours can be estimated from the formula

$$\text{Temp. of brine F} = 32 - \left[\frac{583.3\,a^2}{c \times w}\right]$$

where a = thickness of ice block, in. (11-in. standard)
c = number of cans being frozen in 24 hours
w = weight of block per can (300-lb standard)

This formula is offered in the ASHRAE "Guide and Data Book," and is not suitable for cans other than the standard. By using it in appropriate plants, the troubleshooter can determine quickly whether the brine temperature is too high or unnecessarily low.

6. Circulation of the brine through the freezing tank may be designed to maintain a brine temperature with no more than ½ deg of temperature variation. However, a variation of 1 deg is generally considered to be acceptable.

7. Storage rooms for the ice cakes should provide for at least 3 days supply of ice making, with dimensions based on 13 sq ft of floor area per ton. Where ceiling coils of the direct expansion type are used, allow 2 ft of clearance under the coils and above the ice for air circulation.

8. Temperatures in the ice storage room are usually between 24 and 28 F.

9. Three types of brine are in use. Sodium chloride is suitable down to 5 F, calcium chloride to −40 F, and propylene glycol to −50 F. Corrosion is no problem with the last-named, but is a hazard with the other two unless these brines have been chromate-treated.

(b) *Complaint.* The ice is cracking in the cans during the freezing process.

Causes

1. The metal of which the cans are fabricated is light, probably #16 BWG or thinner. These cans bulge outward when full of water and standing in the air and bulge inward when immersed in the brine. The net result is a cracked ice cake.

2. Freezing of the ice is being accomplished at a temperature of 10 F or colder.

Recommended Actions

1. Replace the cans with units made of #12 BWG steel.

2. Raise the freezing temperature to at least 15 F; a 20 F is better to ensure a perfect cake.

(c) *Complaint.* Ice cakes have cloudy spots and may be a little off-color.

Cause

Various salts have been trapped in the cake. These may be magnesium or calcium originally in solution in the water.

Recommended Actions

1. Check filters in the water supply to the cans. If filters are dirty, clean them with heavy soda ash solutions. If no filters have been installed, provide them.

2. For off-color complaints, the advice of a water treatment expert should be obtained for methods of correction.

(d) *Complaint.* There appears to be some malfunction of the compressors.

Causes and Recommended Actions

Refer to Sec. 9-1(c)–(h). See also Table 9-5.

(e) *Complaint.* There appears to be some malfunction of the ammonia absorption refrigeration system.

Causes and Recommended Actions

Refer to Sec. 9-4(a)–(c). See also Table 9-6.

(f) *Complaint.* Water temperature and flow through the condensing system is unsatisfactory.

Causes

1. Algae and slime are affecting heat transfer adversely in the water tower or spray pond.

2. Water circulating pumps are operating at less than their rated capacity and head.

3. Control devices are out of adjustment.

4. Foreign or corrosive material has accumulated in the ammonia condenser.

Recommended Actions

1. For algae and slime, refer to Secs. 9-7(a)–(d) and 9-2(h) and Table 9-5.

2. For pump trouble, refer to Sec. 9-2(h) and Table 9-5.

3. Check-operate each of the control devices. Readjust or replace faulty items.

4. Flush out under pressure. Cleaning the interior of pipe coils can sometimes be effected by disconnecting the coil and blowing through it with steam. This usually weakens and breaks free a substantial part of the scale and sludge. After steam, wash out with fresh water continuously for 10 minutes. Tap here and there with a hammer during the water-washing cycle for the freeing of additional scale.

The outside of tubes can be cleaned with a 50-percent solution of muriatic acid.

(g) ***Complaint.*** The ice plant is failing to meet its rated capacity, although all components appear to be operating normally.

Causes

1. Too much time is being spent on core pumping. The core pump with a capacity up to 15 gpm, depending on the size of the cans and their number, should handle 36 cans in 20 minutes, including refilling time.

2. Brine temperature is too high.

3. If can freezing is nonuniform as well as too slow, the brine temperature is not uniform either. This may be the result of sluggish operation of the brine circulator.

4. Rate of evaporation is too low, caused by frost on coils, by excessive back pressure, or by condenser malfunction.

5. The density of the brine is either too low or too high. In case of the former, it may freeze or at least deposit frost on the ammonia evaporating coils, thus reducing the rate of heat transfer. If the density is too great, the brine loses some of its heat-conducting ability, and uneven freezing of the ice cakes follows.

Temperatures and densities of the calcium chloride and sodium chloride brines are given in Tables 10-3 and 10-4.

For estimating the volume of ammonia and brines required for plants of various capacities, see Table 10-5.

The arrangement of a freezing tank is shown in Fig. 10-1 and of a can dump in Fig. 10-2.

Table 10-3. Properties of Calcium Brine ($CaCl_2$)

Specific gravity at 60 F	Degrees on salinometer	Pounds $CaCl_2$ per gal °	Freezing point, F
1.09	46	1	22
1.14	70	1.5	10
1.18	90	2.0	−3
1.22	106	2.5	−23
1.27	124	3.0	−46

° Approximate. May vary slightly up or down.

Table 10-4. Properties of Sodium Brine (NaCl)

Specific gravity at 60 F	Degrees on salinometer	Pounds NaCl per gal °	Freezing point, F
1.04	22.7	0.5	24
1.09	45.2	1.0	16
1.12	60.6	1.5	11
1.16	79.1	2.0	5
1.19	94.7	2.5	0

° Approximate. May vary slightly up or down.

Table 10-5. Average Charge, Ammonia and Brines

Rated tons of ice or cans	Ammonia, lb	Sodium ° chloride, lb	Calcium ° chloride, lb
3 tons	330	3,400	4,550
6 tons	545	5,200	6,900
11 cans, 100 lb each	1,005	8,800	11,800
11 cans, 300 lb each	905	11,000	14,800
16 cans, 100 lb each	1,705	14,400	19,300
16 cans, 300 lb each	1,400	15,700	21,000
23 cans, 100 lb each	2,280	16,500	22,200
23 cans, 300 lb each	1,735	21,600	29,000
34 tons	2,415	33,600	45,000
55 tons	3,780	45,600	61,200

° Weight of salt, dry. For weights of brines, see Tables 10-3 and 10-4.

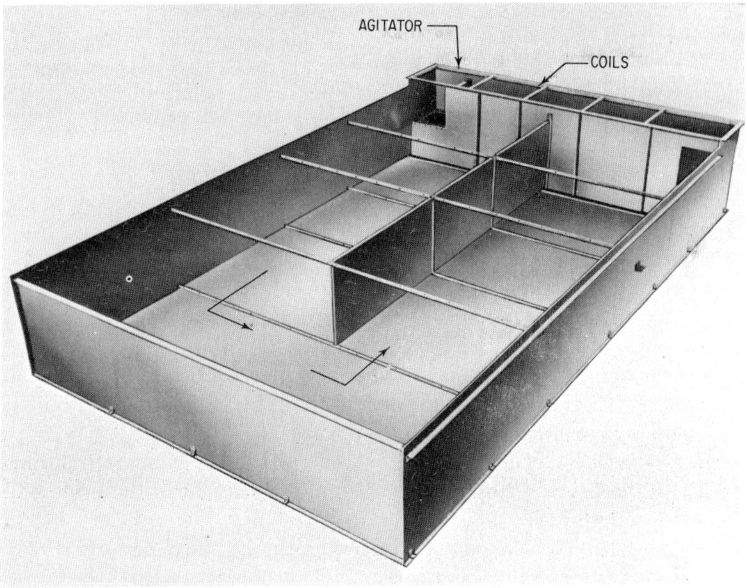

FIG. 10–1. Freezing tank for ice plant. (*Frick Company.*)

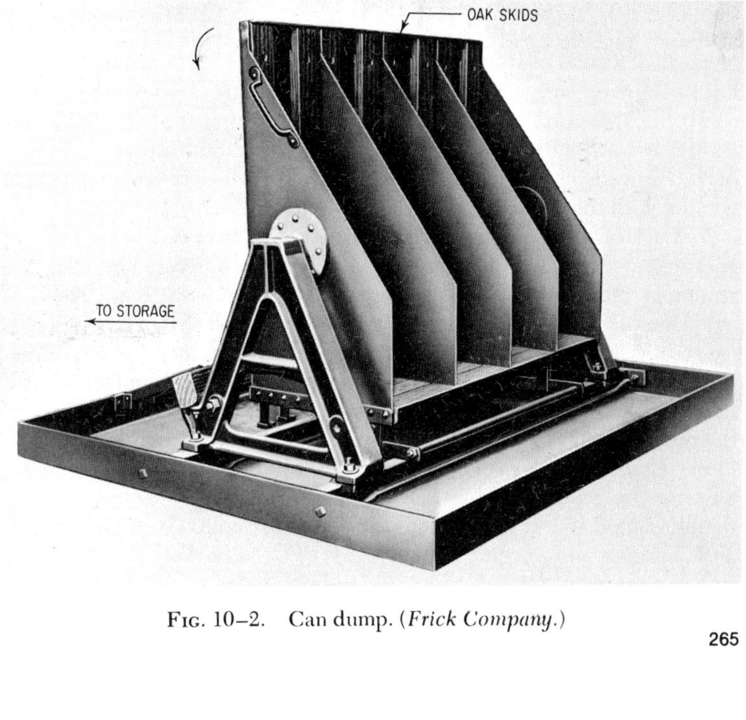

FIG. 10–2. Can dump. (*Frick Company.*)

Recommended Actions

1. Time-check the core-pumping operations. Rearrange procedure if time is excessive. If trouble is in a slow pump-out, examine pump for adequate capacity and performance.

2. Lower the brine temperature by resetting controls.

3. Take thermometer readings on the brine in each section of the freeze tank. If varying by more than 1 deg F, verify speed and head performance of the circulator.

4. Look for ammonia leaks into the brine from the brine coils as well as elsewhere on the ammonia circuits. Ammonia leaks may be located in these ways:

For leaks in the condenser, submerge a piece of litmus paper in the circulating water system of the condenser. The paper will turn blue if any ammonia whatever is present in this water.

For leaks in the evaporator coils submerged in brine, immerse litmus paper in the brine. Any degree of ammonia present in the brine will turn the paper blue.

For leaks in dry coils or pipes, dampen the litmus paper and slowly pass it along the coil or pipe surface. If it contacts a leak, the paper turns blue.

Alternatively for leaks in dry coils or pipes, search out the leak with a lighted sulfur stick. Make the sulfur stick in this way:

In a metal cup or open vessel, melt a chunk of sulfur to the fluid state. Next, take a strip of thin wood or cardboard and dip this strip into the fluid sulfur, turning and redipping the strip as necessary to build up a sulfur thickness on the strip of at least $1/16$ in. After the sulfur stick has cooled and solidified, it may be lighted and poked into suspicious areas of the dry coils or ammonia pipeline. Any presence of ammonia gas in contact with the burning sulfur fumes creates a thick white smoke.

5. Lack of ammonia in the system is signalled by a lack of frost on the compressor suction pipe, together with an inability to readjust the ammonia regulating valve to produce frost on the suction. Another evidence of the lack of refrigerant is a whistling in the regulator valve caused by the passage of gas only. Liquid refrigerant generates a hissing rather than a whistling. For low refrigerant, add ammonia until frost appears on the compressor suction line or until the gage glass on the receiver shows a satisfactory level.

(h) ***Cautions When Correcting Leaks.*** Ammonia valves generally may be repacked while the valve is wide open if these are of the double-seated type. Exceptions are the expansion valves, gage cocks, and charging and purging valves which must be packed with the valve tightly shut.

In order to avoid developing any grade of emergency due to the escape of ammonia, troubleshooters should study the system carefully, including inspection of system diagrams if available, before undertaking the correction of any leak. Closure of *all* the right valves is necessary, and the overlooking of just one valve can precipitate a crisis.

10-4. COLD STORAGE PLANTS

(a) ***General Notes.*** The refrigeration system in a cold storage plant may incorporate any one of these four arrangements:

1. Brine coils mounted on the insulated walls and ceiling.
2. Direct expansion coils, using ammonia or one of the other primary refrigerants, such as Freon.
3. Unit coolers with direct expansion coils over which air is circulated by a fan.
4. Unit coolers having coils, brine spray banks, and fans for air circulation. In operation, the primary refrigerant is in the coils, and the brine spray is recirculated over the coils, thus preventing ice formation and the need for defrosting. The brine used is generally sodium chloride with soda added to improve the pH of the solution. Alternatively, propylene glycol is somewhat favored also.

Cold storage areas may be flanked by chilling rooms in which the material to be stored, such as meat, is cooled at a rate to suit the kind of carcass it is; for example, with beef, 40 deg F in 20 hours, while pork may be satisfactorily chilled in little more than half that time. Most of the refrigeration load occurs in the chill room. Temperatures in the holding or cold storage room generally range from 30 to 34 F.

In a storage warehouse that provides rooms for various materials, each with a different holding temperature, the controls will include back-pressure valves unless each class of storage has its exclusive refrigerating machinery. The back-pressure valves in a cold storage plant might be compared with the balance valves on the returns of a hydronic heating system; each monitors the performance of the circuit in which it is mounted.

Another name for the back-pressure valve is suction-pressure regulator, since it maintains the suction pressure best suited to the room conditions desired. From time to time, the troubleshooter may find the back-pressure regulator to be acting also with a solenoid that is controlled by a thermostat and closes the regulator automatically for periods of defrosting or excess cold.

(b) ***Complaint.*** Control valves are operating erratically. For several consecutive days all may seem normal, then temperatures run out of control, either up or down.

Causes

1. The strainers in the lines connecting to the control valves are either dirty or missing.
2. A thermostat is defective.

3. A suction pressure regulator is out of adjustment.

Recommended Actions

1. Clean or replace strainers as necessary.
2. Verify operating health of the thermostat that controls conditions in the unsatisfactory room.
3. Check adjustment of suction-pressure regulator and correct as necessary.

(c) ***Complaint.*** Temperatures in certain rooms are falling short of the design requirements.

Causes

1. Same as described in Sec. 10-4(b) preceding.
2. Examine inside surfaces of the faulty rooms for breaks in the insulation. In some plants, the insulation for the cold areas may be anchored in the building walls for strength, and subsequent differences of thermal movement between the insulation and the structure can produce breaks detrimental to the maintenance of low temperatures.
3. If brine sprays are being used, check the brine density. The sprays become diluted with the atmospheric moisture they pick up and may be depositing frost or ice crystals on the primary coils.
4. Determine whether the brine circulation is meeting the designed head and capacity of the circulator.
5. If all other possible causes are found negative, check the temperatures of the materials being brought in for storage in the concerned rooms. These may be higher than originally handled, and consequently the room cooling equipment is undersized while this condition lasts. Weights also may be greater than originally intended.

Recommended Actions

1. See recommendations in Sec. 10-4(b).
2. Repair all breaks in the room insulation.
3. Correct brine density as required.
4. If circulator has lost speed and discharge head, as shown by tachometer and pressure gage, repair or replace the pump and its driver. Verify the suitability of the pump suction conditions.
5. Refer to Table 10-6 for specific heats of various materials which may be used for calculating the presence of either an overload or an underload in the rooms where trouble exists.
6. Look for a room with impaired air circulation resulting from an incorrect loading. Air channels may be found blocked or partially blocked by too close packing. Twelve-inch air channels should be considered the minimum.
7. By means of a velometer or anemometer, measure air discharge from the coil fans. If below design, check fan speed and balance, correcting or replacing as found necessary. A simple cleaning of the

Table 10-6. Specific Heats of Various Foods

Food	Temperature of freezing, deg F	*Specific heat*		Latent heat of fusion, Btu
		above freezing	below freezing	
Beef	28	0.74	0.30	98
Fish	28	0.82	0.41	109
Lamb	25	0.67	0.40	73
Mutton	25	0.81	0.67	95
Pork	28	0.51	0.38	66
Poultry	27	0.80	0.42	105
Veal	29	0.70	0.39	90
Beans	30	0.92	0.47	128
Cabbage	31	0.93	0.43	129
Corn	29	0.80	0.43	108
Lettuce	31	0.95	0.48	135
Onions	30	0.90	0.46	127
Peas	30	0.80	0.42	108
Potatoes	29	0.78	0.48	105
Tomatoes	30	0.95	0.48	135
Apples	29	0.86	0.45	120
Lemons	28	0.90	0.39	126
Oranges	28	0.92	0.39	125
Peaches	29	0.90	0.46	125
Strawberries	30	0.92	0.47	130
Milk	31	0.90	0.47	127
Eggs	28	0.78		
Beer	29	0.90		

blades can be expected to improve the fan operation, particularly if it has been in service for more than 1 year.

10-5. LOCKER PLANTS

(a) ***General Notes.*** All temperatures in a locker plant are not held at the conventional 0 F characteristic of the locker area. Chill rooms, aging and curing rooms, and quick-freezing rooms precede the ultimate storage of the meats in the lockers. A schematic diagram of the product movement, from the time it enters the plant until it is taken out, appears in Fig. 10-3.

This diagram does not include all the parallel activities of curing, smoking, and preparing items, such as fruits and vegetables, for special treatment. The purpose of the diagram is to impress on the trouble-

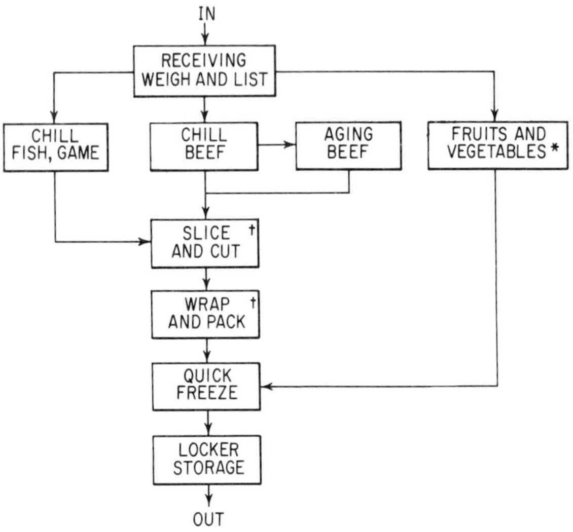

TEMPERATURES:

CHILL ROOMS	34-36 F
AGING OR CURING	38-40 F
FREEZE	-10 F OR -15 F
LOCKERS	0 F

* FRUITS AND VEGETABLES, CHILL TO 40 F ONLY IF TO BE HELD OVERNIGHT
† ROOMS NOT REFRIGERATED

FIG. 10–3. Product movement in locker plant.

shooter that distress calls may originate in various conditions. In short, locker plants are more likely to be complex than simple, especially if each particular condition is served by its own complete refrigerating system.

In addition to arrangements that provide a separate refrigerating unit for each room, two other designs will be encountered. Early designs employed a single condensing system for the entire plant, each room being controlled by a suction-pressure regulator, or by a thermostatic bulb. Presently, practice appears to favor arrangements under which rooms of the same approximate temperature are served by a condensing unit common to the group. This is more economical of equipment investment than is the one room–one unit scheme, and less subject to control troubles than serving all rooms from one central unit.

Condensing equipment for locker plants includes water towers, evaporative condensers, spray ponds, and air-cooled condensers. Some of the air-cooled condensers are designed with multiple refrigerant circuits, each circuit serving an individual compressor. In this arrangement, interruption of the flow by reason of a control action affects only that particular condenser circuit, the condenser continuing to operate for the benefit of the other coils.

The use of a multiple coil condenser, where several separate circuits can be served by the one equipment, offers some economic advantages for consideration by the designer. It also narrows down the trouble-shooter's search for sources of trouble.

(**b**) ***Complaint.*** A given room is too cold, while others on the same compressor are too warm.

Causes

1. The suction-pressure regulators are out of balance.

2. The thermostat in the over-cold room is faulty.

3. Balancing tubing between the expansion valve on a direct expansion installation and its thermostatic valve mounted on the coil outlet has been omitted by the installer. This condition will be found, if it exists, on the coils in the warm rooms.

Recommended Actions

1. Check pressures and temperatures of the suction side of the coil in each room. Adjust the regulators as necessary to establish the design conditions.

2. Check and correct or replace the faulty thermostat.

3. Install the balance tubing between the expansion valve and the thermostatic bulb.

4. Determine whether the too warm rooms are overloaded. Obtain information on the poundage in each.

(**c**) ***Complaint.*** Foodstuffs, when taken from the lockers, sometimes show deterioration in color and weight.

Causes

1. Food is too warm coming into the receiving room and then is not held long enough in the chill room.

2. Poultry, fish, and meats other than beef and game must be sent to the quick freeze as soon as possible for best results. Delay in reaching a refrigerated atmosphere results in a deteriorated product.

Recommended Actions

1. Observe operations over 1 or 2 days, noting the time the product spends in the chill room and quick freeze. Also record weights and temperatures of the food materials as received. Reset the speed of the various operations to correct the unsatisfactory conditions.

2. Verify the correct temperatures in each room, adjusting as necessary.

3. Examine the conditions of air circulation in the various rooms.

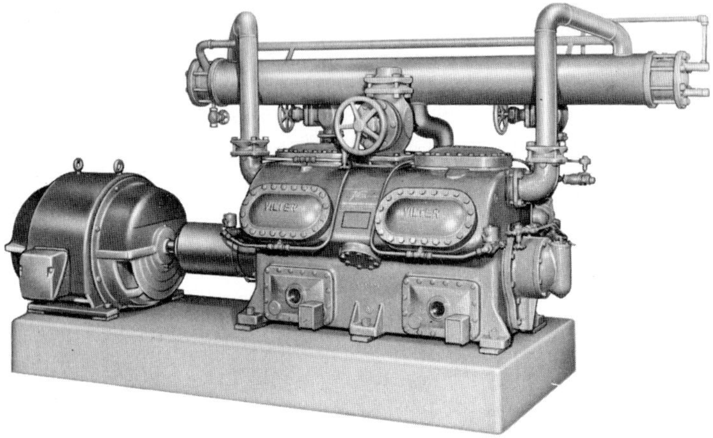

FIG. 10–4. Refrigeration compressor, 12-cylinder, 2-stage with intercooler. (*Vilter Mfg. Corp.*)

Products denied free air circulation are likely to develop heat and spoilage. Repack the rooms if necessary to establish adequate circulation.

(d) ***Compressor Troubles.*** Refer to Sec. 9-1(c)–(h) and Table 9-5.

(e) ***Absorption System Troubles.*** Refer to Sec. 9-4(a)–(c) and Table 9-6.

(f) ***Water-cooling Tower Troubles.*** Refer to Sec. 10-3(f). Give particular attention to Sec. 9-7(a)–(d), concerning algae and slime accumulations.

(g) ***Air Condenser Troubles.*** Refer to Sec. 9-9(a).

(h) ***Evaporative Condenser Malfunctions.*** Refer to Sec. 9-2(h) and Table 9-5.

(i) ***Complaint.*** In a cold storage plant for fruits and vegetables, the magnetic starter for the 7½-hp refrigeration compressor developed welded contacts. The subsequent continuous operation of the compressor pulled down the temperatures so far below freezing that all contents of the storage spaces were spoiled. Loss amounted to over $8,000.

Cause

The design and arrangement of the refrigeration system did not include any safety temperature alarms whatever. All the safety devices were intended to break the electrical circuit through the magnetic contactors, and since these had become welded together, there was no way for the compressor to be stopped automatically. The reason for the welding of the contactors was not known exactly, but knowing that it could happen should have dictated the installation of an emergency alarm outside the refrigerated areas.

Recommended Actions

1. Install both audible and visual alarms outside refrigerated storage areas, these alarms to be actuated by the development of abnormal temperatures, either too warm or too cold.

2. Inspect starter contactors periodically, at least once a month.

10-6. SERVICING HERMETICALLY SEALED MOTORS

(a) ***General Notes.*** The Refrigeration Service Engineers Society has published a manual for dealing with hermetic burnouts, which covers the subject thoroughly. This manual is recommended to all servicemen; copies may be obtained from the offices of the society at 433 North Waller Avenue, Chicago, Illinois.

Because of the importance of a thorough cleaning of the system when a burnout occurs, the society recommends that procedures for cleaning the equipment be obtained from its manufacturer.

Burnouts are classed as either mild or severe, the details of the cleaning up being dependent on which class of casualty has occurred. The procedures for investigation are as follows:

(b) ***Steps in the Investigation***

1. Attempt to start the compressor-motor. If it will not start, check possible reasons as given in Secs. 9-1(f) and 9-2(f).

2. Evaluate the odor of the refrigerant. Is it normal or acrid? If acrid, and the oil is black or heavy with sludge, a severe burnout has occurred, and the refrigerant must be wasted.

3. If the odor is acrid, but the oil is clean and of its normal color, the burnout is mild. The refrigerant in this condition may be saved and reused.

4. If there is no odor and the oil is clean, no burnout has occurred. The compressor-motor unit is suffering from a mechanical fault and must be replaced.

5. In sniffing refrigerant for odor, the fumes may be poisonous, especially if a burnout has occurred. Before undertaking this step, be sure to wear goggles and rubber gloves.

(c) ***Cleanup of Mild Burnouts***

1. See that suction line adjacent to compressor is clean. Install a filter-drier in the suction line.

2. Install a *new* oversize drier (ARI Standards) in the liquid line.

3. Install replacement compressor as required.

4. Charge system and place in normal operation. After 24 to 48 hours, reinspect oil for color and odor. If change of color has devel-

oped, change the oil and replace the drier in the liquid line. Suction drier may be continued in service if pressure drop through it is normal.

5. If possible, check condition of oil again after 14 days of operation.

(d) ***Cleanup for Severe Burnouts.*** If the filter-drier procedure as given in Sec. 10-6(c) above is not used, a flushout with pressurized refrigerant may be adopted. This is done as follows:

1. Install jumper hoses in place of all driers, valves, capillaries, and other restrictive devices. These hoses may be of anything but natural rubber, since the latter may be affected by, or affect, the refrigerant.

2. The circulating pump for flushing refrigerant under pressure through the system is preferably of the reciprocating or pulsating type, the idea being that the surges of the liquid will loosen scale, burnout debris, or any other foreign materials present in the system.

3. Arrange the pump to obtain its supply of refrigerant from a receiver that is so located as to provide a positive suction head on the pump at all times.

4. After the flushing has been completed, remove the excess fluid. This may be accomplished by the introduction of nitrogen under pressure, the pressure being limited by a pressure control valve on the nitrogen flask.

5. When flushing the system this way, all areas must be filled with liquid refrigerant. Calculate the total volume of the system in cubic

Table 10-7. Volume of Copper Tubing

Tubing size, O.D. in.	Cu ft of volume per lin ft of tubing	Lb refrigerant 11 per lin ft *
1/4	0.00020	0.018
3/8	0.00053	0.049
1/2	0.0010	0.092
5/8	0.0017	0.16
3/4	0.0025	0.23
7/8	0.0034	0.31
1 1/8	0.0057	0.53
1 3/8	0.0087	0.80
1 5/8	0.0124	1.14
2 1/8	0.0215	1.98
2 5/8	0.0332	3.06

* Weights given for Refrigerant 11 may be used for other refrigerants by correcting for the relative density of the other refrigerant. For example, density of Refrigerant 11 is 87.6 lb per cu ft at 130 F, and density of Refrigerant 21 is only 80.6 under the same conditions. The latter is therefore only 92% as heavy as Refrigerant 11, and 1 lin ft of 2 5/8 in. copper tubing would contain 3.06 lb × 0.92 = 2.82 lb of Refrigerant 21.

feet or gallons; then add 20 percent to be sure of enough and flood the system accordingly.

6. As a time guide, the flushing procedure should be accomplished in 4 hours or less. By inspecting the recirculated refrigerant for color and the presence of solid materials, the troubleshooter can determine the degree of cleanliness being obtained.

To aid the troubleshooter in estimating the weight of refrigerant necessary for flushing a system, Table 10-7 shows the volume of copper tubing in various standard sizes.

10-7. QUESTIONS AND ANSWERS

QUESTION 1. How does cryogenics fit into the refrigeration picture?
ANSWER 1. Cryogenics, in the simplest meaning of the word, is the art and science of producing cold in the range of −250 F to absolute zero. In the more comprehensive sense, cryogenics deals not only with these temperatures, but is concerned with the discovery, development, research, and practical production of the equipment required to achieve these notably low temperatures. Basic approaches are through the liquefication of various gases, such as helium at −452 F, hydrogen at −423 F, nitrogen at −346 F, oxygen at −362 F, and carbon dioxide at −314 F.

Troubleshooting in the field of cryogenics is the pursuit of specialists in that field. The conventional systems, although some aspects of their function may reach down to temperatures of −100 F or somewhat more, do not incorporate those problems of super-low temperatures for which solutions, in some instances, are still to be found.

In short, cryogenics as applied to commercial refrigeration has not yet been and may never be adapted to the family deep freeze, the neighborhood locker plant, or the ice plant.

QUESTION 2. How much refrigeration is necessary to destroy moths?
ANSWER 2. The following table indicates the relationship between temperature and time:

Temperature, F	Days to kill eggs	Days to kill larvae
0–5	1	2
6–10	2	21
11–15	4	
20–30	21	

Recommended procedure, according to the U.S. Department of Agriculture, for killing moths in furs and fabrics is to (a) store at 20 F for 4 days, (b) raise temperature to 50 F for 2 days in order to hatch the larvae,

(c) drop temperature to 10 F for 2 days, and (d) maintain a steady 35 to 40 F for the remainder of the storage period.

QUESTION 3. How important is humidity in cold storage operations?

ANSWER 3. Next to temperature, nothing is more so. The reason is that the presence or absence of water vapor in the air affects the value and appearance of the foodstuffs being stored. For example, relative humidities above 85 percent can promote the growth of mold on eggs. However, if the humidity is much less than this, the eggs suffer shrinkage and so depreciate in market value. Fruits also can be touchy regarding their humidity of storage. Bananas, for one, do not ripen satisfactorily unless the environment humidity is 90 percent or a bit higher.

QUESTION 4. In addition to the 300-lb ice cakes, what other sizes are in common practice?

ANSWER 4. Many plants, especially those of small capacity, use 100-lb cans. Dimensions of these cans are 8 in. wide by 16 in. by 32 in. deep. Construction is usually all welded, with 16-gage galvanized sides and 14-gage bottoms.

DIVISION 11

Ventilation and Ventilating Systems

11-1. VENTILATION PURPOSES AND METHODS

(a) ***The ASHRAE Definition.*** Ventilation is "the process of supplying or removing air by mechanical or natural means, to or from a given space. Such air may or may not have been conditioned."

The purpose of supplying or removing air, or of the performance of both functions, is to dilute or eliminate odors, to carry away surplus heat, and in some instances of favorable outside temperatures, to bring comfort to otherwise uncomfortable spaces. The troubleshooter's first duty is to ascertain which of these purposes the ventilation facilities have been intended to serve. The area of correction is then apparent.

(b) ***The Four Forms.*** Any given establishment, school, factory, commercial, or whatever, will be served by one of the following arrangements:

1. Natural air movement exclusively.
2. Natural air supply and a mechanical exhaust.
3. Mechanical supply and natural exhaust.
4. Mechanical or forced air movement exclusively.

Whatever the system, for every pound of air that is introduced into a space, an equivalent quantity must leave. Otherwise the new air just cannot enter, and the ventilation then fails in its purpose.

(c) ***Natural Ventilation.*** Natural ventilation describes air movement that results from differences in temperature, wind forces, or both. A common illustration of natural ventilation from temperature differences is in the movement of smoke up and out of a chimney. The hot gases are pushed upward by the cooler and denser air entering the furnace to support combustion. Similar air movements are frequently observed in tall buildings and are termed the "stack effect."

Natural draft may be computed from the formula:

$$D = 0.52 \times H \times P\left[\frac{1}{T} - \frac{1}{T_1}\right]$$

where D = natural draft expressed in inches, water gage
H = height of stack or structure, ft.
P = atmospheric pressure, lb per sq in. (At sea level this is 14.7 lb, less at altitudes.)
T = absolute temperature of ambient air
T_1 = absolute temperature of warm air or gas. Absolute temperature in all cases is the thermometer temperature plus 460 deg.

Obviously, the greater the height between the ground and the point of warm air escape, the greater the amount of draft generated. One-story buildings, especially those with low roofs, tend to produce a relatively weak natural air movement; consequently, these structures are more likely to adopt a mechanical ventilation, rather than natural means which may fluctuate from satisfactory to zero.

Very high temperatures at ground levels, such as may be met in a foundry, forge shop, or any heat-treating industry, sometimes overcome the effects of a one-story structure by substituting a large temperature differential for a large height differential.

Wind as an adjunct to natural ventilation is of uncertain value. Blowing across roof ventilators, it may aspirate space air from below the ventilator, or it may position a rotary ventilator, thus giving strong support to the natural ventilation system. Conversely, as the result of structures and trees in the area, the wind may produce a downdraft, thereby blocking or even reversing the movement of the intended natural ventilation.

Another potential troublemaker for some natural-ventilation installations is the back draft. The back draft can result from openings at elevations above the roof ventilator, when both these openings and the ventilator are in a common building. Figure 11-1 illustrates such an arrangement. The motive power for the back draft is the stack effect of the taller structure. Correction calls for the relocation of the roof ventilator to the higher roof or the addition of a ventilator on this roof.

Figure 11-2 shows the three types of natural-draft ventilator most often encountered by the troubleshooter. The weather vane appendage on the rotary ventilator provides an avoidance of downdrafts from winds; the turbine unit aspirates the warmer air from below when it revolves as the result of wind.

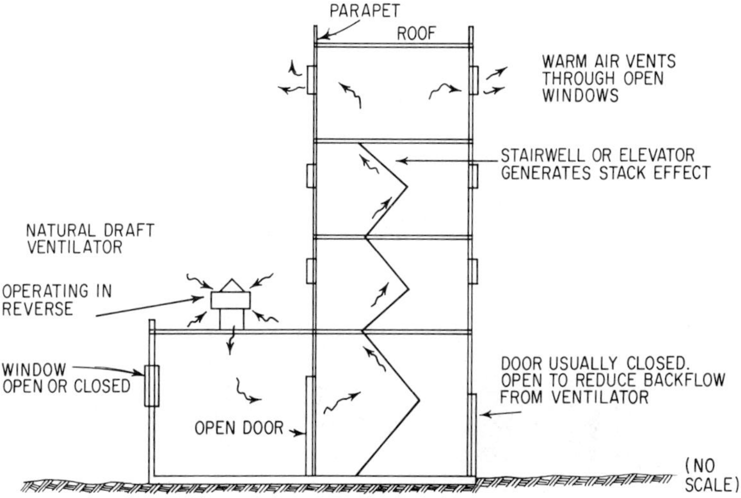

FIG. 11–1. Ventilation backflow (no scale).

(d) ***Mechanical Ventilation.*** Mechanical ventilation aims at freeing the ventilated space from the baneful effects of downdrafts and back drafts and, at the same time, provides a positive and continuing exchange of new air for old. Any desired number of air changes per hour can be obtained by mechanical ventilation equipment; not a single

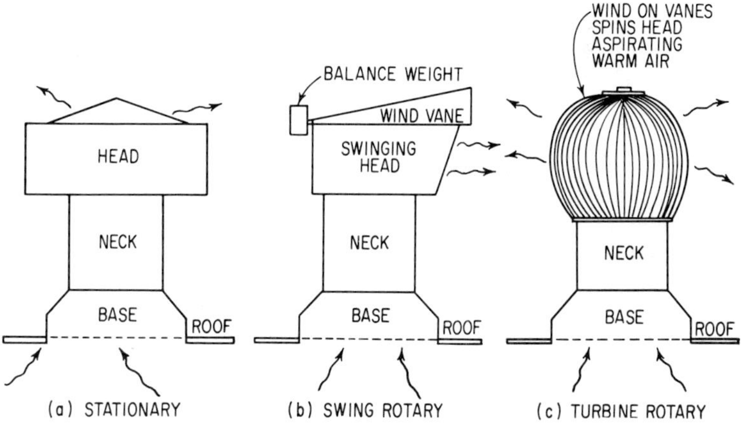

FIG. 11–2. Natural-draft ventilators (no scale).

FIG. 11–3. Portable fabric ducts for ventilation of enclosed spaces.

exchange can be guaranteed for natural draft except under certain stipulated conditions.

The heart of the mechanical-draft system is the fan or blower. It may be arranged to deliver fresh air to the ventilated space or to exhaust it, and its position may be in the wall, on the roof, or as a unit in a duct system. Duct installations include the hoods in kitchens and those mounted above fuming or steaming vats.

Portable mechanical ventilation equipment may be used for establishing a safe working atmosphere in areas of danger, such as underground tunnels, the interiors of tanks, the holds of ships. Welding operations in a generally closed space call for ventilation as a fundamental requirement.

Figure 11-3 shows portable fabric ducts in service delivering and removing air from a closed and overheated space. The blowers for these ducts are not shown, being approximately 100 ft distant.

Since the ventilation troubleshooter may find himself facing emergency conditions, Table 11-1 provides data on gases and vapors; Table 11-2 serves a similar purpose for fumes, dusts, and mists, and Table 11-3 lists the densities of common gases.

Table 11-4 offers a scale of relative odor values, as developed by C. P.

Yaglou in ASHRAE Research Report No. 1524 titled "Ventilation Requirements for Cigarette Smoke." This simple scale may be applied to intensity estimates for any undesirable smell.

Fume removal systems are illustrated by the unit ventilating system

Table 11-1. Industrial Hygiene Standards for Gases and Vapors*

Gas or vapor	Parts per million by volume	Odor †
Acrolein	0.5	Burning fat
Acrylonitrile	20.0	
Ammonia	100.0	Pungent, choking
Benzine or benzol	35.0	Light, frail
Bromine	1.0	Suffocating, irritating
Carbon dioxide	5,000.0	No odor
Carbon monoxide	100.0	No odor
Carbon tetrachloride	50.0	
Chlorine	1.0	Pungent, irritating
Chloroform	100.0	Sweet
Ethyl alcohol	1,000.0	Penetrating
Formaldehyde	5.0	
Freon 12, 21, 114	1,000.0	No odor
Gasoline	500.00	
Hydrogen chloride	5.0	Penetrating, acid
Hydrogen cyanide	10.00	Bitter almonds
Hydrogen fluoride	3.0	Pungent, irritating
Hydrogen sulfide	20.0	Rotting eggs
Methyl chloride	100.0	Burning fat
Naptha, coal tar derivative	200.0	Pungent
Naptha, petroleum derivative	500.0	Pungent
Nitrogen oxides	25.0	Pungent, sweet
Nitroglycerine	0.5	
Ozone	1.0	Pungent, irritating
Phosgene	1.0	Freshly cut hay
Sulfur chloride	1.0	
Sulfur dioxide	10.0	Pungent, irritating
Toluene	200.0	
Turpentine	100.0	Penetrating
Vinyl chloride	500.0	
Xylene	200.0	Musk, animal

* Based on an acceptable tolerance for an 8-hour working day, repeated over a 5-day working week.

† Differences in human response affect the description of an odor, so that a common agreement is not possible. Any odor, unless instantly identified as harmless, must be regarded as indicating a hazard until proved otherwise.

Table 11-2. Industrial Hygiene Standards for Fumes, Mists, and Dusts *

Fumes, mist, or dust	Mg per cu m †
Antimony	0.5
Arsenic	0.5
Barium	0.5
Cadmium	0.1
Cyanide	5.0
Iron oxide as a fume	15.0
Lead	0.15
Magnesium oxide as a fume	15.0
Manganese	6.0
Mercury	0.1
Phosphorous	0.1
Selenium	0.1
Sulfuric acid	1.0
Tellurium	0.1
Zinc oxide as a fume	15.0

* Based on an acceptable tolerance for an 8-hour working day, repeated over a 5-day working week.

† One mg = 0.00003527 oz. One cu m = 35.31 cu ft. Fumes, dusts, and mists are described in terms of weight.

Table 11-3. Relative Densities

Gas, vapor, or fume	Density (air = 1)
Ammonia *	0.0597
Butane	2.07
Chlorine *	2.44
Coal gas	0.47
Carbon dioxide	1.529
Carbon monoxide	0.0967
Hydrogen cyanide (prussic acid) *	0.93
Hydrogen fluoride *	0.71
Hydrogen sulfide *	1.19
Natural gas	0.56–0.70
Nitrogen oxides *	1.59
Oil gas (Pacific)	0.47
Ozone	1.655
Phosgene	3.45
Producer gas	0.86
Propane	1.55
Sewer gas	0.79
Sulfur dioxide *	2.26

* Water soluble

Table 11-4.

Grade 0	Imperceptible
Grade 1	Not objectionable
Grade 2	Acceptable
Grade 3	Objectionable
Grade 4	Endurable
Grade 5	Intolerable

shown in Fig. 11-4. Each of the fume sources is vented by a local "sucker" that may be attached to the source by means of a magnetic contactor, if so desired. The attachment pictured is to grinder hoods.

(e) ***Volumes and Velocities.*** Where fumes and other odoriferous air handling is concerned, a primary action of the troubleshooter is to check on volumes and velocities. Two methods are in common use. The first is to establish the volume of the space being ventilated and then to determine the number of air changes per minute in order to properly neutralize the atmospheric contamination.

The second method is to start from a fairly standardized air exchange rate of 4 cfm per sq ft of area. This rate is used by the General Services Administration and is called for by some municipal codes, particularly as these apply to commercial kitchens.

Velocity is calculated from the moving volume which is divided by

FIG. 11–4. Unit fume removal systems. Air is decontaminated in wall separator and returned to space. (*Aget Manufacturing Company.*)

the square foot area of the exit channel or duct. General practice indicates a maximum velocity of 2,000 ft per minute through the ducts, if ducts are part of the system.

Discharge of the supply air must be limited in velocity in order to avoid drafts and uncomfortable blows on personnel. The following figures may be considered a safe guide:

If point of discharge is less than 8 ft above floor, 120 ft per minute.

From 8 to 12 ft above the floor	250 fpm
From 12 to 18 ft above the floor	500 fpm
Over 18 ft above the floor	1,000 fpm

Points of air entry should be placed in all cases at such a distance from exhaust openings as to avoid short-circuiting from entry to exit. The same is true of venting appliances, vats, fuming tanks, furnaces, and other consumers of air. An ill-chosen location for the points of air entry may serve nothing but some form of venting equipment, leaving the main space unventilated and uncomfortable. These air-using units should be provided with their individual ventilating systems, as in the grinders in Fig. 11-4.

As a further guide to recommended velocities in kitchen hoods, consider the following figures:

Central hood, 4 exposed sides	150 fpm across face of hood
Wall-hung hood, 3 exposed sides	100 fpm across face of hood
Corner-hung hood, 2 exposed sides	85 fpm across face of hood
Hood with aprons on 3 sides, 1 exposed	85 fpm across face of hood

These velocities also may be considered as a starting point for the removal of fumes and odors, but of course they are subject to modification if they are found insufficient to accomplish the ventilating purpose.

(f) ***Handling Fumes and Odors.*** Conventionally, the designer has a choice of five methods for handling objectionable atmospheres.

Method No. 1 is simply to flush out contaminated spaces with enough clear air to replace the foul, or at least to so dilute it that the mixture establishes a tolerable condition. This is the simplest and, in most instances, the cheapest technique available. However, antipollution ordinances in some localities may bar its use. See Table 11-5.

Method No. 2 is to absorb the objectionable odor or fume, thus creating a mixture that can be disposed of as waste. The expense of this method is in the constant drainage of the absorbing material. Also there may develop an equally nettling problem of where to dump the mixture without fouling some other space or irritating the neighbors. Water sprays and scrubbers are used with soluble gases.

Method No. 3 permits the fume or gas to find a resting place in the

Table 11-5. Exhaust Air Required for Welding and Burning in Enclosed Spaces *

Distance, exhaust nozzle to arc or torch, ft	*Minimum air flow per operator, cfm*		
	Space volume over 50,000 cu ft	Space volume 5,000–50,000 cu ft	Volume under 5,000 cu ft
8 or less	200	225	250
8–10	250	275	300
10–12	300	350	375
12–15	350	400	450
Fan capacity per welder, cfm	2,500	1,850	1,000
Time for complete change, minutes	20	15	4

* Data from U.S. Department of Labor, Bulletin 304, "Maritime Safety Data for Shipyards."

body of a host material called an "adsorbent." The chemical makeup of the adsorbent is not affected by the foul visitor and merely loads up with it temporarily. When fully loaded, the adsorbent is reactivated by being heated, the contaminants are driven off, and the adsorbent host is then ready to repeat the cycle. A roll call of adsorbents includes activated carbon, silica gel, and activated alumina. This system uses the adsorbent as a transfer station, collecting the contaminants from the air, and dumping them inoffensively where they will bother nobody.

Method No. 4 employs a catalyst. The catalyst is formed into a mat of metal wires or ribbons, using a platinum alloy, and operates at a temperature between 500 and 600 F. This temperature is below that of visible flame; consequently, the destruction of combustible fumes and gases is carried out without flame. Flue gases leaving the catalyst are higher in temperature than they were on entering it, and there is likely to be considerable heat thus made available for a waste-heat boiler, an absorption refrigeration system, or some other useful application. This salvaging of heat helps pay for the catalyst.

Method No. 5 concentrates on destroying objectionable gases and fumes by means of an afterburner. The most familiar installation of the afterburner probably is found in incinerators. The idea is simply that materials escaping destruction in a primary fire or furnace will not be so lucky if obliged to transit a second furnace where combustion is maintained at high temperatures. Afterburners are discussed in detail in Division 17.

The ventilating troubleshooter may find himself faced with complaints originating in any one of these five methods for contaminant destruction. He should be prepared accordingly.

(g) ***Fans and Blowers.*** These units customarily are mounted on the roof or in a wall when performing exhauster duty. In these positions, they handle contaminated air, which is sometimes corrosive; in other instances, the air is heavy with dust or grease or oily vapors that in time may deposit enough foreign material on the fan blades to create an imbalance.

As evidence of deposit accumulations, the motor load shows an increase, speed diminishes, and the effectiveness of the fan is noticeably decreased. All exhaust fans and blowers should be checked periodically for a correction of developing overloads.

Figure 11-5 shows a form of wall fan; Fig. 11-6, two forms of roof ventilator, both power-driven. Drive may be direct-connected or through a belt, depending on the form selected.

When answering a trouble call that appears to have originated in the fan, note first of all whether the fan or blower is fabricated of plastic or other corrosion-resistant material. Either way determine the formula; the material will have a bearing on the correction of the trouble and particularly on the replacement of defective parts.

FIG. 11–5. Flush-type wall exhauster. This is a direct-connected fan, but belt-driven units are available and similarly mounted inside the wall box. Access is through the removal of the aluminum louver. (*Swartwout, Inc.*)

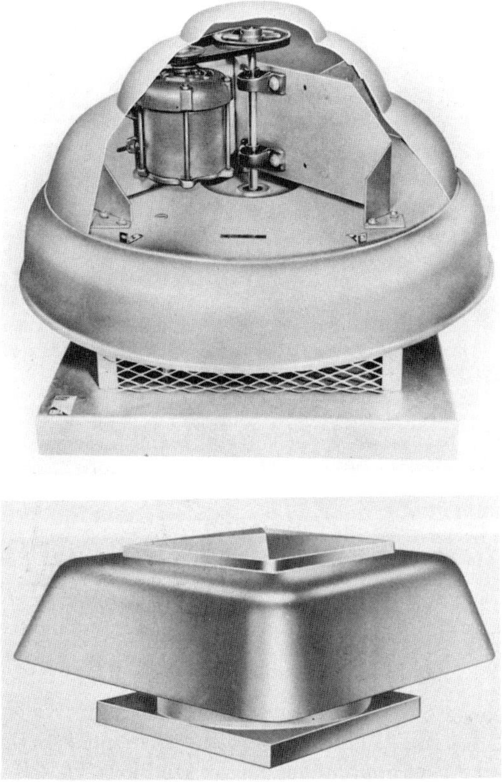

FIG. 11–6. Two forms of roof ventilator, both power-driven. (*Ilg Industries, Inc.*)

Another important inspection procedure is to examine carefully for evidence of water entry. A wetting could be disastrous to the motor.

Figure 11-7 illustrates a V belt–driven blower of the type frequently used for the removal of dusts and the delivery of contaminant-loaded air to some form of separator. It is a component of a ventilating system using ducts.

(h) ***Filters.*** In general, three forms of filters may be encountered in ventilation systems: dry filters; wet filters, such as scrubbers and spray filters; and electrostatic filters. Since all of these introduce an element of friction and, consequently, pressure loss in the air passing through, their practical application is limited to installations incorporating a fan or blower. Also they are likely to be mounted in the exhaust rather

FIG. 11–7. Belt-driven exhauster fan suitable for service with duct systems. (*Northern Blower Division, Buell Engineering Company, Inc.*)

than in the supply, in order to monitor the contaminants leaving the ventilated spaces. Occasionally, recovery of these dusts and particles has economic value.

Dry filters and electrostatic filters are both common in air-conditioning systems. Troubles with these components therefore may be referred to the appropriate sections of Divisions 7 and 9.

Air washers using water sprays may be subject to considerable corrosion both from the water and from the contaminants being absorbed. Complaints will be examined in the later sections of this Division, as they apply to this equipment.

(i) ***Ventilating Air Requirements.*** The troubleshooter on ventilating complaints can expect to become involved in problems of air volumes, points of air entry, and locations of the air-removal openings.

Table 11-6, "Ventilating Air Volumes for Structures," is presented as a guide and assistant in solving these problems. By use of an anemometer or velometer, the flow of air in either direction can be measured, and a comparison of this measurement with Table 11-6 will give some idea of the adequacy of the flow.

This table should not be treated as a standard; it is simply the result of experience by many persons, and for any given ventilated space, its values may be excessive or deficient. For specific applications, if the anemometer readings approximate the table values but the ventilation is unsatisfactory, the course of correction must consider an increase in the air volume. By similar reasoning, if the unsatisfactory ventilation is accompanied by a less-than-table-value of the moving air, certainly this deficiency calls for an improvement that may clear up the situation without further search for trouble.

In another situation the air movement substantially exceeds the table values, but defective ventilation exists nonetheless. This evidence suggests trouble in some area of operation other than volume; the troubleshooter then looks for short-circuiting, faulty circulation resulting from unbalanced dampers, external windage, or variations from the original conditions for which the system was designed.

Treatment of these difficulties in detail appears in subsequent sections.

Table 11-6. Ventilating Air Volumes for Structures *

Structure	*Air removal, cu ft per hour per sq ft floor*	
	Average contamination	Bad contamination
Churches	90–100	
School classrooms	90–110	125
Dairy industry	120	300 or more
Dry cleaning establishments	100	more for heat removal
Offices	60	
Forge shops	100–200	double or more
Foundries	300 for high roofs	500–600
Repair garages	80–120	200 or more
Light manufacturing	.100–110	200
Gymnasiums	80–90	100–120
Hotel corridors	60	80–100
Kitchens, commercial	240	400
Kitchens, domestic	200	300
Dining rooms	60–90	125
Toilet rooms	120	200
Laundries	200	300 or more
Assembly rooms (with smoking)	120	160
Shops, machine and woodworking	75–100	125–200
Paint spray rooms	180–200	250–300
Retail stores	80–100	200
Theaters	80	200 (projection booth)

* Volumes are approximations and have been compiled from several sources and the author's experience. They should be used as check values only. Every job has peculiarities affecting its requirements.

11-2. TROUBLES WITH NATURAL VENTILATION SYSTEMS

(a) ***Complaint.*** Shop structure with monitor roof and clerestory type windows, Fig. 11-8, depends on natural ventilation. Unless a wind is blowing and entering through the ground-level windows, there is no noticeable movement of air upward to the monitor with attendant venting through the clerestory windows. Ventilation is unsatisfactory.

Causes

1. The space immediately below the monitor roof and above the side roofs forms a heat trap. This condition is aggravated on sunny days, the solar effect on the monitor creating a temperature inversion relative to temperatures in the shop at ground level. Since warm air at ground level cannot move upward rapidly because of the still-warmer air in the heat trap, ventilation at ground level is ineffective.

2. A contributing cause is that the clerestory windows are more likely to be closed than open as a protection against the entrance of rain. Even if open, these windows require a horizontal movement of warm air to vent through them, usually resulting in an ooze rather than a flow.

Recommended Actions

1. If natural ventilation is to continue, install roof ventilators, using one of the types shown in Fig. 11-2, and keep the clerestory windows closed.

2. Provide positive air movement by installing motor-driven roof exhausters as in Fig. 11-6.

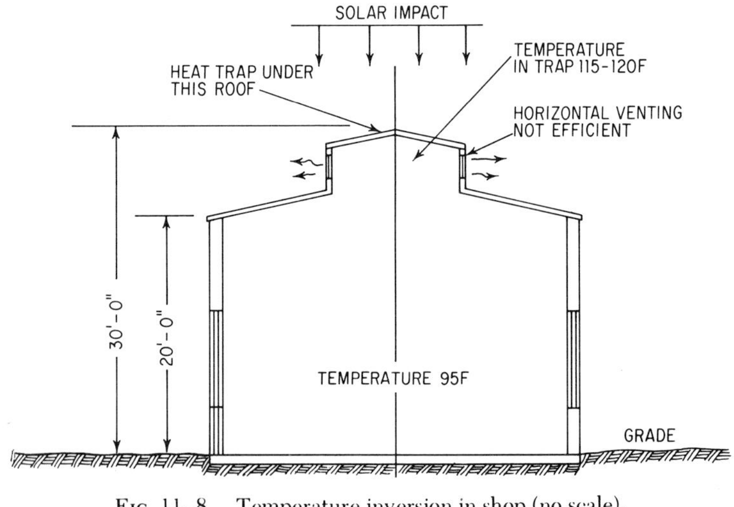

FIG. 11–8. Temperature inversion in shop (no scale).

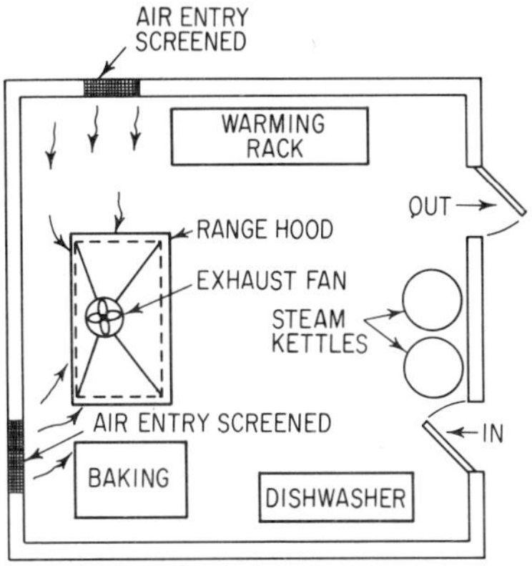

FIG. 11–9. Faulty kitchen ventilation (no scale).

3. Augment the effect of the monitor gravity exhausters, Recommended Action No. 1, with auxiliary roof gravity units mounted on the wing roofs.

(b) ***Complaint.*** Restaurant kitchen, Fig. 11-9, remains hot and odorous, despite the existence of ventilating openings in the walls.

Causes

1. During operation of the exhauster fan above the range, the outside air is short-circuited, moving directly from the inlet openings to the range hood.

2. When the range is not in operation, and when the exhauster fan is not running, natural air movement through the openings is opposed by the resistance of the insect screening across the openings. Unless very high temperatures are developed in the kitchen, enough to overcome the frictional resistance of the screens, fresh air from the outside does not enter, and the kitchen atmosphere remains stagnant.

Recommended Actions

1. Convert the ventilation system from natural to power-operated by installing at least one power-driven blower to provide positive entry of air to the kitchen.

2. Install a pressure-actuated shutter over the high wall opening. This shutter should be designed to open when the air pressure in the kitchen becomes slightly above normal, such as when the range exhauster fan is not operating.

3. If possible, locate the new power fan close to floor level and across the kitchen, full width or length, relative to the range hood. In this position, the sweep of fresh air gives the most benefit to persons working in the kitchen.

4. In some instances, fresh air from above the roof is brought down to a ground-level kitchen by means of a duct, the blower being mounted on the roof. In this arrangement, there may be some danger of the blower picking up the discharge from the range hood exhauster, at least during certain directional sources of the wind. Locate the fresh air inlet to avoid this danger.

(c) ***Complaint.*** Despite several windows in the upper walls and a natural-draft ventilator on the roof, an automobile garage and repair shop is consistently heavy of atmosphere. By the end of the working day, some of the shop people have headaches.

Cause

The gases and vapors coincident with the operation of internal combustion motors are generally heavier than air; consequently, high-set windows are of small value in venting gas accumulations close to the floor. See Table 11-3 for comparative densities. Gasoline, not included in the chart, is heavier than air to start but gradually dilutes, and some percentage eventually may rise. The danger here is mainly pyrotechnic; a spark may cause it to flash.

All these gases will create physiological conditions and symptoms if breathed for a sufficient number of consecutive hours.

Recommended Actions

1. Gas and vapor collection systems must operate from collection points close to the floor. In this complaint case, install ducts with their openings about 16 in. above the floor. This arrangement is illustrated in Fig. 11-10, a commercial garage.

2. Alternatively, patented collection systems employ undersurface ducts to move the gases and vapors first down, then out.

3. Because of the density of these gases and vapors, natural ventilation is unlikely to be satisfactory. Positive evacuation is essential.

4. The urgency of the required action can be estimated from an oxygen deficiency check of the atmosphere at working level. At least 17 percent oxygen must be present for working conditions that are reasonably safe.

(d) ***Complaint.*** The show windows in a retail store develop frost and condensation during many of the winter days.

Cause

The inside air, in contact with the cold surfaces of the window glass, is chilled below its dew point and deposits moisture on the glass. If the outside temperature is low enough, this condensate freezes and forms frost, thus impairing vision through the window. Either way, the condensate becomes a problem.

Recommended Actions

1. Treatment of the condition may take one of two forms. The first

FIG. 11–10. Garage ventilation duct system.

is to reduce the relative humidity of the air in contact with the window. This may be accomplished by introducing cold, low-humidity outside air through small grilles set in the outside wall *beneath* the window. This air, gradually rising and expanding as it picks up heat from the interior conditions, may be absorbed in the store ventilation system or vented outdoors through a high wall exit in one of the store walls, preferably opposite the window. Natural ventilation may be expected to function satisfactorily with this plan. See Fig. 11-11a.

The second scheme serves a closed-in window, as noted in Fig. 11-11b. It also requires a small circulating fan to discharge rising and warmed air to the outside atmosphere through a grille set *above* the glass window.

2. By means of drying equipment, maintain the entire store atmosphere at a noncondensing humidity. This is a costly operation and conducive to respiratory membrane troubles from excessive dryness.

11-3. TROUBLES WITH MECHANICAL VENTILATION SYSTEMS

(a) ***Preliminary Discussion.*** Ventilation is an integral part of most air-conditioning systems, accomplishing its purpose by means of blowers, fans, and dampers. For correcting complaints in this area of

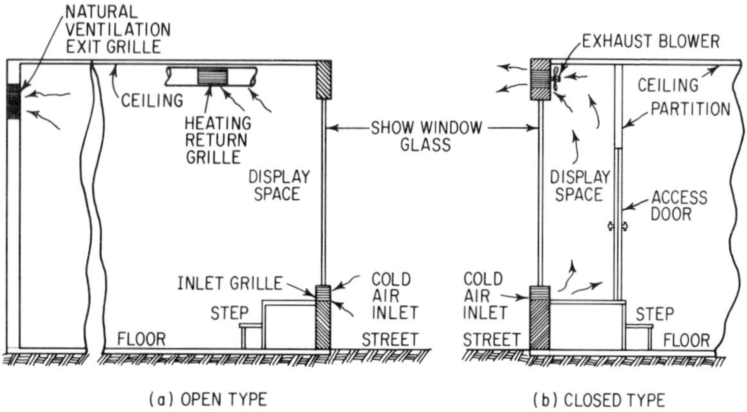

NOTE:
FOR SIZING INLET AND EXHAUST AREAS, PROVIDE AT LEAST 3 SQ IN. OF EACH AREA FOR EVERY LINEAR FOOT OF WINDOW GLASS MEASURED ACROSS THE FRONT.

FIG. 11–11. Show window ventilation (no scale). Two arrangements.

application, the troubleshooter is referred to the appropriate sections in Division 9.

Mechanical ventilation that is ventilation solely, and not a factor in some other treatment of the atmosphere, is usually encountered in harsh industrial applications, such as foundries, forge shops, paint-spray operations, handling corrosive and toxic fumes, and dust collection and transfer, as well as in tunnels, tanks, and mines.

Complaints may be separated into two categories: first, complaints of inadequate performance, and second, complaints involving the sudden breakdown or development of defective operation in the ventilating machinery itself. Differences in the approaches to correction will become clear as the individual cases are examined in the following sections. In many instances, these differences are caused by peculiarities of the areas being ventilated and of course by the nature of the air being handled.

(b) ***Complaint.*** Motor-driven roof exhausters develop excessive vibration.

Causes

1. Fan blades have accumulated deposits of adhering materials from the air being handled. This has thrown them out of balance.

2. Because of slight shifting or settling of the supporting structure, the relationship of motor and fan has become misaligned.

Recommended Actions

1. Clean fan blades as a basic principle.

2. Tap each blade lightly with a hammer to determine if it is loose or if it contains a crack.

3. Check alignment of fan and motor.

4. If unable to identify source of vibration, remove to the shop and check complete unit for balance.

5. As routine in field examination, check firmness of mounting.

(c) ***Complaint.*** Shop areas are divided into hot spots and regular caves of the winds. Worker efficiency is diminished by both conditions.

Causes

1. Velocity of discharge is too high, or supply fan is too near the working area, thus producing an undesirable draft.

2. The hot spots are suffering from a lack of air change, resulting from exhaust openings that are too distant from the hot areas or are undersized, and are unable to handle the volumes required for satisfactory ventilation.

3. High heat-producing equipment, such as heat-treating ovens or furnaces, has not been provided with separate ventilating facilities to counteract their extreme environments.

Recommended Actions

1. Measure the outputs of the supply fans in cfm for volume and fpm for velocity.

2. Check obtained measurements against the figures in Sec. 11-1(e) and Table 11-5.

3. To counteract the excessive draft, consider relocating the supply fan. If this is not practical or desirable, try setting up a pedestal fan in such a location that its discharge serves as an air baffle to the ventilating fan producing a turbulence rather than a draft. Figure 11-12 illustrates the relationship.

4. Ventilate the high-heat-producing unit individually by the placement of an exhauster designed to pull ambient air over the high-heat maker and vent it directly outdoors.

5. For hot spots not presently receiving any benefit from the general ventilation, or not enough benefit, install an additional ventilator, either roof or wall, as appropriate. Item for caution: the new ventilator may upset the balance elsewhere if the supply fans are not sized to handle the additional flow of air. Treatment, therefore, may involve increasing the capacity of the supply fan. Item for hope: the new ventilator in the hot spot may neutralize the undesirable draft in an adjacent area.

(d) ***Complaint.*** In a multistory office building having toilets on

FIG. 11–12. Countering air blasts (no scale).

each floor and exhauster fans on the roof to vent them, the removal of odors in the toilet rooms is unsatisfactory.

Cause

Ventilating air for toilet rooms is expected to originate in the access corridors of the building, entering the toilet rooms whenever the doors are opened. If these doors are tight-fitting and opened infrequently, no movement through the toilet ventilating system is possible since there is no makeup for air that would be removed by the exhaust system.

Recommended Action

Provide a small no-sight louver in the bottom panel of the toilet room door. This will permit a continuing ventilation of the room.

(e) ***Complaint.*** In a battery-charging room, the exhaust fan is discharging through a side wall. Despite the use of filters, there have been complaints in the area from the fumes.

Causes

1. These fumes are difficult to separate from the air by the conventional filters. Fine-spun mineral wool has been effective in some instances.

2. A low volume of ventilating air will produce high temperatures in the battery room and a relatively high concentration of the fumes. If this is the case here, more ventilating air must be circulated through the room.

Recommended Actions

1. Measure the volume of air being handled by the exhaust fan. The approved volume, for the average battery room, is 75 cu ft per hour per sq ft of floor area. If the room is congested with equipment, allow 100 to 150 cu ft, as necessary, to maintain temperatures at safe levels.

2. In addition to providing sufficient volume of air to suit conditions, consider shifting the exhaust fan to the roof. From a roof discharge, the fumes may be expected to undergo quicker diffusion than with a wall outlet. Also this diffusion will take place at an elevation above the usual working areas.

3. All of the exhaust components should be of corrosion-resistant materials. Some plastics are excellent in this application, others unsatisfactory. Obtain a guarantee from the manufacturer, based on the plastic's chemical reaction to the fumes, before adopting a plastic exhaust system.

(f) ***Complaint.*** Figure 11-13 shows the side elevation of a church. Blowers located at each side of the sanctuary and in compartments approximately 14 ft above the floor are sized to throw chilled air in summer and warm air in winter the full length of the church, right up to the narthex partition. When neither heating nor cooling is needed, the blowers operate simply as ventilating equipment. Operation of these blowers sets up strong draft conditions throughout the year and results in frequent complaints from the congregation.

Causes

1. The velocity necessary to carry the air 70 to 80 ft to the back of the church is high, and drafty conditions at various areas along the line of throw are inevitable with the arrangement as shown.

2. Blower speed is too high, and the volume of air discharge is larger than needed.

Recommended Actions

1. Measure the volume of air leaving each of the two blowers. If the sum of the two in cubic feet per minute is enough to produce a com-

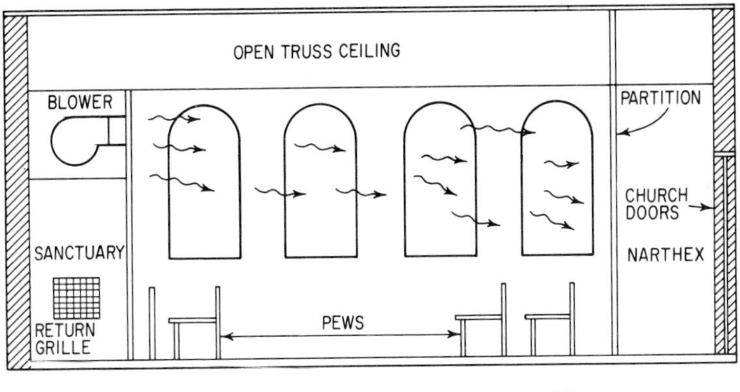

FIG. 11-13. Church ventilating (no scale).

plete change oftener than four times an hour, there is too much air being discharged. Reduce the volume accordingly.

2. Erect a counterblow outlet to face each main blast, as shown in Fig. 11-14. These counterblows, recirculating air from the church auditorium, convert much of the velocity to turbulence. The effect is that of having four blowers instead of two, thus requiring relatively low velocity from each.

(g) ***Complaint.*** The below-street-level storerooms of a retail women's shop remain so damp that full use of these rooms is practically impossible.

Causes

1. Because the storerooms are below the street level, a degree of surface dampness is carried through the foundations and into these rooms by capillary movement. This moisture, being isolated from free air contact, evaporates very slowly in the ground, but the evaporation process speeds up when the moisture appears on the surfaces of the basement walls and floor. This speedup reduces the vapor pressure in the basement to some value lower than that in the exterior ground, thereby promoting the inflow of new moisture.

2. Because of a lack of ventilation in the average basement, the humidity gradually rises until near saturation produces the complaint described.

Recommended Actions

1. Paint the walls and floor of the basement with a moisture-sealing paint. This treatment has been successful in some areas, less than satisfactory in others. Check with the sealant manufacturer for reports on its success in the geographical area of the complaint.

2. Install drying equipment of types that are electrically reactivated.

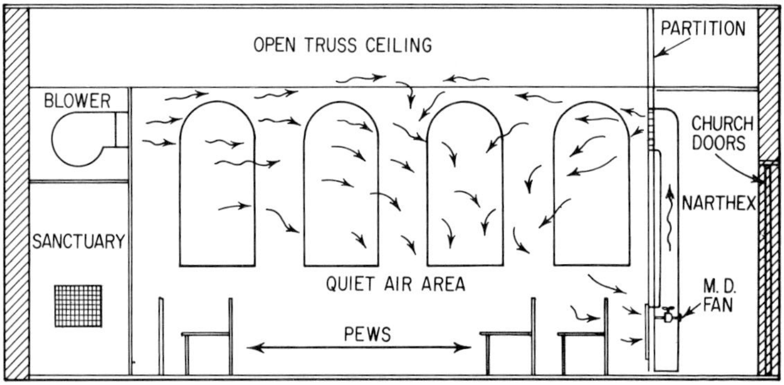

FIG. 11–14. Recirculating fan and duct (no scale).

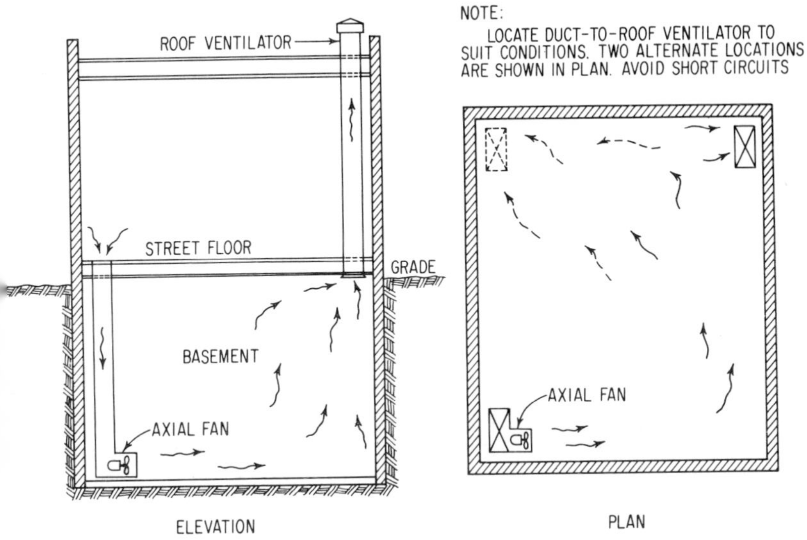

FIG. 11–15. Basement ventilation (no scale).

Dual units will be needed, one to maintain the drying cycle while the other is being reactivated.

3. Ventilate the basement as suggested by the sketch in Fig. 11-15. In order to escape spotty drying, easy movement of the ventilating air is necessary; consequently, circulation must not be blocked by stored merchandise. Actuate the fan by humidistat.

4. Depending on weather conditions and the nature of the soil in the area, attaining a dry state for the basement may extend over two or three months regardless of what system is used.

(h) ***Complaint.*** The nozzles in an air washer corrode badly and must be replaced at frequent intervals.

Causes

1. The water supply carries some chemical that reacts with the metal of the nozzles.

2. The nozzles are being eroded rather than corroded by gritty material in suspension in the water supply.

Recommended Actions

1. Neutralize the corrosive element in the water. To do this successfully requires an expert in water treatment. Communicate with a water treatment firm.

2. If neutralization is not practical or is considered too expensive, replace the nozzles with ones of stainless steel. This is a costly change,

but the increased service from the nozzles may well justify the investment.

3. Introduce filters, a settling basin, or both into the water supply to the nozzles in order to remove the gritty matter and thus protect the nozzles.

(i) ***Complaint.*** Water sprays in an air washer are steadily diminishing in density and distance.

Causes

1. Pressure to the sprays is falling off. If on city water pressure, the pressures may well fluctuate and diminish as the area grows and the demand for water increases. If water is furnished by a pump, pump wear could be a cause.

2. Filtering equipment is sufficiently fouled to affect water flow.

Recommended Actions

1. Check pressure. Compare gage readings with pressure in the area known to be standard with the city system. Compare gage readings also with the design discharge pressure of the pump.

2. If city pressure cannot be bettered, install a booster pump.

3. If pump is at fault, repair and replace parts as necessary.

4. Clean or replace filters as found to be needed.

5. Analyze water supply for excessive amount of foreign material which may be clogging filters or expediting wear of pump parts. Identify the reason for excessive particulates and act as required to improve the conditions.

Table 11-7. Troubleshooter's Field Guide for Ventilation

Possible causes	How to check for cause	Recommended ways to correct
COMPLAINT: Shop structure with monitor roof has insufficient ventilation.		
REFERENCE: Sec. 11-2(a).		
1. Space under monitor forms a heat trap.	*a.* Compare temperature under monitor with temperature at work level.	*a.* Install roof ventilators on monitor. Install power-driven exhausters on top of monitor. Add auxiliary exhausters on wing roofs.
COMPLAINT: Restaurant kitchen is hot and odorous, using natural ventilation.		
REFERENCE: Sec. 11-2(b).		
1. Range fan is short-circuiting air makeup to kitchen.	*b.* Show air movements in kitchen by smoke test.	*b.* Rearrange fresh air inlets to sweep kitchen. Convert gravity

Table 11-7. Troubleshooter's Field Guide for Ventilation (*Continued*)

Possible cause	How to check for cause	Recommended ways to correct
		ventilation to power-driven type. See Sec. 11-2(b) for details.
COMPLAINT: Atmosphere in automobile garage and repair shop is causing headaches in workers. REFERENCE: Sec. 11-2(c).		
1. Present ventilating openings are high, but most hazardous gases are as heavy or heavier than air.	*c*. Check percentage of oxygen in air at working level. Must have 17 percent minimum.	*c*. Ventilate from close to floor. See Sec. 11-2(c) for details.
COMPLAINT: Show windows in a retail store show frost and condensate on numerous winter days. REFERENCE: Sec. 11-2(d).		
1. Air in contact with glass is chilled below dew point.	*d*. Note temperature and percent relative humidity.	*d*. See Sec. 11-2(d) for details.
COMPLAINT: Motor-driven roof exhausters develop excessive vibration. REFERENCE: Sec. 11-3(b).		
1. Fan blades are out of balance from material adhesions.	*e*. Inspect the blades. Scrape with knife. Look for nicks in blades.	*e*. Clean blades.
2. Fan and motor are out of alignment.	*f*. Check alignment with dial indicator or other measurements.	*f*. Correct misalignment if found, in shop if necessary.
3. Mounting is loose.	*g*. Shake or stress unit for tightness of bolts.	*g*. Tighten bolts as necessary.
4. One or more fan blades is cracked.	*h*. Tap with hammer in addition to examination.	*h*. Replace blade or entire fan.
COMPLAINT: Areas in a large shop alternate between over hot and over drafty. REFERENCE: Sec. 11-3(c).		
1. Velocity of air discharge is too great in	*i*. Measure the velocity with anemometer.	*i*. Slow fan or install louvers in discharge.

Table 11-7. Troubleshooter's Field Guide for Ventilation (*Continued*)

Possible cause	How to check for cause	Recommended ways to correct
drafty areas, or fan is too close to workers.		Set up counteracting fan.
2. Hot spots are not getting adequate air change.	*j.* Take temperatures. Check air movement with test smoke.	*j.* Provide new and more ventilation.
3. High - heat - making equipment is not effectively ventilated by area air movement.	*k.* Take temperatures beside equipment, all levels.	*k.* See Sec. 11-3(c) for details.
COMPLAINT: Toilets in a multistory office building contain persistent offensive odors.		
REFERENCE: Sec. 11-3(d).		
1. The movement of air into the toilets is so restricted that exhauster fans are unable to vent the rooms.	*l.* Check air movement with test smoke.	*l.* Provide louvered openings in bottom panel of room door.
COMPLAINT: Fumes from a battery-charging room are a neighborhood annoyance.		
REFERENCE: Sec. 11-3(e).		
1. Air filters are not suitable for these particular fumes.	*m.* Examine cleanliness of filters. Check with manufacturer for this service.	*m.* Install filters of more suitable design as recommended by filter makers.
2. Inadequate ventilation of battery room is resulting in an excessive concentration of fumes in the exhausting air.	*n.* With anemometer, note volume of air being discharged. Compare this with calculated volume.	*n.* Increase ventilation by speeding up fan or replacing with larger fan. Consider shifting point of discharge to roof.
COMPLAINT: Drafts in church are so bad that many in the congregation stay away and others complain.		
REFERENCE: Sec. 11-3(f).		
1. Velocity of air discharge is too great to meet required throw.	*o.* Check velocity of discharge with anemometer.	*o.* Reduce velocity by change in fan speed or reset the outlet

Table 11-7. Troubleshooter's Field Guide for Ventilation (*Continued*)

Possible cause	How to check for cause	Recommended ways to correct
		louvers. Set up a counterblow, as in Fig. 11-14.
COMPLAINT: Storage room in retail shop basement is almost unusable because of high level of dampness. REFERENCE: Sec. 11-3(g).		
1. Lack of ventilating air movement in basement permits dampness from ground water to develop.	*p.* Check relative humidity percent, and temperature. With test smoke, note direction of air movement, if any.	*p.* Seal walls and floor of basement with special paint for this purpose. Install electric drying equipment. Ventilate as shown in Fig. 11-15.
COMPLAINT: High incidence of corrosion is being experienced with certain air washer spray nozzles. REFERENCE: Sec. 11-3(h).		
1. Water is chemically contaminated.	*q.* Obtain a water analysis.	*q.* Neutralize by water treatment.
2. Nozzle destruction is from erosion.	*r.* Look for grit in water to sprays.	*r.* Install filters, settling basin, or both. Consider stainless nozzles and pipes.
COMPLAINT: Output and effectiveness of sprays in air washer show progressive reduction. REFERENCE: Sec. 11-3(i).		
1. If sprays are on municipal system, system pressure is falling.	*s.* Check pressure at washer against the published city pressure.	*s.* Notify city water department. If increase not possible, install booster pump.
2. Water is being furnished by private pump, and pump needs overhaul.	*t.* Check pump discharge pressure against design pressure. Measure output in gpm against design head-pressure curves.	*t.* Overhaul pump.

Table 11-7. Troubleshooter's Field Guide for Ventilation (*Continued*)

Possible cause	How to check for cause	Recommended ways to correct
3. Water filters are fouled excessively, introducing high flow resistance.	*a*. Inspect filters.	*a*. Clean or replace filters as needed.

DIVISION 12

Humidity Control

12-1. RELATIVE HUMIDITY, ITS MEANING AND MEASUREMENT

(a) *Definitions.* "Humidity" is the name given to water vapor present in the air. The Romans called it "humidus," which is the Latin for moist.

"Relative humidity," quoted as a percentage, indicates the amount of moisture present compared with the maximum amount that the air can hold at the existing temperature. Temperature is very important to the capacity of the air for absorbing moisture. The warmer the air, the greater its capacity for holding moisture.

When the air contains all the moisture it can at the existing temperature, it is said to be "saturated." It is also said to be of 100 percent relative humidity, saturation and 100 percent relative humidity having identical meanings.

Saturated air can be made unsaturated simply by raising its temperature. It also can be made over-saturated with a precipitation of the excess moisture by lowering the temperature. The temperature at which this precipitation begins is called the "dew point." Dew point is of top importance in air-conditioning installations, since by exercising the dew point, the relative humidity of the air entering a conditioned space can be established at whatever level may be desired for the maximum comfort.

Moisture also may be removed from the air, without changing the temperature, by passing the air over a moisture-adsorbing chemical, such as silica gel or activated carbon. To increase the moisture without changing the air temperature, water vapor must be added in the form of sprays or by evaporation. Steam jets are used frequently for this purpose in industrial problems.

A knowledge of these relations of moisture, temperature, and dew point is important to the troubleshooter. In them he will usually find the solution to his problem.

"Absolute humidity" expresses the amount of water vapor present, not in a percentage as with relative humidity, but as grains per pound of dry air. This weight of moisture tells us how much water we must remove from a given condition of temperature and humidity in order to establish a desired condition. To illustrate, if we have a starting condition of 90 F dry bulb and 40 percent relative humidity for every pound of dry air, 102 grains of water vapor are present. Now if we want to

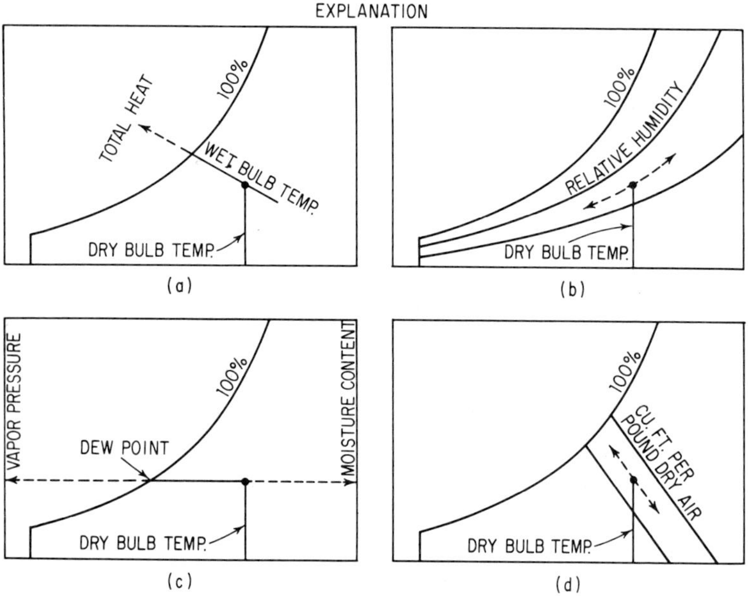

Fig. 12–1. Psychrometric chart. (*General Electric Co.*) If two properties of air are known, all properties may be found as follows:

Dry-bulb Temperature is read directly by following vertically down to the bottom scale.

Wet-bulb Temperature (a) is read directly at the intersection of the wet-bulb line with the 100 percent relative humidity line (saturation curve). The scale is marked along the 100 percent line.

Relative Humidity (b) is read directly from the curved lines marked Relative Humidity. For a point between the lines, estimate by distance.

Moisture Content, or absolute humidity (c), is read directly from horizontal lines with scales to the right and left of the chart, and is the weight of water vapor contained in a quantity of air and water-vapor mixture which would weigh 1 lb if all water vapor were extracted.

change this to 80 F and 50 percent relative humidity, this desired state will contain 82 grains to the pound, indicating that we must do something to get rid of 20 grains of moisture. Incidentally, there are 7,000 grains in a pound, and if the conditioned space is large enough, we will have plenty of pounds of water to dispose of.

The psychrometric chart in Fig. 12-1 shows all these values for the benefit of the troubleshooter, as well as for the designer.

(b) ***Discomfort Index.*** Although this term has been largely superseded by a more scientific expression, it is of high significance to the troubleshooter. By calculating this index for any given space, he can evaluate the true worth of many summer complaints originating in the said space. What makes some persons uncomfortable does not bother others. For example, an index of 70 disturbs only a few; half of those

Dew-point Temperature (a) is read at the intersection of a horizontal line of given moisture content with the 100 percent relative humidity line.

Enthalpy (Total Heat) (a) is read directly by following the wet-bulb line to the scale marked Total Heat. Total Heat refers to a quantity of air and water-vapor mixture which would weigh 1 lb if all water vapor were extracted, and includes the heat of the water vapor.

Specific Volume (d) is read directly from the lines marked Cu ft per Lb of Dry Air. For points between lines, estimate by distance. Specific volume is the volume occupied by a quantity of air and water-vapor mixture which would weigh 1 lb if all water vapor were extracted.

Vapor Pressure (c) corresponding to a given moisture content is read directly from the left-hand scale marked Pressure of Water Vapor.

EXAMPLES:

A. Air at 80 deg dry bulb and 65 deg wet bulb.

Relative Humidity	45 percent
Moisture Content	68.5 gr per lb
Dew Point	56.8 F
Total Heat (includes heat of 1 lb of dry air and heat of 68.5 grains of water vapor)	30.0 Btu per lb
Specific Volume	13.81 cu ft per lb
(Actual weight of 13.81 cu ft. = 1.0098 lb; 1 lb for the dry air and 68.5 grains for the moisture content, 7000 grains = 1 lb)	
Vapor Pressure	0.228 lb per sq in.

B. Air at 75 deg wet bulb and 67 deg dew point.

Dry-bulb Temperature	94.7 F
Relative Humidity	40.0 percent
Moisture Content	99.2 gr per lb
Total Heat	38.5 Btu per lb
Specific Volume	14.28 cu ft per lb
Vapor Pressure	0.328 lb per sq in.

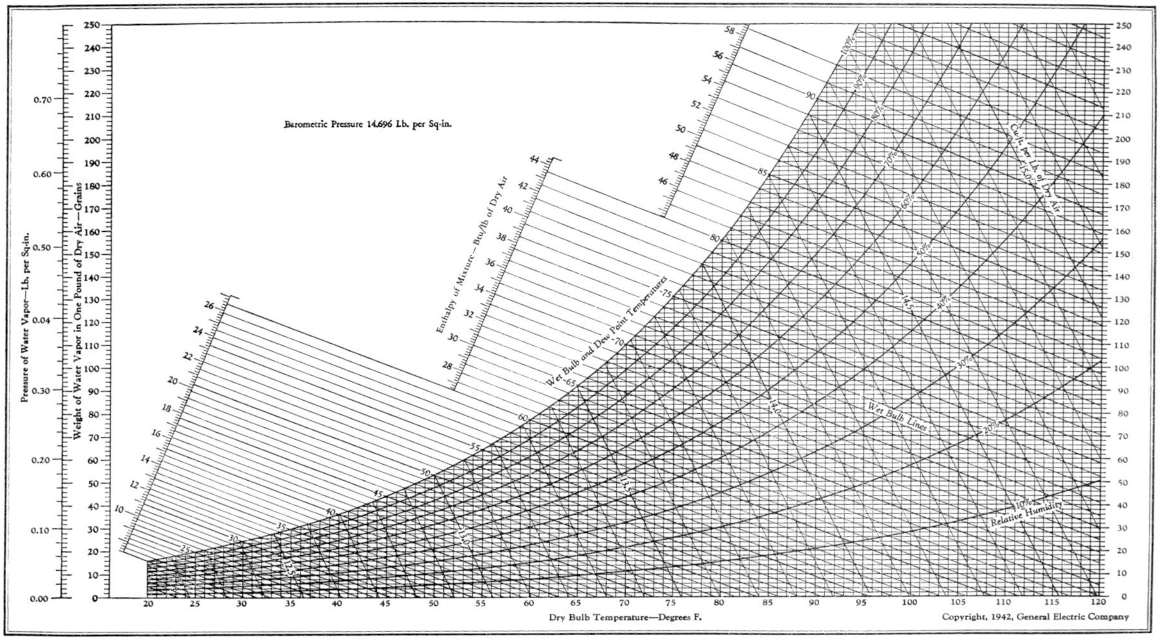

FIG. 12-1. Continued.

present may complain at 75 and by the time it reaches 80, just about everybody will be looking for relief.

The troubleshooter has two approaches to determination of the index. The first uses the following formula, for which the wet-bulb reading must be available:

$$DI = 0.4\,(t_d + t_w) + 15$$

where DI = Discomfort Index
t_d = dry-bulb temperature
t_w = wet-bulb temperature

The second formula is applicable where we know the relative humidity.

$$DI = t_d - [(0.55 - 0.55 \times rh)(t_d - 58)]$$

where t_d = dry-bulb temperature
rh = relative humidity, percent
0.55 = a constant
58 = a constant

Both of these formulas incorporate both temperature (dry bulb) and humidity (wet bulb) or use humidity expressed as a percentage.

Occasionally, a Discomfort Index as high as 90 may be reported, and the highest of a published value has been 92.

If an installed air-conditioning system cannot hold down the Discomfort Index to at least 85, probabilities are it has been designed for more optimistic conditions than actually exist. The troubleshooter meeting this situation, finding everything working to capacity, may be obliged to look toward new units as the only permanent cure.

12-2. SYMPTOMS OF UNSATISFACTORY HUMIDITY

(a) ***When the Relative Humidity Is Too Low.*** This condition is common during the heating season, a time in which outdoor humidity is very low, frequently less than 10 percent. Because of the difference in vapor pressure between an inside and an outside environment, moisture released into the inside atmosphere tends to leak out, and inside conditions show these signs of undesirable dryness:

Woodwork shrinks. Glued furniture comes apart at the joints. Floors shrink and develop openings at joints. Mucous membranes in the nose and throat feel parched and inflamed. There is in some individuals an increased tendency to colds. (*Note:* At relative humidities of between 50 and 55 percent, airborne bacteria are least likely to initiate an infection.)

The troubleshooter, seeking to evaluate a complaint of excessive atmospheric dryness, will need a humidistat, a sling psychrometer, or combination of wet and dry bulb with which to determine the existing relative humidity. Simply to know that this condition exists is only the start of the program. The decision on how to correct it, and by what kind of equipment, depends on finding out how much increase in humidity is necessary.

(b) ***When the Relative Humidity Is Too High.*** Excessive humidity is not only physically uncomfortable but promotes mold, peeling of paint, rotting of wood, and moisture streaks on cold walls as the water vapor condenses on the surface, and is believed to provide a congenial environment for the support of termites.

In a residence, excessive humidity may result from a tightly insulated building with vapor barriers to prevent the escape to the outdoors of moisture entering the inside atmosphere from cooking, showering, laundry operations, and bathing. An often unsuspected source of moisture is the ground beneath the building when it is separated from the structure by a crawl space. When the building is set on a concrete slab, ground moisture will rise through the slab and appear as a damp surface on top of the slab, unless the slab has been set on an effective waterproof membrane.

12-3. CORRECTING LOW HUMIDITY

(a) ***Complaint.*** All normal activities are carried on in the house by the two adults and three children in the family, but the dryness of the air is ruining the furniture, making the draperies brittle, and causing chronic conditions of irritated throat.

Causes

1. The house is not vapor barriered.

2. Considerable traffic in and out of the house, particularly by the children, allows super-dry, outside winter air to enter, and because of its greater density, it displaces an equal volume of warm inside air.

Recommended Actions

1. Introduce moisture into the inside atmosphere. This may be done by boiling water on the stove, running the bath showers with the bathroom doors open, doing laundry work where the vapors may diffuse through the house.

2. For a duct heating system, provide a humidifier that permits a water pickup by the heated air leaving the furnace.

3. For a steam radiator system, install humidifying vents on radiators as necessary. These vents inject steam vapor into the atmosphere. Alternatively, install an evaporative humidifier on at least one radiator in each room. Also alternatively, mount a steam jet humidifying nozzle in the hot air duct leaving the steam coil on those systems that

heat air by means of a steam coil. These are predominantly industrial and commercial arrangements.

4. In one or two rooms, a portable type of "slinger" humidifier, using plug-in electrical service for the motor, may produce all the humidity desired. These are small, inexpensive units and may be obtained with either automatic or manual control.

Figures 12-2 and 12-3 illustrate commercial equipment of this type. The latter is intended primarily for walk-in refrigerators, where maintenance of a normal humidity is essential in preserving the quality of the eggs, fruits, vegetables, and meats stored there. However, this unit may be suitable to many industrial environments.

Table 12-1 provides data that will enable the troubleshooter to evaluate conditions as he finds them, compared with what they should be for satisfactory performance.

Figure 12-4 (a) and (b) shows a form of evaporative humidifier that may be mounted within radiator cabinets where a heating coil, usually served by a hot water circulating system, is part of the arrange-

FIG. 12–2. Console cabinet form of humidifier. (*Herrmidifier Company, Inc.*)

FIG. 12–3. Humidifier for refrigerated storage room. (*Herrmidifier Company, Inc.*)

ment. This evaporative unit is also suitable for those self-contained air conditioners that have through-the-wall connections for condensing air and room-ventilating air.

(b) ***Pneumatic Humidification for Industry.*** Figure 12-5 shows a piping arrangement for injecting water vapor into either an ambient atmosphere or a moving air stream, as required. Operation is controlled by a humidistat, and when the humidity falls below a desired value, the humidistat opens a solenoid valve in the air supply line. If the humidifying system is entirely self-contained with its own compressor, the humidistat starts the compressor at the same time it opens the solenoid.

A valve in the water supply line is next opened when the air pressure reaches its designed value. Automatic shutdown occurs when the humidistat is satisfied.

(c) ***Complaint.*** In a textile mill, the level of humidity is too low. There appears to be nothing wrong with the functioning of the pneumatically operated system.

Causes

1. The system capacity has been underdesigned for the humidity demands of the mill.

2. Atomizing nozzles are partially clogged.

3. Air pressure is lower than designed.

4. Water supply is reduced by corroded areas in the supply piping.

5. Air, water valves, or both are not opening to their designed values because of excessive resistance. This resistance may be originating in a slightly deformed spindle or from accumulations of foreign materials.

Recommended Actions

1. Make a field calculation of the amount of humidifying moisture, either grains or pounds, that must be introduced to achieve the desired level of humidity. Compare this with the rating of the humidifier. If unit is undersized, consider ways of increasing its capacity, such as raising air or water pressure. As an ultimate expedient, replace with units

Table 12-1. Optimum Industrial Temperatures and Humidities

Industry	Temperature, F	Percent relative humidity
Paper binding	70	50–65
Paper folding	77	55–65
Paper printing	75	45–65
Textiles:		
Cotton carding	75–80	50–55
Cotton spinning	60–80	50–70
Cotton weaving	68–75	85
Rayon spinning	70	85
Silk processing	75–80	60–70
Wool carding	75–80	65–70
Wool spinning	75–80	55–60
Wool weaving	75–80	50–55
Food storage:		
Apples	31–34	75–85
Citrus fruit	32	80
Eggs	30	80
Grains	60	30–45
Meat, ripening	40	80
Sugar	80	35
Flour	60–75	55–65
Yeast	32–45	60–75
Candy	60–68	50–65
Cigar, cigarette making	70–75	55–75
Libraries	65–70	38–70

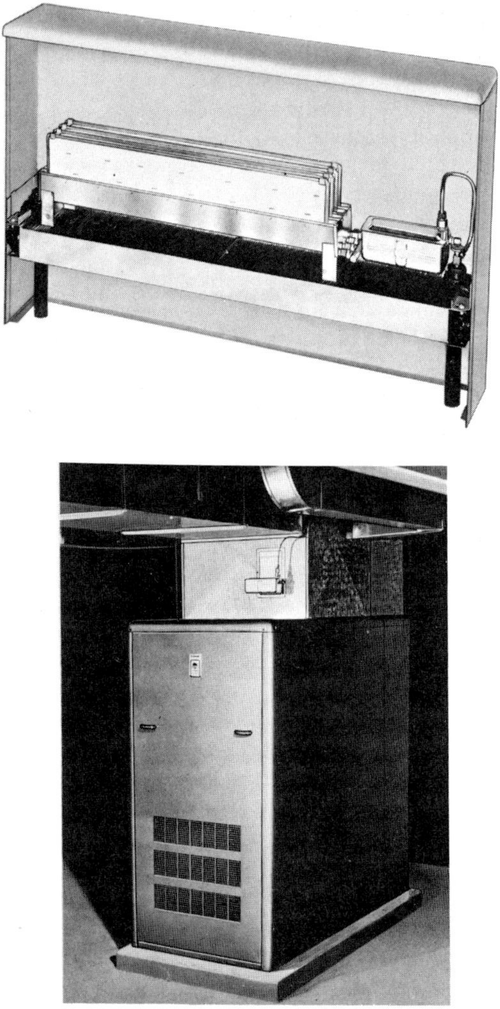

FIG. 12–4. (a) Evaporative humidifier in convector heating cabinet. (b) Evaporative humidifier in furnace plenum. (*Maid-O'-Mist.*)

of greater capacity. Also, determine if the humidifying demands have been increased since the time of the original rating and installation.

2. Note cleanliness of nozzles. Clean if necessary. Note especially whether deterioration is present from corrosion. Nozzle replacement may be required.

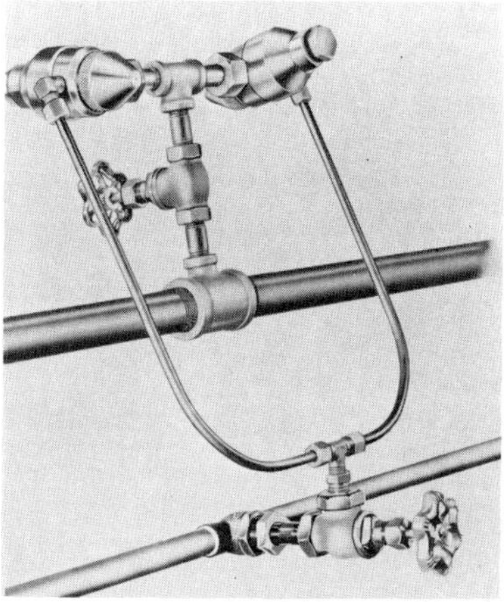

FIG. 12–5. Pneumatic water spray nozzles for space humidification, showing connections to compressed-air and water lines. (*Herrmidifier Company, Inc.*)

3. Compare existing air and water pressures with those on which the rating of the system was originally based. Step up as required, if possible, to satisfy the rating.

4. Observe and time the opening of all solenoid or motor-operated valves. Note the distance of travel. If obstructions or excessive friction appear to be present, open the valve, locate the obstruction or area of excessive friction, and correct. Removal of the valve to the shop for the corrective work may be necessary.

5. Note whether periods of inadequate humidity are occurring at certain times of the day only. This would indicate a recurring drop in water pressure due to peak demands elsewhere or the development of a temporary condition in the general atmospheric environment over those particular hours.

6. Verify the operation of all humidistat controls.

(d) ***Complaint.*** This is the same as the complaint in Sec. 12-3(c), except that humidity is being produced by a steam jet rather than by compressed air and water. See Fig. 12-6.

Causes

1. System is underdesigned for capacity.
2. Atomizing nozzles are eroded.
3. Steam pressure is below the required level.

Recommended Actions

1. Check the required amount of humidifying moisture with weight of water vapor, in grains or pounds, actually being delivered. If deliv-

FIG. 12–6. Steam jet humidifier serving an air duct. (*Armstrong Machine Works.*)

ery is short, consider ways of increasing it, such as an increase in steam pressure or the installation of additional nozzles.

2. Steam jet nozzles are vulnerable to high-velocity erosion and should be examined carefully. Replace those found to be in deteriorated condition.

3. Check the operation of the humidistat-controlled steam valves. Make sure the valves are opening at rated speed and to the rated orifice. To regrind seats and disks, removal of valves to the shop may be required.

4. Verify accuracy and performance of the humidistats. Make sure their control settings are the ones desired. As a rule, the control element is a human hair or animal membrane which is hygroscopic in its nature and shrinks with growing dryness of the air and expands or stretches as the humidity increases.

(e) ***Complaint.*** During the winter, moisture loss from the interior spaces of a building through the walls appears to be beyond control.

Causes

1. The travel of moisture through the building construction is the result of the inside vapor pressure being much higher than that of the outside winter air. Just as water flows downhill, the higher inside pressure forces the atmospheric moisture into the lower pressure of the outdoors.

2. There is no vapor barrier in the walls.

Recommended Actions

1. Obtain a good rubber-base wall paint and go over the entire surface of the inside walls. This kind of paint interposes an effective, if not perfect, obstruction to the outward travel of the inside moisture.
2. Increase the output of whatever humidifiers are presently in use.

12-4. CORRECTING EXCESSIVE HUMIDITY

(a) ***Discussion.*** Excessive humidity conditions may be corrected by:

1. Installing chemical dryers, such as activated carbon or silica gel.
2. Chilling the humid air below its dew point, thereby "squeezing out" the excess moisture, followed by a reheating which restores it to the desired temperature.
3. Ventilating areas in which humidity is being created by a chemical process, cooking procedures, laundry operations, and the like.
4. Preventing the emission of water vapor from the ground underneath a given building. This is accomplished by covering the ground with a suitable material, such as roofing, which is laid on top of the bare ground with joints overlapping for a minimum of 4 in.
5. Ventilating attic spaces.
6. Ventilating crawl spaces.
7. Laying a vapor barrier before pouring a concrete slab. The barrier prevents moisture migration through the concrete.

The amount of moisture evaporated into the air inside a residence is shown for various points of origin in Table 12-2. Undesirable conditions resulting from excessive humidity are treated in the following complaints.

(b) ***Complaint.*** The white paint on the outside walls of a clapboard-sided house is peeling, although it has been on less than two years.

Causes

1. At the time the paint was applied, the wood was wet. Paint will not adhere satisfactorily to wet wood.
2. The development of moisture conditions within the siding structure, for example, the rising of water vapor from the ground into the spaces between the studs, has penetrated the wood and produced a wet surface beneath the paint. The combination of wet surface and the swelling of the wood can be expected to dislodge the paint.

Recommended Actions

1. The peeling paint must be removed completely and the wood surface prepared for accepting a suitable primer; then after adequate drying, the final coats may be applied.
2. Determine source of moisture getting into walls. If rising from

the ground in a crawl space, cover the ground surface with roofing, allowing lap joints of at least 4 in. If there is some possibility of wind lifting the roofing, ballast the roofing with sand.

3. Make sure that crawl space is ventilated properly. The formula for adequate ventilation of a crawl space is 2 sq ft per 100 ft of building perimeter plus ½ of 1 percent of the building area. For example: a building 60 × 70 ft has a perimeter of 260 ft × 2 sq ft per hundred for ventilation = 2 × 260/100 = 5.2 sq ft of ventilator. To this add 0.005 × (60 × 70) = 21 sq ft for a total of 26.2 sq ft. Provide at least four openings for cross drafts.

(c) ***Complaint.*** Insulation was blown into the walls of an old mansion finished on the outside with clapboard siding. Within a year, the clapboards at the bottoms of several walls began to bulge.

Cause

The insulating material was hygroscopic. In absorbing moisture, it expanded, exerted pressure, and bulged the siding.

Recommended Actions

1. Remove the insulation in its entirety. After the walls and siding have become dry, reinsulate, being careful to choose an insulation that will not absorb water.

2. Determine sources of moisture entering the walls. Take corrective steps as previously noted to stop water vapor emissions from the ground where appropriate. Also ventilate the crawl space, basement, and attic.

3. Reduce vapor migration through the walls from the interior spaces of the house by painting the walls on the inside with a good rubber-base paint.

(d) ***Complaint.*** In cold weather, condensation on the windows of a house becomes so bad that it runs over the window sills, destroying the paint on them, and trickles down the wall beneath the window, leaving streaks. The window sash is of aluminum.

Causes

1. The surface temperature of the glass is lower than the dew point of the air in the room.

2. Aluminum is an excellent conductor of heat; consequently, heat loss through the metal sash is expanding the temperature difference in the area.

Recommended Actions

1. Reduce the relative humidity of the room air by an exhauster fan above the cooking stove; do the same for the laundry and bathrooms.

2. When weather permits, leave a 1-in. opening for each of the condensing windows. This will lower the relative humidity in contact with the glass and aluminum by mixing with the cold dry air from outside.

3. Reduce the relative humidity by means of a chemical dryer. These are commercially available in cabinet form.

(e) ***Complaint.*** Condensation and frost formation are nuisances that appear on commercial show windows.

For a discussion, together with correction drawings, refer to Sec. 11-2(b) and Fig. 11-11a and b.

(f) ***Complaint.*** During the summer, relative humidity is so high and contaminated with industrial acids that it pits silver and other metals.

Cause

Airborne acid fumes, when mixed with high relative humidity, may create a corrosive atmosphere. This is also true of salt air.

Recommended Actions

1. If not presently in place, install an air-conditioning system. By dropping the coil temperature below the dew point, the excess humidity is condensed and drained off from the coil.

2. If air conditioning is in fact in operation, and there is pitting notwithstanding, the indications are that the fume must be removed from the air by special equipment. See Sec. 11-1(f) for details of this equipment and its manner of use.

(g) ***Complaint.*** Since connecting a new, gas-fired boiler to an existing brick chimney, water stains have appeared on the plaster finish of the chimney, where the chimney passes through a room in the house. In several spots, the plaster is actually crumbling.

Causes

1. Indications are that moisture leaving the boiler as a product of combustion is condensing on the cold inside surfaces of the chimney, then soaking through the brick and into the plaster.

2. The chimney is oversize. This condition extends the warming up time whenever the boiler lights off after a shutdown.

3. The chimney may be lacking a liner.

4. The top opening may be particularly exposed to an entry of rainwater.

5. The percentage of water vapor released by the combustion of a gas fuel is greater than from either oil or coal fuel. Combined with one or more of the other causes, such as the lack of a liner, water penetration through the plaster and into the room becomes a natural expectation.

Recommended Actions

1. Examine the inside of the chimney. If it has no liner and the installation of a liner does not reduce its cross-sectional area unacceptedly, insert one.

2. Provide an exhauster fan to remove the high humidity products of combustion as promptly as possible.

3. See that a suitable rain-exclusion cap is mounted, if the field exam-

ination shows such a cap to be necessary. Make sure that the ultimate flue-gas escape openings under the cap provide a total escape area at least four times the cross-sectional area of the chimney.

Table 12-2. Sources and Quantities of Water Vapor in a House

Source	Quantities and units of time in which released
Bathing by shower	½ lb (3500 grains) added per shower taken
Bathing by tub bath	⅛ lb (875 grains) added per tub bath taken
Laundry washing and drying	26 lb (182,000 grains) per week *
Mopping the floor	3 lb (21,000 grains) per 100 sq ft of floor
Cooking by electricity	2 lb (14,000 grains) per day *
Cooking by gas	4.7 lb (32,900 grains) per day *
Dish washing	1.0 lb (7,000 grains) per day *
Unvented gas appliances	0.1 lb (700 grains) per cu ft of gas
Personnel emission, breath, skin †	0.12 lb (840 grains) per person per hour †
Personnel emission, breath, skin	26 lb (182,000 grains) per person, per day †
Ground emission by test	100 lb (700,000 grains) per 24 hours, average

* Based on an average family of four persons.

† Personnel emissions are based on normal physical activity in the house. Greater activity, such as dancing and physical work, can be expected to increase the water vapor emitted per person from 2 to 2½ times those values shown.

(h) ***Complaint.*** Droplets of water emerge from ceiling registers and creep over the ceiling leaving streaks.

For causes and recommended actions concerning this condition, see Sec. 9-8(c), Division 9.

(i) ***Complaint.*** During summer months, the floor and walls of a basement become damp to the point of actual wetness.

Causes

1. Most of this moisture is condensation from the basement air. It happens when the dew point of the air is above the temperature of the floor and walls.

Table 12-3. Troubleshooter's Field Guide for Humidity Control

Possible causes	How to check for cause	Recommended ways to correct
COMPLAINT: Dryness of the air is causing throat irritations and damage to furniture and draperies.		
REFERENCE: Sec. 12-3(a).		
1. Structure is not vapor-barriered.	*a.* Examine structure.	*a.* Inject mist into the inside air.

Table 12-3. Troubleshooter's Field Guide for Humidity Control (*Cont.*)

Possible cause	How to check for cause	Recommended ways to correct
2. Super dry outside air is entering through the frequent opening of doors, also by leakage around windows and doors.	*b.* Note traffic in and out. Examine for weather stripping.	*b.* Inject mist into the inside air. Weather stripping may help.
COMPLAINT: Inadequate level of humidity is found in textile mill which operates a pneumatic atomizing system. REFERENCE: Sec. 12-3(c).		
1. System is undersized.	*c.* Compare injection rate with design rate.	*c.* Step up mist diffusion by adjustment. Install new unit.
2. Air pressure is lower than design calls for.	*d.* Check pressure at nozzle.	*d.* Increase pressure by adjustment.
3. Water supply is inadequate in quantity or pressure.	*e.* Check pressure. Compare with design.	*e.* Step up pressure, volume, or both.
4. Nozzles are clogged.	*f.* Inspect.	*f.* Clean or replace.
5. Valve operation is sluggish or limited.	*g.* Test operation of each valve for time and distance.	*g.* Replace or repair defective valves as indicated by condition.
COMPLAINT: Same as above but using a steam-atomizing system. REFERENCE: Sec. 12-3(d).		
1. Steam pressure is below requirements.	*h.* Check pressure to nozzles.	*h.* Step up pressure at boiler or at pressure-reducing valve.
2. Atomizing nozzles are eroded.	*i.* Inspect.	*i.* Replace.
COMPLAINT: Loss of moisture through building walls appears to be excessive. Atmosphere is too dry for best results in certain industrial operations. REFERENCE: Sec. 12-3(e).		
1. No vapor barrier in the walls.	*j.* Inspect. Examine design drawings.	*j.* Paint inside wall surfaces with rubber-base paint.

Table 12-3. Troubleshooter's Field Guide for Humidity Control (*Cont.*)

Possible cause	How to check for cause	Recommended ways to correct
2. Insufficient humidity being generated.	*k.* Compare humidity available with the requirements of the area.	*k.* Step up the production of water vapor. Add new humidifiers if necessary.
COMPLAINT: Paint on the siding of a clapboard frame house is peeling from moisture in the wood beneath it. REFERENCE: Sec. 12-4(b).		
1. Paint was originally applied on wet wood.	*l.* No way to determine this.	*l.* Remove paint, dry the wood, and repaint on new primer.
2. Water vapor has entered the wood. Vapor may be from ground or inside sources.	*m.* Check all possible sources of moisture.	*m.* Reduce moisture generation. See Sec. 12-4(b) for details.
COMPLAINT: Insulation blown into building walls has become water-soaked and, on expanding, bulged the outside wall surfaces. REFERENCE: Sec. 12-4(c).		
1. Insulating material is hygroscopic.	*n.* Remove some of the insulation and weigh for water content.	*n.* Replace insulation with nonhygroscopic material.
COMPLAINT: Condensation from windows and metal sash is ruining window sills and floors beneath the windows. REFERENCE: Sec. 12-4(d).		
1. Relative humidity maintained within the structure is excessive.	*o.* Check relative humidity against window glass temperature and the existing dew point.	*o.* Reduce the humidity to lower the dew point below the glass temperature. See Sec. 12-4(d) for details.
COMPLAINT: Condensation and frost on commercial show windows stop visibility. REFERENCE: Sec. 12-4(e). For recommended actions, see also Sec. 11-2(b).		
COMPLAINT: High atmospheric humidity is mingled with industrial contaminants and is corroding metals. REFERENCE: Sec. 12-4(f).		
1. Geographic location.	*p.* Note distance to the industrial plants, also	*p.* Install an air-conditioning and air-

Table 12-3. Troubleshooter's Field Guide for Humidity Control (*Cont.*)

Possible cause	How to check for cause	Recommended ways to correct
	prevailing wind direction.	filtering system. See Sec. 11-1(f) for special treatments.
COMPLAINT: Water stains have appeared in the plaster finish on a chimney since the firing methods have been changed to gas fuel. REFERENCE: Sec. 12-4(g).		
1. Chimney is oversize.	*q*. Calculate cross-sectional area of chimney flue and compare with area needed.	*q*. Insert a tile chimney liner if none is in place.
2. Rainwater is making a too-easy entry into the chimney.	*r*. Check size of chimney top opening. Note whether any type of cap is in place.	*r*. Provide cap. See Sec. 12-4(g) for details.
COMPLAINT: Drops of water are falling from ceiling registers. REFERENCE: Sec. 12-4(h). See Sec. 9-8(c) for details of the causes and handling of this complaint.		
COMPLAINT: Basement in summer is wet on both floor and walls. REFERENCE: Sec. 12-4(i).		
1. Temperature of floor and walls is lower than dew point of contacting air.	*s*. Take reading of ambient relative humidity and temperature of floor. Check dew point on psychrometric chart of air condition and compare with floor temperature.	*s*. Heat basement to drop dew point below floor and wall temperatures. Install a dehumidifier. Ventilate basement with small exhaust fan.
2. Moisture is migrating from ground and entering through concrete.	*t*. When air temperature in basement is made dry by dehumidification ventilating or any other way, note whether moisture spots still appear on floor or wall.	*t*. Paint inside surfaces with a commercial sealing paint.

2. Some of this moisture is migrating into the basement by movement through the concrete from the ground.

Recommended Actions

1. Heat the basement space sufficiently to lower the relative humidity to a value having a dew point below the surface temperature of the floor and walls.

2. Place a dehumidifier in the basement.

3. Ventilate the basement with a small exhaust fan.

4. Paint the walls and floor with a sealing paint, presently available from commercial sources. This paint will stop ground migration into the basement but will leave the relative-humidity level probably high enough to require ventilation or dehumidification.

DIVISION 13

Compressed-air Equipment and Systems

13-1. BASIC FACTS FOR THE TROUBLESHOOTER

(a) *The Common Forms of Air Compressors*

1. *Horizontal Reciprocating:* Mostly of one or two stages, this type is well favored by industrial activities. The one-stage designs are equipped with an aftercooler; the two-stage units have a between-the-stages cooler called the intercooler, plus the aftercooler.

The purpose of the coolers is to reduce the volume of the compressed air, thereby permitting the use of smaller pipe on the final discharge and increasing the efficiency of the high-pressure cylinder in the two-stage units. They also eliminate water from the air.

2. *Vertical Reciprocating:* In this form, the movement of the piston is up and down instead of horizontally from left to right and back again. A saving in floor space is claimed for the vertical units, plus a little less wear on the cylinder liners, since the weight of the piston is not imposed on the liners. The incorporation of coolers between and after the stages is a standard practice here also. Some of these compressors take the form of a Y setting for the cylinders, thus making them a modified vertical design.

3. *Centrifugal Compressors:* Traveling at a tip speed of 1,000 ft per second, these units develop high pressures from relatively small machines. Cooling is as essential with centrifugals as with any other form of compressors. In some multistage machines, each stage has its own cooler section contained within the general casing. A fairly new variation is the diaphragm design, in which each impeller is followed by a diaphragm, both being shrunk on the common shaft. These are designed on the module plan so that more capacity can be obtained by bolting on additional impeller-diaphragm sections.

4. *Other Forms:* Less frequently encountered will be the rotary and axial type of compressors. The former are positive displacement machines, moving air just as rotary pumps move water.

The axial flow compressor is being coupled in some instances with a centrifugal unit. The axial section can be compared with the blower of the gas turbine, when of a multistage design. Between adjacent rows of rotating vanes, stationary vanes guide the air smoothly from stage to stage in its journey through the machine. There is no interstage cooling in the axial design, and aftercooling may or may not be added, depending on the conditions of service. Single-stage units are often used where considerable volume at moderate pressure is needed.

(b) *Compressor Drives*

1. Steam cylinders built into the compressor as integral parts have long been the driving source for reciprocating compressors. For the troubleshooter, this arrangement involves investigations into steam

FIG. 13–1. Angle type of compressor. (*Pennsylvania Pump and Compressor Company, Division of Cooper-Bessemer Corporation.*)

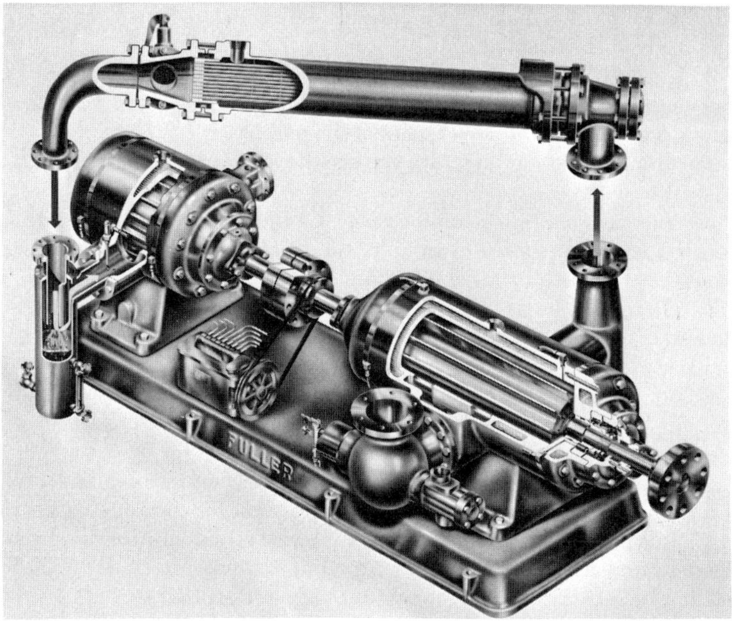

FIG. 13–2. Two-stage axial compressor. (*Fuller Company.*)

pressure, quality of steam, steam and exhaust valves, length of stroke, cylinder lubrication, governor performance, and all other inherent aspects of steam engine operation.

Alternatively, the steam engine drive may be a separate unit, applying the necessary power through a belt. The same problems face the troubleshooter here, plus those that originate in the belt.

Steam turbines are used with centrifugal and axial compressors, being well suited to the high-speed requirements of these compressors. Refer to the appropriate sections of Division 14 for troubleshooting guides for steam engines and steam turbines.

2. Electric motor drives tend to use squirrel-cage forms in sizes of 125 horsepower and smaller. However, if an improvement of the power factor is required, synchronous motors are a common selection. The synchronous motor is also favored for the larger, direct-connected drives, since these motors show high efficiency at partial loads, and the elimination of the belt removes a source of trouble and maintenance.

The troubleshooter should obtain the full operating data on the driving motor, usually furnished to the owner by the compressor manufacturer, to be used in making comparisons with the data he collects in his field study of the complaint.

3. Internal-combustion engines are furnished with some compressors, either belted or direct-connected, depending on the manufacturer's practice and the preferences of the purchaser.

For troubleshooting these drivers, see the appropriate sections of Division 14.

(c) ***Lubrication of Reciprocating Compressors.*** Since lubrication is a frequent source of trouble, the troubleshooter should understand these characteristics of the oiling system:

1. There may be as many as three types of oil required for a single machine, such as one type for the compressor cylinder, one for the crankcase and bearings, and one for the power cylinder.

2. For a single-acting, trunk piston compressor, the oil must be suitable for both cylinder and crankcase lubrication. It is termed "compressor-cylinder oil."

3. If compressor is of the double-acting form, compressor-cylinder oil is to be used for the cylinder, and crankcase oil for the running gear.

4. Steam cylinders are to be lubricated with oil suitable for the steam temperatures it must withstand. An oil for temperatures below 550 F will not be satisfactory for temperatures higher than that.

5. If an engine is using cylinder oil having a *high* pour point and discharging into a surface condenser, trouble from a fouled condenser tube surface can be expected to develop. Therefore, an engine operating with a surface condenser should select a cylinder lubricant with a low pour point and thus keep the oil in a fluid state to help it pass through and out of the condenser.

6. Lubricating recommendations are customarily furnished to a compressor owner by the machine's manufacturer. With these in hand, the troubleshooter can evaluate the oil actually being used. Items to check include:

Flash point, F
Viscosity, SSU at 100 F
Viscosity, SSU at 210 F
Pour point, F
Conradson carbon residue, %
Fire point, F

(d) ***Lubrication of Centrifugal and Axial Compressors.*** There is no lubrication involved here beyond that which is needed for the bearings. Obtain specifications for the oil from the compressor manufacturer. Compare with the characteristics of the oil actually being used.

(e) ***Compressed-air Coolers.*** Compressor cooling may be achieved by circulating water through a jacket or by blowing air across finned cylinder surfaces designed for the purpose.

The intercooler between stages and the aftercooler on the discharge from the final stage perform a twin purpose: they reduce the air volume by lowering its temperature, and they condense water and oil mists that otherwise could be carried into the distribution system.

The water enters the compressor as humidity in the air being compressed, and the oil is from the cylinder lubricant. Both are potentially dangerous. If not removed before entering the distribution system, water in the compressed air can create a water hammer and adversely affect the performance of pneumatic tools and control systems. The oil, if heated sufficiently, can precipitate a destructive explosion. As in a diesel engine cylinder, this oil, when mixed with air under high pressure, needs only a sympathetic temperature in order to explode. This sympathetic temperature may already exist in the uncooled compressed air, and in this situation, the richness of the oil-air mixture determines whether or not there will be an explosion.

The coolers are generally of the shell and tube form, the air passing through the tubes, the cooling water around them. As rule-of-thumb information, the amount of cooling water circulated through an aftercooler should approximate 1 gal for every 100 cu ft of free air that are taken into the compressor. The objective is to cool the compressed air to a temperature level about 15 deg above that of the entering water. A larger differential tends to leave the compressed air too wet for satisfactory use; a smaller differential may be uneconomical to provide and maintain. Each job must be evaluated on its requirements.

(f) ***Ratings and Volumetric Efficiency.*** The capacity, or actual delivery, of a compressor is expressed in cfm of the entering air at the specified conditions of total temperature, total pressure, and relative humidity.

Volumetric efficiency is determined by the ratio between the capacity of the compressor and the displacement. This term applies to reciprocating compressors only. Displacement can be calculated from the following formula, expressed in cfm:

$$\frac{D = (\text{stroke, in.}) \times (\text{cylinder bore, sq in.}) \times (2\ \text{rpm})}{1{,}728} = \text{cfm}$$

The displacement is the volume that the cylinder theoretically can handle, but actually the delivery is less than this because of losses through valves and piston clearance. The actual volume of the air delivered at the outlet of the compressor, divided by the displacement, equals volumetric efficiency.

The troubleshooter can readily figure the compressor displacement, but the determination of the actual delivery is a fairly complicated procedure and is described in the ASME Test Code for compressors. Where the complaint concerns the efficiency of the compressor, the troubleshooter's checking procedure should follow the ASME Code.

Sometimes a compressor capacity will be referred to as cfm of "free air." Since free air is simply the ambient atmosphere, the term has no significance unless the temperature, pressure, and humidity are stated and, moreover, apply at the entrance to the first stage of the compressor.

Another term, "standard air," has been developed by the ASME and refers to air at 68 F, absolute pressure of 14.7 psi, and relative humidity of 36 percent. Density under these conditions is 0.0750 lb per cu ft.

In the compression of gases other than air, it is customary to use a temperature of 60 F, with the characteristics of the particular gas being given as related to that temperature.

(g) ***Unloaders.*** In order to protect driving motors of reciprocating compressors, the machines are arranged to start on a nonload condition. Refrigeration and gas compressors are often protected similarly.

A starting control that functions as a response to speed is possibly the most favored expedient. When the compressor stops, the load is dumped and stays dumped until the machine is brought back to speed.

This is accomplished in some instances by rotating weights, comparable to the flyball governor of steam engines. When the weights are swinging wide from centrifugal force, a ball valve closes, the air in the compressor can no longer escape to the atmosphere or crankcase by way of the valve, and pressure builds up immediately.

Other devices will be found for accomplishing the same purpose.

(h) ***Intake Filters.*** The first requirement of good filter practice is to size the filter so that air flow is not restricted.

The second requirement, applying to wet filters, is to use the same type of wetting oil that is used as a lubricant in the cylinders. For machines that are nonlubricated in their compressing parts, the filter must be dry. Trouble can result from injecting an oil mist into the air stream, as noted previously.

(i) ***Air Silencers.*** These are commercial devices, available on the market or as an extra from the compressor manufacturer. In instances of noise complaints, the troubleshooter's work may consist of correcting a deficiency of the silencer, or if no silencer is in use, the problem can be resolved either by eliminating the noise or by containing it. The adoption of a silencer is another alternative.

This writer has had satisfactory results from containment in the form of a room designed to contain a battery of air compressors. Intake air from the outside atmosphere reached the compressors by strategic baffling in this room, thus counteracting both suction and unloading noises.

As a suggestion, thin lead sheets might be considered for a containment room of light construction. Lead has excellent sound-deadening characteristics and has been used successfully in various situations.

(j) ***Leak Checking.*** Where pressures have fallen off and a leak is suspected, the troubleshooter should be prepared to follow out the distribution system in this manner:

1. After noting the discharge pressure from the compressor, proceed to each tool or point of use and, by means of a hypodermic needle gage, note the pressure at that point while the tool or other application is in service. If the needle gage shows an excessive drop in the supply from the compressor, a leak in that particular line can be suspected.

2. Scan the suspected supply line with a soapy water solution, daubing it on wherever a leak seems possible or likely.

3. As an alternative to the soapy water treatment, sweep the pipe with the flame from a lighted candle. If the leak is sizable, the flame will signal it. This method of leak detection must be undertaken with care, so that there is no danger from a combustible atmosphere or from contact with combustible material.

4. Inject an odorizer into the air stream. This may betray the area of a leak to a person present there. Wintergreen is reported to have been used successfully for this purpose. Actually, any aromatic injection will do if it is nontoxic and nonflammable in any concentration.

5. Leaks in the shell and tube coolers can be identified by the appearance of air in the discharge of the circulating water, provided the air pressure is higher than the water pressure and consequently leaks into the water. Higher water pressure reverses the direction of leakage into the compressed air, and excessive water in the latter must be considered suspicious. If coloring material injected into the circulating water appears in the water separated from the air, we have definite proof of leakage in the cooler.

6. Instead of a soapy water solution for leak checking, it is possible to use a proprietary compound that is nonflammable, noncorrosive, and noninjurious. It is conveniently squirted on the area of suspected leak, and it indicates a leak by forming a foam of white bubbles. This is probably neater to handle than the soapy water technique.

13-2. COMPRESSED-AIR COMPLAINTS AND THEIR TREATMENT

(a) *Some Significant Symptoms*

1. If the pressure in the intercooler in psia is much in excess of the design pressure, it points to trouble in the high-pressure cylinder of a

kind that obstructs the normal movement of the air from the lower pressure to the higher pressure cylinder. The probable situation is one of leaky or damaged intake valves on the high-pressure cylinder.

2. An abnormally low pressure in the intercooler points to some inadequate performance of the low-pressure cylinder: the air is simply not being delivered as it should be. This, too, could mean a defective valve.

3. Damaged intake valves also may be indicated by abnormally hot intake valve covers. Whenever this condition is observed, shut down the machine at once; otherwise serious physical damage can result.

4. Temperature of water leaving a water-jacketed compressor at an abnormally hot level suggests either a restricted flow of water or a jacket that is fouled with mud or other foreign material.

5. If the air receiver is being drained once a day as it should be but the water level in the receiver goes high, there is a possibility of leakage in one of the coolers.

6. Pulsation or surging in an axial or centrifugal compressor may signify that the machine is operating at its lower limit of stable performance.

7. A falling off in pressure which is not accompanied by a reduction in speed suggests the possibility of worn shaft seals and an air leakage therefrom.

(b) ***Complaint.*** Compressor is operating with a distinct knock.

Causes

1. If knock is in harmony with the rpm, the knock is probably originating in a main bearing.

2. If the knock occurs at a frequency of twice the rpm, look for something loose in the reciprocating parts.

3. For a knocking when the machine is unloaded, look for worn piston rings or a piston that is loose on the rod.

Recommended Action

Shut down the machine, open it up, and make a check of each cause possibility. Replace worn parts, tighten loose items, and do whatever must be done to correct the bad condition.

(c) ***Complaint.*** Compressor runs abnormally warm, and a sort of sliding squeak is noticeable.

Causes

1. Cylinder is receiving insufficient lubrication.

2. Valves are defective or poorly fitted.

Recommended Actions

1. Increase the oil feed for cylinder lubrication.

2. Shut down the machine, open it up, and examine the valves. Look for scoring or uneven wear of the cylinder walls.

3. Examine particularly the surfaces that show uneven friction wear.

4. Verify the correctness of the lubricating oil analysis as compared with the analysis called for by the compressor manufacturer.

5. Correct valves and change oil as necessary.

6. Check performance of lubricating oil pump. If defective, adjust or replace as situation indicates.

(d) ***Complaint.*** Compressor installation vibrates undesirably.

Causes

1. Compressor is not bolted firmly to its foundations.

2. Air pulsation in the distribution system is being carried back to the compressor.

3. Vibration of other equipment either is being transmitted through the building structure to the compressor or is traveling over the distribution piping.

Recommended Actions

1. Improve the firmness of the anchors if the vibration is rising from this source.

2. For vibrations traveling over the piping, install vibration absorbers in each pipeline, as close to the compressor as may be feasible. In selecting the absorbers, be sure they are suitable to the maximum pressure in the line. Use spring hangers for the piping.

3. Building transmittals call for either eliminating the vibration at its source or isolating the compressor foundations from the building structure.

(e) ***Complaint.*** Compressor repeatedly shuts down when load comes on.

Causes

1. Driver is too small for the load.

2. Excessive pulsation in line develops recurring peaks of horsepower demand.

3. Air pressure at discharge is excessive, thus increasing the horsepower.

4. Voltage is consistently low.

5. Compressor is running hot due to cooling failures.

Recommended Actions

1. Verify the size of the driver for the load. If too small, replace it. Normal pressures with a shutdown can be significant.

2. Check valve operations as contributing to the air pulsation in the line. Evaluate size and performance of the receiver from the viewpoint of reducing or eliminating the pulsation.

3. Adjust compressor settings to bring down the discharge air pressure to the design value.

4. For consistent or recurring low line voltage, determine whether

the condition is originating with the electric power company or is the result of some other item or items of equipment in the plant. Trouble from some sudden heavy demand recurring with machinery on the premises could require correction by some form of demand limiter, possibly a synchronous motor, a change in the operating schedule, or the replacement of an inefficient unit.

5. Prove the operation of safety relief valves. A valve setting that tolerates an excessive pressure without opening not only can stop the compressor but is potentially dangerous. Look for foreign materials in the seat.

6. In winter, ice in the lines can prevent satisfactory operation of the compressor.

(f) ***Complaint.*** Temperature controller on the air discharge indicates the presence of excessive air temperature by shutting down the compressor. It does this not regularly but frequently.

Causes

1. Cooling water through the intercooler or aftercooler or both is not removing the heat in the air passing through.

2. Mud in a jacket cooling system for the compressor could produce a similar symptom.

Recommended Actions

1. Check flow of cooling water through each heat exchanger, whether intercooler, aftercooler, or compressor jacket. If flow is less than the designed volume, look for mud and silt in the water passages; also consider a falloff in the pump discharge or municipal pressure.

2. Clean passages either by flushout or by the use of a proprietary chemical.

3. Temperature check the incoming water for a possible heat contamination prior to the water reaching the exchangers. Correct this situation by either eliminating the contamination or increasing the volume of warmer water being circulated.

(g) ***Complaint.*** Machine shuts down too frequently because of a drop in lubricating oil pressure.

Causes

1. Oil reservoir is at a low level.
2. Oil circulating pump needs adjustment or repair.
3. Oil feed is excessive for the needs of the compressor.
4. Oil is too light for the work.
5. There is leakage somewhere along the oil piping.

Recommended Actions

1. Add oil of a proper weight for a condition of low oil or wrong characteristics introduced in a previous filling.

2. Check over the oil pump, looking for leaky seals. With some

Table 13-1. Weights and Volume of Air at Sea Level *

Temperature, F	Volume, cu ft/lb	Weight, lb/cu ft
32	12.39	0.0807
40	12.59	0.0794
50	12.843	0.0778
60	13.095	0.0764
70	13.347	0.0749
80	13.599	0.0735
90	13.851	0.0722
100	14.103	0.0709
110	14.355	0.0696
120	14.607	0.0685
130	14.859	0.0673
140	15.111	0.0662
150	15.363	0.0651
160	15.615	0.0641
170	15.867	0.0630
180	16.119	0.0621
190	16.371	0.0611
200	16.623	0.0602
212 †	16.925	0.0591
220	17.127	0.0584
240	17.631	0.0567
260	18.135	0.0552
280	18.639	0.0537
300	19.143	0.0523
400	21.663	0.0462
500	24.184	0.0414
600	26.704	0.0375
700	29.224	0.0343
800	31.744	0.0315

* Usually satisfactory up to 1,000 ft above sea level.

† Boiling point of water at atmospheric pressure at sea level.

designs, internal leakage occurs after some years of wear. Consider this possibility.

3. Increase the oil feed if there is also evidence of above-normal heat being generated.

(h) ***Complaint.*** Air is blowing back and out through the inlet air filter.

Cause

Machine has a broken or malfunctioning inlet valve.

Recommended Action

Stop the machine. Open up and replace or repair the damaged valve.

(i) ***Complaint.*** Water is seen from time to time creeping along the piston rod.

Causes

1. The condensate trap on the aftercooler is not functioning.
2. There may be a leak in the head gasket.
3. Water pressure in the intercooler is greater than the air pressure, and water is entering the air side through a leak.

Recommended Actions

1. Check the trap performance. Repair or replace as found necessary.
2. Install a new head gasket and make sure the head bolts are all tight.
3. Check the coolers for leaks after comparing the circulating water pressure with the air pressure. If the former is the higher, a leak is possible.

Table 13-2. Troubleshooter's Field Guide for Compressed-air Equipment and Systems

Possible causes	How to check for cause	Recommended ways to correct
COMPLAINT: Compressor is operating with a knocking sound. REFERENCE: Sec. 13-2(b).		
1. Worn or misaligned main bearing.	*a.* Check frequency of knock against rpm.	*a.* Stop machine; repair or replace bearing.
2. Worn piston rings or a loose connection to the rod.	*b.* Knock shows up mostly when machine is not loaded.	*b.* Renew rings. Tighten up all loose connections.
COMPLAINT: Compressor runs abnormally hot and occasionally squeaks. REFERENCE: Sec. 13-2(c).		
1. Cylinder lubrication is insufficient.	*c.* Test by increasing rate of lubricating oil	*c.* Establish an increased level of lu-

Table 13-2. Troubleshooter's Field Guide for Compressed-air Equipment and Systems (*Continued*)

Possible cause	How to check for cause	Recommended ways to correct
	feed. Note whether there is any reduction in temperature and elimination of squeaking.	bricating oil feed.
2. Valves need adjustment.	*d.* Stop compressor; lift valves and inspect for excessive wear and other defects.	*d.* Renew valves. Do not switch valves unless approved by compressor manufacturer.
3. Oil analysis is not that which is called for by the compressor manufacturer.	*e.* Have sample of lubricative oil analyzed especially for viscosity and critical temperatures.	*e.* Replace incorrect oil with one of proper analysis.
COMPLAINT: Compressor vibrates excessively. REFERENCE: Sec. 13-2(d).		
1. Machine is loose on foundation.	*f.* Check tightness of bolts holding down compressor.	*f.* Make bolts tight.
2. Vibration from other sources is traveling through the building structure on air piping.	*g.* Disconnect motor and run. If vibration is in compressor, look for bent shaft, misalignment, etc.	*g.* If motor vibrates, take to shop. If motor does not vibrate, open compressor. For foreign sources of vibration, isolate compressor with conventional absorbers and isolate foundation from contacting earth or floor.
COMPLAINT: Compressor does not carry the load dependably. REFERENCE: Sec. 13-2(e).		
1. Driver is too small.	*h.* Compare installed driver hp with required hp by calculation.	*h.* Install driver of adequate size.

Table 13-2. Troubleshooter's Field Guide for Compressed-air Equipment and Systems (*Continued*)

Possible cause	How to check for cause	Recommended ways to correct
2. Pulsations in air distribution system create momentary peaks of hp needs that cause overloader to cut out machine.	*i.* Air receiver may be too small to smooth out normal pulsating effect of reciprocating machine.	*i.* Resize the receiver.
3. Air discharge pressure is higher than normal level.	*j.* Check existing pressure with design pressure of discharge.	*j.* Adjust to proper pressure.
4. Voltage is low for driving motor.	*k.* Take readings at compressor over a period of several hours. Note whether energy is bought or generated on premises.	*k.* Notify commercial source of power. For premises generation, look for overloaded generator and realign load, during critical hours.
COMPLAINT: Temperature limiter on air discharge shuts down machine. REFERENCE: Sec. 13-2(f).		
1. The cooling water through coolers is too warm to start or is not in sufficient volume.	*l.* Take temperature of water entering each cooler and leaving. Note if too warm for volume being circulated.	*l.* Step up water flow, obtain a cooler supply, or both.
2. If compressor is water-jacketed, jacket is choked with mud, as are coolers.	*m.* Check for obstructions to flow by taking pressure and temperature of entering water and of the effluent.	*m.* Clean passages by commercial mud remover or flush out with high-pressure water or air.
COMPLAINT: Compressor shuts down too frequently because of low oil pressure. REFERENCE: Sec. 13-2(g).		
1. Oil is low in reservoir.	*n.* Sight oil level. Check rate of feed for wastage.	*n.* Add oil to normal level. Reduce feed if appropriate.

Table 13-2. Troubleshooter's Field Guide for Compressed-air Equipment and Systems (*Continued*)

Possible cause	How to check for cause	Recommended ways to correct
2. Pump is out of adjustment.	*o.* Check discharge rate and head of oil pump.	*o.* Reduce volume and head of discharge if found excessive.
3. Oil is too light.	*p.* Obtain analysis, especially viscosity.	*p.* Replace with oil specified by manufacturer.
4. Oil piping is leaking.	*q.* Trace out the oil lines to locate any leakage.	*q.* Stop leakage.
COMPLAINT: Air is blowing back and out through the compressor intake filter.		
REFERENCE: Sec. 13-2(h).		
1. Machine has broken or malfunctioning intake valve.	*r.* Stop machine. Open it up.	*r.* Replace or repair defective valve.
COMPLAINT: Water is observed creeping along piston rod and dripping therefrom.		
REFERENCE: Sec. 13-2(i).		
1. Leaky head gasket.	*s.* Examine joint.	*s.* Replace gasket.
2. Trap on aftercooler is not operating.	*t.* Inspect trap.	*t.* Repair or adjust trap as necessary.
3. Water is leaking in through cooler.	*u.* Compare pressures of air and circulating water. If latter is higher, a leak is possible.	*u.* Repair cooler leak if found.

DIVISION 14

Engines and Turbines

14-1. STEAM ENGINES, CHARACTERISTICS

(a) *Discussion of Steam Engine Types.* Basically, all steam engines develop mechanical power as the result of steam expanding within a cylinder and against a piston. The movement of the piston delivers the power against whatever work it is called on to do.

General descriptions of steam engines refer to them as horizontal or vertical, overrunning or underrunning, condensing or noncondensing, single-stage or multistage. In marine installations, a common form of steam engine is a vertical, triple expansion (or three-stage unit), operating condensing.

Figure 14-1 illustrates several of these types. The uniflow engine, a particular form of the horizontal, is sketched in Fig. 14-2.

Engines are further described according to their valving. In slide valves, the seat is a smooth plane over which the valve slides back and forth, alternatively opening and closing the ports for the steam admission and exhaust.

Poppet valves rise from the seat, admit or exhaust the steam as appropriate, and then return to their seats.

In piston valves, the piston is a cast-iron plug that moves back and forth across slots in the steam chest, alternatively opening the admission and exhaust ports. In principle, it is similar to the slide valve.

Corliss valves are of the four-port type, two of the ports being for steam admission, two for exhaust. The actuating mechanism is fairly complex; consequently, Corliss valves are best suited to slow-speed engines, usually less than 150 rpm.

Uniflow, as a term used to describe a steam engine, refers to a cylinder design in which the steam always moves in the same direction. For

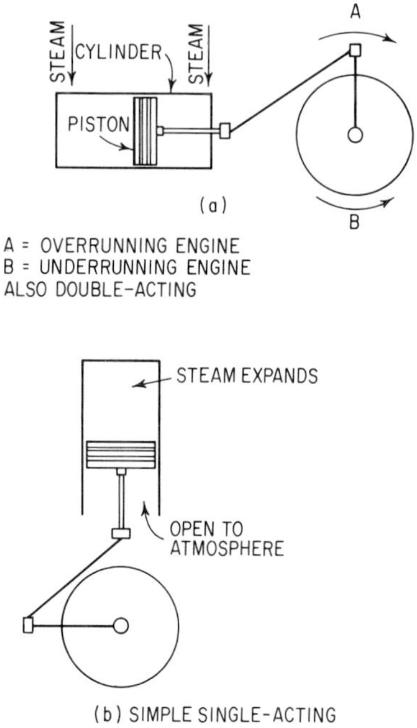

FIG. 14–1. Typical steam engines, vertical and horizontal.

example, steam admission is through valves at the head and frame ends of the cylinder, and the exhaust is through center ports. The advantage claimed for the uniflow design is a reduction in cylinder condensation, since the exhaust temperatures do not get back to and cool the admission areas.

As a further counteraction against cylinder condensation, engine cylinders are frequently steam-jacketed, a phase of vapor cooling.

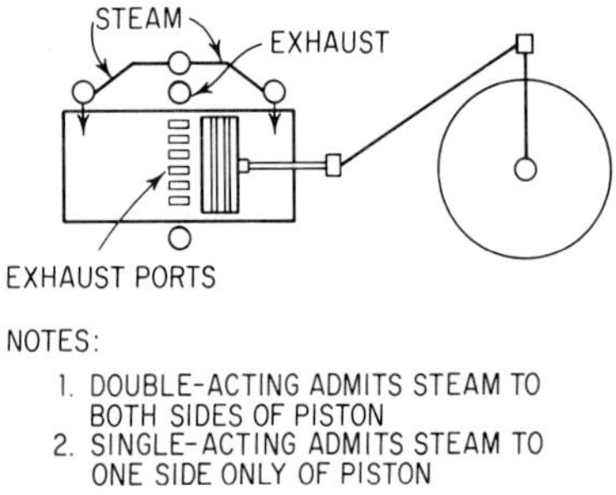

FIG. 14–2. Uniflow engine.

Compound engines are two-cylinder arrangements in which the exhaust steam from the high-pressure cylinder is passed into a low-pressure cylinder for the purpose of getting more useful work out of the steam. Since the two cylinders do not operate in unison, a receiver must be provided between the cylinders to smooth out the pulsations and to receive the high-pressure exhaust until the low-pressure valves open and admit it to the low-pressure cylinder.

When the two cylinders are in horizontal positions parallel to each other, the engine is termed a "cross-compound" design. A steam reheater may be used between the cylinders in addition to the receiver.

A "tandem compound" engine is arranged with the two cylinders in line, and the two pistons are carried on a common piston rod. This enables the engine to use a single connecting rod and crank. No receiver is needed, since both cylinders are timed to operate their valves simultaneously. However, a reheater can be applied to the steam crossover piping if desired.

Compound engines are considered to be relatively economical to operate. However, tandem engines are difficult to inspect and service.

(b) ***The Steam Engine Indicator.*** The troubleshooter seeking to improve engine operation must become familiar with the indicator and its diagram. The indicator diagram tells the operator what conditions are present in the engine at the particular time the diagram is made; for example, a delayed or late admission of steam, an untimely cutoff, or high compression. The diagram also permits a determination of the mean effective pressure (MEP), which is essential in calculating the indicated horsepower.

The formula for indicated horsepower (IHP) is:

$$\text{IHP} = \frac{\text{PLAN}}{33{,}000}$$

where P = mean effective pressure, pounds per square inch
L = length of stroke, feet
A = area of piston, square inches
N = number of strokes per minute counted as rpm $\times$ 2

Because of the internal arrangements of the piston and its cylinder, certain modifications of the formula units become necessary, as follows:

1. The area of the piston (A) must be reduced by the area of the piston rod on the crank end of the piston, since steam cannot occupy an area already given over to the rod. In those engines which have the piston rod extending right through the piston, the area reduction for the rod must be applied to both sides of the piston.

2. The MEP requires taking an indicator diagram for each working end of the cylinder. Slight differences in valve settings can produce differences in the two diagrams.

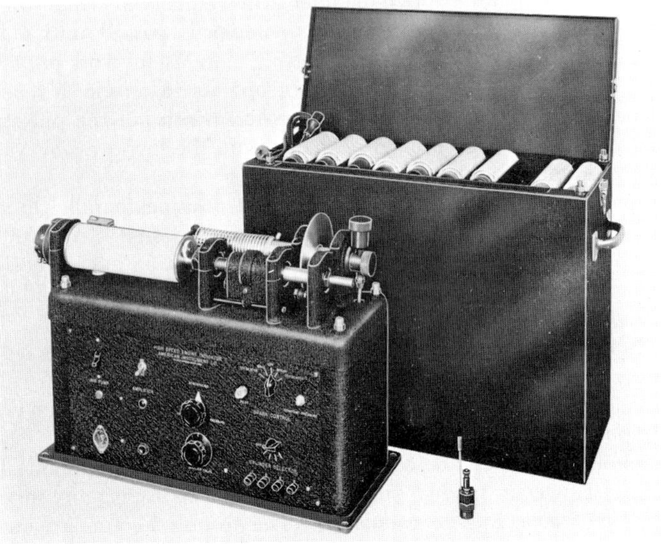

FIG. 14–3. Steam engine indicator with fittings. (*American Instrument Company, Inc.*)

Figure 14-3 shows a modern indicator.

As a rule, the engine owner will have an indicator among his tools and instruments which the troubleshooter can use in his check of the engine performance.

(c) ***Pressure and Back Pressure.*** When a steam engine operates noncondensing, the back pressure becomes highly important in limiting the output of the engine. The difference between the throttle pressure and the exhaust pressure is all the energy the engine has to work on.

In responding to a complaint, the troubleshooter will do well to examine the pressure records over the period that developed the complaint. Any decrease in the former or any increase in the latter can be expected to deteriorate the engine performance. Such findings may explain the trouble and point to troubles outside the engine, as reflected in the engine performance.

14-2. STEAM ENGINE TROUBLES

(a) ***Complaint.*** There appears to be excessive condensation in the engine cylinders.

Causes

1. Steam is wet. Boiler priming may be throwing water into the steam mains.

2. Defective or inadequate steam line insulation is producing condensation in the supply piping.

Recommended Actions

1. Check boiler water for contamination from oil, grease, or other coagulant. If found, shut down boiler and clean internally.

2. Check velocity of steam through the supply main. If main is undersized for the weight of steam passing through, the velocity may be high enough to pull boiler water into the piping. If so found, increase size of piping.

3. Install a separator in the main supply line ahead of the engine. Trap the main as necessary.

4. Increase the insulation on the steam piping.

5. If the trouble is in a compound engine, consider providing reheating between the high- and low-pressure cylinders.

6. If boiler is of the Scotch or fire-tube type and is operating at top capacity, consider installing a superheater.

(**b**) ***Complaint.*** Cylinder head is breathing or flexing with each stroke of the piston.

Shut down the engine immediately. Otherwise run, do not walk, to the nearest exit. This is a dangerous condition.

Causes

1. Examine the cylinder head for a crack. Sometimes these are so fine as to be almost invisible, yet they result in a flexible performance of the head.

2. Engine is operating at a pressure higher than that for which it is designed.

3. Excessive water in the cylinder is building up a pressure plug between the piston and the cylinder head.

Recommended Actions

1. If found defective, replace cylinder head with perfect part.

2. Check steam and exhaust pressures. Adjust to suit rating of engine.

3. Take steps to eliminate excess water as described in Sec. 14-2(a).

(**c**) ***Complaint.*** Gland packing will not remain tight for any substantial length of time. Leakage from high-pressure cylinder becomes an annoyance in the engine room. (Gland also may be called the stuffing box.)

Causes

1. There is probably a misalignment of the crosshead guide.

2. Piston rings are worn, allowing steam blow-by.

3. Cylinder or cylinder liner is worn out-of-round, producing uneven wear on the gland packing.

Recommended Actions

1. Correct the crosshead alignment by resetting the shoes ("slippers"

or "gibs") on the crosshead. This can be done by adjusting the wedges. Do no set up so tight as to cause heating.

2. Consider lowering the crosshead on horizontal engines in an amount equal to the uneven wear, or drop, of the piston. This keeps the piston rod in line.

3. Stop blow-by by replacing worn piston rings.

4. If practical, consider reducing exhaust pressure to relieve pressure differential across the glands.

(d) ***Complaint.*** Engine fails to meet the power requirements for which it was installed.

Causes

1. Engine was sized too small.

2. Steam supply pressure is too low, exhaust back pressure is too high, or both conditions exist.

Recommended Actions

1. Check the engine operation by means of the indicator. This will point to the necessary adjustments in timing, valve settings, etc.

2. Adjust valves as much as the indicator diagram shows to be needed for improved performance. This adjustment will increase the economy of operation also.

3. Investigate the economy and practicality of adding a reheater between stages to increase the power output from the low-pressure cylinder.

Table 14-1. Troubleshooter's Field Guide for Steam Engines

Possible causes	How to check for cause	Recommended ways to correct
COMPLAINT: There is excessive water in engine cylinders.		
REFERENCE: Sec. 14-2(a).		
1. Wet steam.	*a.* Make calorimeter quality check of steam at throttle.	*a.* Install separator in steam supply. Trap low points of steam line.
2. Inadequate insulation on steam mains.	*b.* Take temperature ot surface of insulation. Note if insulation is off in spots.	*b.* Add insulation as necessary.
3. Compound engine is using saturated steam.	*c.* Take a calorimeter quality check at inlet to low-pressure cylinder.	*c.* Consider adding reheater between stages.

Table 14-1. Troubleshooter's Field Guide for Steam Engines (*Cont.*)

Possible causes	How to check for cause	Recommended ways to correct
COMPLAINT: Cylinder head is breathing with each compression stroke of piston.		
REFERENCE: Sec. 14-2(b).		
1. Cylinder head has fine cracks.	*d.* Inspect with magnifying lens.	*d.* Replace head. Do not repair cast iron.
2. Check engine operating pressure with designed pressure.	*e.* If pressure is too high, reduce.	*e.* Note exhaust back pressure. Improve this.
3. Water of condensation is building up a plug between piston and cylinder head.	*f.* Check for wet steam.	*f.* Correct same as for wet steam.
COMPLAINT: Gland, or stuffing box, persists in leaking.		
REFERENCE: Sec. 14-2(c).		
1. Misalignment of the crosshead guide.	*g.* Check alignment of piston with crosshead.	*g.* Reset the slippers or gibs to restore piston alignment with crosshead.
2. Piston rings and/or piston are worn out-of-true.	*h.* Check piston for loose play in cylinder.	*h.* Renew rings or piston. For horizontal engines, lower crosshead to meet drop of piston in cylinder.
COMPLAINT: Engine will not provide the power output for which it is rated.		
REFERENCE: Sec. 14-2(d).		
1. Engine is too small for the load.	*i.* Calculate the load and compare with the engine rating.	*i.* Replace with engine of proper size.
2. Differential pressure between steam inlet and exhaust is less than the design calls for.	*j.* Take pressure readings at throttle and at exhaust.	*j.* Elevate supply steam pressure, or decrease back pressure, or both.

14-3. DIESEL ENGINES, CHARACTERISTICS

(a) *Discussion of Basic Factors.* A diesel engine is distinguished by its dependence on the heat of compression for ignition of the fuel which has been injected into its cylinders. The instrument of compression is the piston traveling back and forth in the cylinder.

Diesels are more amenable to multicylinder construction than are steam engines. Eight- and sixteen-cylinder arrangements are not rare, and they are all of a size. The cylinders may be set in line, in a V formation, and in opposed pairs. "Pancake" or flat arrangements, with all cylinders horizontal, are available for special applications, as are radial diesels. This variety of possibilities makes the diesel the most versatile power producer presently available.

Diesel cylinders are designed for either two-cycle or four-cycle operation. The sequence of events in the four-cycle engine is (1) the admission of air into the cylinder; (2) the compression of the air in the cylinder by the movement of the piston; (3) the injection of the fuel into the hot compressed air and the ignition of the fuel from the heat, followed by the power stroke of the piston actuated by the expanding gases of the burning fuel-air mixture; and (4) the exhaust of the products of combustion.

Four strokes of the piston are needed to complete this cycle: first, drawing in the air; second, compression of the air; third, the power stroke of the piston; and fourth, the exhaust stroke which ejects the burned gases from the cylinder.

The same sequences are carried out in the two-cycle diesel, but it is all accomplished in just two strokes of the piston. On the first or power stroke an exhaust port is uncovered as the piston nears the end of its travel. Reversing its direction of movement, the piston now opens an air intake, and working on the air charge so obtained, completes its second or compression stroke.

Many diesels incorporate a supercharger in order to increase the power output. A supercharger is simply an air compressor driven as a blower by the engine exhaust, or it can be a reciprocating, rotary or other form of compressor driven mechanically from the engine.

By precompressing the combustion air, more of it can be packed into the fixed volume of the engine cylinder, and of course the more air available, the more fuel it can burn. Reports give supercharging credit for output increases up to 20 percent more than the same engine without supercharging.

The troubleshooter should understand something about the different fuel-injection methods. In solid or airless injection, the fuel oil is projected into the cylinders as fine jets at high velocity. Pressures behind the fuel may range all the way from 4,000 psig to double that.

In the common rail system, oil is delivered to the individual cylinder spray valves from a common oil header, hence the name. One pump serves all.

Another arrangement provides a pump for each cylinder. The amount of oil delivery is controlled by a governor that varies in design with the engine using it. The spray valves are like the others.

In a system that uses air injection, air at a pressure considerably higher than the approximately 500 psig common in a cylinder at the end of the compression stroke carries the fuel into each cylinder.

Fuel pumps may be of the constant or variable stroke design. The former is regulated through the operation of a bypass that opens to reduce the amount of fuel being fed into the system. In the variable-stroke unit, the stroke may be lengthened or shortened to meet the conditions established by the engine.

Temperatures in a diesel cylinder at the moment of ignition may be expected to range from 950 to 1100 F, depending on the level of compression. Both the temperatures and pressures are higher in a diesel than in a comparable gasoline engine.

(b) ***Auxiliary Systems, Cooling.*** Three methods of engine cooling are presently in use. Air-cooled cylinders are largely limited to small, relatively slow-speed engines. For this method, the cylinders are provided with fins over which the cooling air may pass, delivered by a fan or blower.

Water cooling, with the water circulated through cylinder jackets, is probably the most often encountered. Outlet temperatures vary according to the manufacturers' specifications, and a 190-F temperature recommended for a given engine might be 30 deg higher than the manufacturer of another engine would approve. Thus for any complaint that appears related to cooling water temperatures, the starting point for corrective action is in the designated temperature for the water leaving the jacket.

The third method, and the one that elevates the operating efficiency of the engine most significantly, is called "vapor phase cooling." What it means is that the cooling water is maintained at boiling or above boiling temperatures in the jacket and is permitted to flash into steam after leaving the engine.

The advantages of vapor phase cooling are these: first, the temperature of the cylinder liner is kept more nearly uniform from top to bottom than with other methods. Eliminating the otherwise large temperature differential also eliminates conflicting expansion strains, reduces or eliminates cylinder condensation, and lengthens the life of the liners by removing a prime cause of liner cracks.

The second benefit is that the low-pressure steam generated in the flash tank after the water leaves the engine becomes a useful source of heat, often in much demand for process work. If there is no application for this steam, it simply is condensed and returned to the cooling circuit. The amount of steam generated is reported to range from 2 to 3 lb per brake horsepower hour.

Vapor phase cooling has been known to increase engine efficiency from 3 to 4 percent at full load, and much more than this at partial loads.

Removing the heat from a water-cooling system or providing condensing water for a vapor phase installation is accomplished by the use of water towers, spray ponds, or evaporative coolers. Troubleshooting a diesel complaint, therefore, may well involve these auxiliary items of the power system.

Since the higher jacket temperatures in a vapor phase cooling arrangement usually result in some increase of temperature for the lubricating oil, particularly in the cylinders, an auxiliary oil cooler may be required. If the use of lubricating oil is observed to be excessive, and especially if the increase has been sudden or recent, the troubleshooter will do well to check out the oil cooler for leaks or for extreme fouling of the heat-exchange surfaces.

(c) ***Auxiliary Systems, Waste Heat Recovery.*** Since the exhaust gases from a diesel may range up to 1100 F and moreover may be of large volume, the recovery of this heat can be an economy factor of importance.

Methods of recovery include passing the exhaust gases through a waste heat type of steam boiler for the generation of steam or hot water for process or industrial use. Such boilers are standard items with many of the boiler building firms and may be of either the water-tube or the fire-tube type. With some models, auxiliary firing is incorporated into the design in order to continue a steady production of steam during periods of diesel shutdowns.

For trouble involving these boilers, refer to the appropriate sections of Division 2.

For the smaller volumes of exhaust gases, the use of a conventional shell and tube heat exchanger may meet all requirements for heat recovery. See Sec. 6-2(d) for a complaint on a gas-to-water heat exchanger. Because of gas pulsations from the engines, the tubes of a heat exchanger on the exhaust system may tend to leak earlier than on a line containing gases at stable pressures.

(d) ***Auxiliary Systems, Silencing.*** Noise in the exhaust pipe of a diesel engine is the result of pulsations and vibrations set up by the operation of the engine exhaust valves. Here is what happens when the exhaust port of a two-cycle diesel opens and releases gas into the exhaust pipe: first, a high-pressure impulse starts down the exhaust pipe at the velocity of sound or a little better, say 1,400 ft per second. By the time it reaches the end of the pipe, some of it is starting back in the relatively low-pressure wake following the wave. If the exhaust port is closed, as it should be to prevent reentry of the wave into the cylinder, the wave is reflected once more toward the end of the exhaust pipe, and this back-and-forth oscillation continues until the wave is dampened out and finally lost in the ambient atmosphere. Long exhaust pipes, without a silencer near the engine, are particularly bad.

Silencers, or snubbers as they are sometimes called, usually are most effectively located close to the engine. From this position, there is very little if any oscillation in the exhaust gases on their way to the atmosphere.

In addition to silencing the exhaust to bearable levels, these devices may incorporate a spark-arresting facility. This can be quite important in an area where discharged sparks from the engine could light off industrial or forest fires. Another aspect of the silencer may be the inclusion of heat recovery surfaces, the heat being absorbed by circulating water. Figure 14-4 illustrates the internals of a spark-arresting design of silencer.

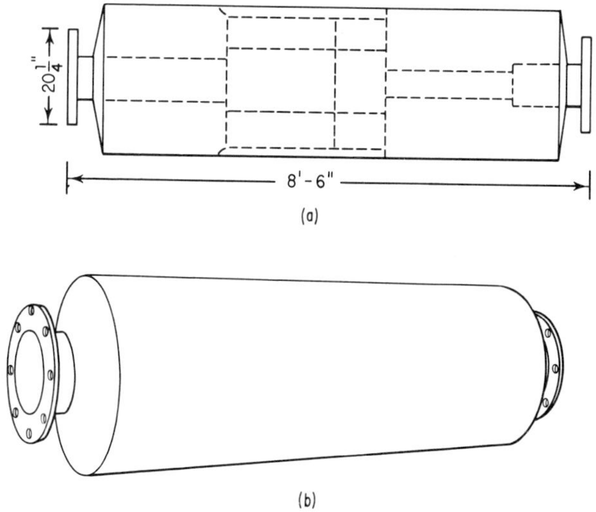

FIG. 14–4. Exhaust silencer. (a) Internal view; (b) external view. *Note:* Dimensions show proportions only. Actual sizes range widely. (*AMF Beard, Inc., a subsidiary of American Machine & Foundry Company.*)

14-4. DIESEL ENGINE TROUBLES

(a) ***Complaint.*** The engine is smoking to an extent that annoys the neighborhood. Also it is of a density that is forbidden by the local anti–air pollution ordinance.

Causes

1. Fuel-air ratio is out of adjustment, resulting in incomplete combustion.
2. There may be a leaky head gasket on one or more of the cylinders.
3. Valves are not seating tightly and are drooling.
4. Fuel oil may be of a poor grade or not suited to this particular engine.
5. The fuel spray nozzles are dirty.
6. There is a malfunction in the fuel pump, possibly in the fuel delivery valve or in the bypass if it is that kind of pump.

Recommended Actions

1. Adjust the fuel-oil injection timing to correct for presently being too early or too late.
2. Replace leaky gaskets.
3. Regrind and refit poorly seating valves.
4. Clean the spray nozzles.
5. Replace present fuel with fuel as recommended by the engine manufacturer.
6. Readjust or repair fuel pump and valves as necessary.

(b) ***Complaint.*** The engine knocks.

Causes

1. If knock is in a single cylinder, the spray valve may be sticking from dirt or advanced corrosion.
2. Spray valve spring is broken.
3. Fuel delivery pump is malfunctioning, probably from some form of valve trouble.
4. A valve is not seating, or the head gasket is leaking.

Recommended Actions

1. Inspect spray valve and clean it up.
2. Inspect and correct, as found necessary, the engine valves and fuel pump valves.
3. Replace all leaky gaskets.

(c) ***Complaint.*** Knocking is in all cylinders and is fairly steady.

Causes

1. Fuel oil is not suited to the engine.
2. Dirt, water, or corroded matter is causing valve sticking.
3. Engine is feeling effects of overworn bearings or piston pins. Also trouble may be in badly worn liners, allowing pistons to thump.

4. Crosshead pin is badly worn.

Recommended Actions

1. Clean up the fuel oil or replace it with fresh clean oil of the proper analysis for this engine.

2. Check condition of all bearings, pins, and liners, including the flywheel, and replace, renew, or tighten as necessary.

(d) ***Complaint.*** Engine is generally noisy in operation.

Causes

1. These are generally similar to those enumerated in the preceding Sec. 14-4(b).

2. Also the valve operating complex may be worn and loose. This may be extended to include the timing gears.

Recommended Actions

1. Check and correct as given in Sec. 14-4(b).

2. Inspect and repair or renew worn parts in the valve mechanism and gears.

(e) ***Complaint.*** The engine overheats and loses some power as a result.

Causes

1. There is excessive friction or binding of the pistons in the cylinder liners. This could be the result of uneven wear with either the piston or the liner or both being out of round.

2. Crankshaft bearings are too tight and possibly scored.

3. Scale or other contamination has accumulated in water jacket, and the flow of cooling water is obstructed.

4. Circulation of cooling water is restricted from wrongly set valve, pump trouble, or a scarcity of the general water supply.

5. Water is not being cooled adequately in tower or spray pond.

6. Lubricating oil system is failing to deliver adequate amount of oil.

7. The timing is incorrectly set for injection of fuel.

Recommended Actions

1. Check water temperatures both to and from the engine. If trouble is found here, correct it in whatever way is necessary.

2. From the cooling water, go to the lubricating oil system and note whether temperatures are normal entering the bearings. If too high, check oil cooler and amount of lubricating oil being circulated. The cooler may leak, and piping obstruction may be choking down oil flow. Clean system if trouble is here.

3. Examine for overheated bearings indicating local difficulties. Stop engine and ease up on bearings as needed. Burned or scored bearings should be replaced in total.

4. Take down pistons at first opportunity and inspect rings and liners. Overhaul as needed.

5. Flush out water jacket, heat exchanger water passages, or both, using clean water with a chemical solvent. A 50-percent muriatic acid solution is effective in some instances, but do not inject any acid type of cleaner until it has been approved by the engine manufacturer.

(f) ***Complaint.*** Engine starting is uncertain. Sometimes it takes considerable trouble to get it going.

Causes

1. Compression is low because of a leaky head gasket.
2. One or more valves may be warped and pitted.
3. Carbon is clinging to valve seats preventing complete closing of the valve.
4. Air filter may be dirty and clogged.
5. Broken valve spring is preventing operation of valve.
6. Fuel pump fails to function, or filter is plugged up with sediment.
7. Too much water is in the fuel oil.
8. Air binding in fuel system is preventing delivery of fuel to cylinders.

Recommended Actions

1. Check compression. If low, look for leaky gasket and replace.
2. Clean carbon from valves. Grind in valves with reseating tool and replace weak or broken springs.
3. Inspect air and fuel filters, cleaning and renewing as needed.
4. Test operation of fuel pump and inspect lines for air pockets.
5. Bleed off bottom of fuel oil tanks and analyze for water content. Pump out the water.

(g) ***Complaint.*** The formation of carbon on the piston and in the valves is both frequent and heavy. Engine requires shutdown and cleaning abnormally often.

Causes

1. Fuel is of the wrong characteristics for this engine.
2. Fuel injector plungers are sticking.
3. There is an after dribble of fuel oil after the injector valve closes.
4. Exhaust gas passages are fouled, or the exhaust is meeting other causes for back pressure.

Recommended Actions

1. Check the fuel oil analysis with the engine builder's approved characteristics.
2. Dress up the injector parts, removing carbon, foreign matter, and sludge.
3. Check the silencer and exhaust piping for obstructions.
4. Check for low compression and, if found, correct as suggested in the preceding Sec. 14-4(f).

Table 14-2. Troubleshooter's Field Guide for Diesel Engines

Possible causes	How to check for cause	Recommended ways to correct
COMPLAINT: Engine is noisy and smoking.		
REFERENCE: Sec. 14-4(a).		
1. Fuel-air ratio is incorrect for good combustion.	*a.* Take Orsat test of exhaust for percent CO.	*a.* Increase air to cylinders as needed to eliminate CO.
2. A leaky head gasket is losing compression.	*b.* Take pressure of compression, if leak is not visible otherwise.	*b.* Tighten up head bolts. If not enough, replace gasket.
3. Valves need work, and some drooling is present.	*c.* Inspect each valve and injector for carbonization or other fouling.	*c.* Regrind and reseat valves after cleaning.
4. Fuel is unsuited to the engine.	*d.* Match fuel being used with engine maker's instructions.	*d.* Replace present fuel with a more suitable grade.
COMPLAINT: The engine operation is accompanied by knocking sounds.		
REFERENCE: Sec. 14-4(b).		
1. Spray valve is sticking from dirt or corrosion.	*e.* Inspect valve internals.	*e.* Clean up. Replace parts in poor condition.
2. Injection timing is poor.	*f.* Check timing for early or late injection for each cylinder.	*f.* Reset injection sequence to correct values.
3. Engine bearings, pins, or both are worn beyond good operating tolerance.	*g.* Listen to bearings and rods with stethoscope, if necessary, to locate knocks.	*g.* Refit bearings and pins to bring back correct clearances.
COMPLAINT: Engine overheats and loses power.		
REFERENCE: Sec. 14-4(d).		
1. Binding of the piston in the cylinders.	*h.* With engine running, listen for excessive rubbing sounds.	*h.* Remove piston and free up.
2. Circulation of cooling water is being obstructed.	*i.* Take water temperatures in and out of jacket. Evaluate for excessive rise.	*i.* Clean water system with pressure and chemical solvent.

Table 14-2. Troubleshooter's Field Guide for Diesel Engines (*Cont.*)

Possible causes	How to check for cause	Recommended ways to correct
3. Engine injection is badly timed.	*j*. Look for incomplete combustion as smoke. Listen for timing knocks.	*j*. Correct injection timing sequence.
4. Temperature of cooling-water supply is higher than desired.	*k*. Inspect the cooling units, such as tower, spray pond, and exchanger, for subrated performance.	*k*. Correct water-cooling auxiliaries as found necessary.
5. Lubricating oil system is providing inadequate lubrication, especially in cylinders.	*l*. Take lubricating oil temperatures. Look for a pressure buildup that could mean a pipeline obstruction.	*l*. Increase the flow in minimum steps and watch effect on engine temperatures. Clean lubricating system.
COMPLAINT: Engine starting is undependable. REFERENCE: Sec. 14-4(e).		
1. Low compression resulting from leaky head gasket.	*m*. Search out leaky gasket.	*m*. Tighten up on head bolts. Replace gasket if tightening does not suffice.
2. Valves are not functioning effectively.	*n*. Inspect valves for warping and pitting, deposits of carbon, or a broken spring.	*n*. Clean, regrind, and reseat valves as necessary.
3. Fuel pump is ineffective from sediments, air binding, etc.	*o*. Check volumetric output of pump. Look for a dirty filter or air binding in the line.	*o*. Restore pump to proper performance. Vent air binding. Clean filter.
4. Water in fuel oil.	*p*. Pull sample of oil and check for water.	*p*. Drain bottom of fuel-oil tank. Run off water.

Table 14-2. Troubleshooter's Field Guide for Diesel Engines (*Cont.*)

Possible causes	How to check for cause	Recommended ways to correct
COMPLAINT: Engine performance is affected by recurring carbon formations on pistons and valves.		
REFERENCE: Sec. 14-4(f).		
1. Engine is burning the wrong-formula fuel.	*q*. Compare characteristics of present fuel with recommendations by engine builder.	*q*. Discontinue present fuel and substitute the correct analysis.
2. Fuel injector plungers are sticking.	*r*. Observe operation of injectors with particular reference to plungers.	*r*. Clean up plungers.
3. Back pressure on exhaust is abnormally high.	*s*. Take pressure reading of exhaust. If too high, inspect passages for carbon accumulations and other obstructions.	*s*. Clean exhaust system. Give special attention to silencer.
4. Engine compression is low.	*t*. See preceding complaint for low compression procedures.	*t*. See preceding section for procedures.

14-5. STEAM TURBINES, CHARACTERISTICS

(a) ***Discussion of Turbine Types.*** Basic references to type distinguish impulse and reaction turbines and combinations of the impulse and reaction wheels. The *simple impulse* turbine coordinates a single row of buckets on the circumference of a rotating disk, or wheel, and one or more nozzles that direct the steam in one or more jets against the buckets and thus rotate the wheel. This combination of nozzle and single wheel is also called a *velocity* turbine by some makers.

Multistage impulse turbines have several of these impulse wheels mounted on a common shaft, with additional sets of nozzles between the wheels for the purpose of efficiently directing the steam leaving one set of buckets into a succeeding row on the next wheel. This expansion of the steam through a series of nozzles and rotating buckets continues until the expanding steam reaches the exhaust pressure. See Fig. 14-5.

In the *velocity compounded* or *reaction* turbine the first stage will be

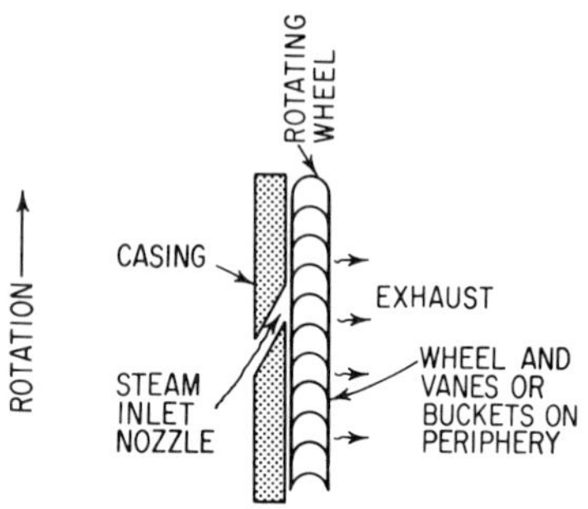

FIG. 14–5. Impulse turbine.

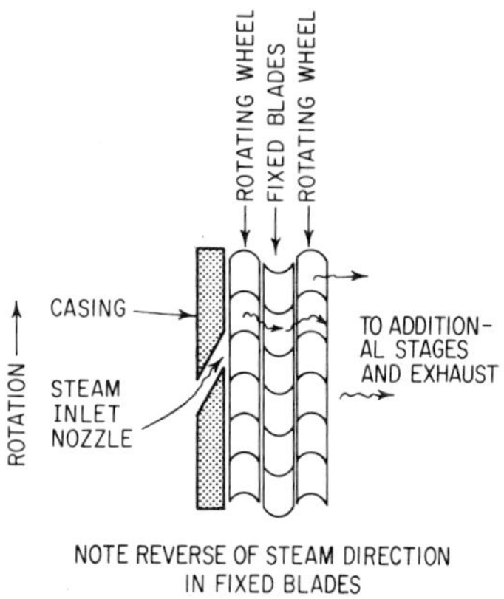

FIG. 14–6. Reaction turbine.

of the impulse design, followed by a row of stationary blades set in a diaphragm that reverse the direction of the steam jet and thus redirect it into a second row of moving blades. This reversing redirection function continues until exhaust pressure is reached as the steam leaves the last rotating wheel. Sometimes termed a *velocity stage* turbine, this design is also familiar to turbine users as the *Curtis stage* turbine.

Further description is likely to note whether the turbine is condensing or noncondensing, whether it provides steam extraction at one or more reduced pressures, whether it employs steam reheating between a high-pressure and low-pressure section, and whether there are any combinations of these characteristics.

Noncondensing turbines are designed to operate at an exhaust back pressure that will provide steam at a desired condition for process or some other service application. Extraction turbines meet the same process needs but not in volumes equivalent to the entire output of the turbine exhaust. Multiple points of extraction are useful in satisfying more than one level of pressure.

In all these applications of used or partially used turbine steam, the turbine, in addition to driving a generator or other mechanical equipment, is performing the function of a steam pressure-reducing valve.

Topping turbines are excellent examples of the reducing-valve concept. In this arrangement, the exhaust from a high-pressure turbine is led to a lower-pressure turbine, in this way increasing the overall economy of a low-pressure system. It is an expedient for lengthening the productive life of some of the older plants.

(b) *Control and Safety Features.* The normal speed of the turbine is controlled by a governor working to (1) throttle down the steam to the first stage nozzles or (2) admit steam through nozzle valves, each nozzle valve monitoring a group of nozzles. In this latter arrangement, all nozzle valves will be wide open except those of the last group. Nozzle-valve control is considered to provide better steam economy when the turbine is operating at a partial load.

Governor forms may be of the familiar centrifugal type, in which the upward and downward movements of the spindle reflecting the speed of the turbine shaft are transmitted to the steam inlet, or governing may be achieved through variations of oil pressure in the discharge from an oil pump attached to, or driven by, the turbine shaft.

Overspeed trips are a part of all governor systems. These are usually set about 10 percent above the normal running speed of the turbine.

Since turbine casings at the exhaust ends of the turbines may not be capable of withstanding full line steam pressure, protection against the sudden development of excessive pressures in that area take the form of (1) an atmospheric relief valve or (2) a blowout, or rupture, diaphragm. Either form is designed to vent the steam to atmosphere any time the pressure in the condenser or turbine exhaust rises 5 percent above normal.

Still another device offers protection for condensing turbines in the form of a condenser pressure sensor that shuts down the main governor whenever the vacuum in the condenser falls significantly below normal. The common critical level for setting this sensor is 10 in. of mercury, compared with an average normal vacuum of 27 to 28 in. of mercury.

(c) ***Turbine Lubrication.*** Bearings for very small turbines may be of the self-oiling design, with oil rings or "slingers" that pick up oil from the reservoir in the bearing shell and splash it over the bearing surface. In the larger designs, oil is fed to the bearings by pump, the pump being either direct-driven from the turbine shaft or set up as an auxiliary piece of equipment.

Lubricating oil coolers are usually of the shell and tube design, with the tubes of copper, Admiralty Metal, or brass. Cooling water to these units is subject to the same inherent troubles that exist with other heat exchangers, that is, deposition of scale or sludge, water too warm or too cold, and the development of leaks.

Opinions as to what is the optimum temperature to be maintained in turbine bearings vary with the turbine builders. A safe range of oil temperatures entering the bearings seems to be between 105 and 140 F. The oil leaving the bearings may be from 130 to 160 F. In the experience of this writer, the 160-F temperature is about the upper limit of acceptance, and 150 F is to be preferred.

A substantial percentage of turbine troubles originate in faulty lubrication; consequently, the troubleshooter will do well to keep close watch on the bearing performance. With the bearings in new turbines, finding oil passages clogged is not uncommon, sometimes as the result of machine turnings that were overlooked and went out to the job with the turbine. Sometimes the bearings fit too tightly against the shaft, thus preventing proper oil penetration, and the probable end is a wiped bearing.

As an auxiliary to a large turbine, or to serve a group of turbines, the troubleshooter will often meet with a centrifuge or centrifugal form of oil separator. The responsibility of the centrifuge is to keep foreign particles out of the lubricating system. Any evidence of lubricating trouble calls for an inspection of the centrifuge as a matter of routine.

14-6. STEAM TURBINE TROUBLES

(a) ***Complaint.*** Turbine vibrates over a fairly wide range of operation instead of being limited to a critical speed.

Causes

1. Turbine shaft is bent or is out of balance. This condition is particularly worth attention if there is no turning motor to keep the turbine revolving when not in service.
2. Turbine bearings are worn and may be partially wiped.
3. There is a misalignment between the turbine and the driven equipment.
4. With the turbine and driven unit disconnected, check each for rotating balance. Trouble is sometimes found in the former by reason of missing buckets in one or more wheels or by axial misalignment of the shaft.

Recommended Actions

1. By means of a dial indicator, check the trueness of the turbine shaft when running at designed normal speed. If a bow is discovered, communicate with the turbine manufacturer for instructions. This is the correct action for imbalance also.
2. Observe the temperature of each bearing. Elevations above the normal of 10 percent or more indicate the presence of trouble either in the bearing or in the oil feed to the bearing. Lift bearing caps and note condition of oil in the grooves and any evidence of burning. Check clearances and correct by machining or by inserting new bearings. In some instances it may be necessary to pour new babbitt before machining. Fit bearings with Prussian blue to indicate contacting surfaces.
3. Measure the shaft for axial position. Do this when the turbine is hot. Adjust as necessary.
4. If source of vibration resolves into wheels or buckets, communicate with the maker. If not corrected, vibration in these parts may end up with failure from fatigue or a burst wheel.

(**b**) ***Complaint.*** The control governor persists in hunting.

Causes

1. There may be a bent valve stem or bent spindle.
2. The steam valve, operated by the governor spindle, is sticking.
3. There is lost motion in the governor parts.

Recommended Action

Open up the governor valve and all actuating mechanisms and check for deteriorated conditions, excessive wear, bent stems or spindles, and excessive friction. Correct as necessary.

(**c**) ***Complaint.*** Turbine either does not come up to speed or does so with extreme slowness.

Causes

1. Governor setting is incorrect and is cutting off steam too soon.
2. Steam pressure reaching nozzles is too low for the machine to carry its designed load.
3. Unit being driven by the turbine is imposing a heavier load than was expected.
4. Foreign material is partially blocking the steam supply, either ahead of the throttle valve or ahead of the nozzles.
5. Back pressure on turbine, whether from noncondensing operation or from condensing operation malperformance by the condenser, is so high as to affect adversely the performance of the turbine.

Recommended Actions

1. Check pressures of steam to throttle valve and at the exhaust. If pressure differential is less than designed, correct whichever is found wrong. This may involve carrying the checking back to the condenser, to the steam generator, and to the process steam system.
2. Compare the actual load with the designed load. If in excess of the turbine rating, reduce the load accordingly. Otherwise, accept the reduction in turbine performance and the increase in steam consumption with its falling off of steam economy.
3. Open the steam line, strainers, and valves, and examine for the presence of foreign material. This writer once found a workman's black glove stuck in the throttle valve of an auxiliary turbine. Occasionally rats, mice, and birds are trapped in pipelines.

(**d**) ***Complaint.*** Turbine bearings are running hot, right up to the maximum temperature before wiping.

Causes

1. Oil feed is inadequate as result of oil pump difficulties or failure or because the oil supply lines are more or less plugged by sludge or foreign material.
2. Supply obstruction is in the drilled oil passages through the bearing pedestal. New turbines are especially susceptible to this trouble because of careless checking in the shop.

3. Clearances in the bearings are too restricted to allow admission of the needed amount of oil.

4. The oil itself is at fault, being too heavy or possibly contaminated, or it might be charged with water or air to form an emulsion.

Recommended Actions

1. Examine the oil pump to determine whether operation is normal. If not, repair or adjust pump to obtain correct delivery of oil.

2. Take oil pressure at pump discharge and at the bearings. Excessive pressure drop suggests a measure of obstruction in the oil feed piping, probably from sludge or corrosion. This condition may require a complete dismantling or replacement of the oil piping in order to correct it.

3. Check clearances and trueness of bearings. Any binding on the shaft will produce an overheated bearing. Trim bearings to increase clearances, if necessary. If bearings have been shimmed into position, see that bearing caps have not been displaced as a result.

4. Take a sample of oil and analyze for viscosity and cleanness. If a centrifuge is being used to clean the oil, check its performance.

5. Inspect the oil grooves for correctness of arrangement and size. In some instances, it may be necessary to rebabbitt the bearings and recut correct oil grooves.

6. Check oil level in reservoir. If low, add oil to proper running level.

(e) ***Complaint.*** Turbine suffers from an accumulation of scale on the blades. This increases steam consumption and lowers the economy of operation.

Causes

1. Turbine is getting the effects of boiler priming or foaming.

2. Boiler water treatment is inadequate so that priming surges carry over calcium and magnesium scale into the turbine.

3. Steam line to throttle valve is not effectively trapped. Alternatively, no separator has been installed in the steam line, although the steam to the turbine is not superheated.

Recommended Actions

1. Look into the cause of the priming or foaming. Sometimes the latter may result from the boiler water treatment when coupled with a heavy pull of steam from the boiler. Draw a sample of the boiler water and analyze it both before and after treatment.

2. Does the turbine draw exceed the normal maximum rating of the boiler? If it does, water carry-over is almost inevitable and ways must be found to reduce the steam demand, or an improvement is essential in separating water from the steam before it reaches the turbine throttle.

3. Consider the economy and practicality of installing a superheater on the boiler.

(f) ***Complaint.*** The capacity and efficiency of the turbine are believed to be falling short of the turbine ratings and guarantees.

Causes

These can be numerous and may be the result of factors in the previously listed complaints, (a) through (e).

Recommended Action

The turbine capacity and efficiency of operation should be tested in accordance with the "ASME Performance Test Code for Steam Turbines" PTC-6 1964 or latest edition.

Table 14-3. Troubleshooter's Field Guide for Steam Turbines

Possible causes	How to check for cause	Recommended ways to correct
COMPLAINT: Turbine vibrates over a fairly wide range of operation.		
REFERENCE: Sec. 14-6(a).		
1. Turbine shaft is bowed or out of balance.	*a.* Set up a dial indicator on the shaft. Rotate shaft and read deflections on indicator.	*a.* Ask turbine maker to recommend a straightening procedure. Do the same for correcting balance.
2. A misalignment of the connecting coupling between turbine and driven machine.	*b.* Alignment may be checked with indicators on each shaft. Deflections should be approximately similar.	*b.* Break coupling and remake the joint.
3. Bearings are too tight.	*c.* Note temperature of bearings. Tight bearings will run hot.	*c.* Refer to Sec. 14-6(d).
4. Bearing has wiped.	*d.* Lift bearing cap and inspect.	*d.* Rebabbitt or replace with new bearing.
COMPLAINT: Control governor persists in hunting. Turbine speed is erratic.		
REFERENCE: Sec. 14-6(b).		
1. Valve stem or governor spindle is bent.	*e.* Inspect for trueness.	*e.* Replace with new stem or spindle to suit.
2. Steam valve is sticking.	*f.* Check steam valve for ease of operation.	*f.* Remove source of excessive friction.
3. There is lost motion in governor parts.	*g.* Check all moving parts for excessive play.	*g.* Tighten up as necessary.

Table 14-3. Troubleshooter's Field Guide for Steam Turbines (*Cont.*)

Possible causes	How to check for cause	Recommended ways to correct
COMPLAINT: Turbine is very slow in reaching speed; sometimes it does not make it. REFERENCE: Sec. 14 6(c).		
1. Governor is set wrongly or is faulty in its operation.	*h.* Note point of governor speed take-over for normal running.	*h.* Reset governor. Replace defective part if found.
2. The steam pressure reaching nozzles is too low.	*i.* Check pressure at boiler and at nozzles. Note amount of drop.	*i.* Increase boiler outlet pressure if practical and within boiler design. Look for some obstruction in steam line.
3. Turbine capacity is too little for the imposed load.	*j.* Compare existing load with rating of turbine.	*j.* Consider ways of increasing turbine capacity with higher steam pressure or reduction of load.
4. Back pressure on turbine is higher than machine is designed for.	*k.* Compare back pressure with design limits.	*k.* Determine ways to reduce back pressure as more vacuum on condenser or lower pressure on process uses of exhaust steam.
COMPLAINT: Turbine bearings run out. Occasionally one wipes. REFERENCE: Sec. 14-6(d).		
1. Oil feed to bearings is inadequate.	*l.* Take oil pressure at bearing of supply. If low, bearing is likely to be oil-starved. Examine pump for malfunctioning.	*l.* Do whatever is necessary to restore pressure to normal level.
2. Bearings are set up too tight. Oil cannot reach contacting surfaces in required quantities.	*m.* Lift bearing caps and measure clearances. Note if bearings are scored.	*m.* Restore clearances to normal.

Table 14-3. Troubleshooter's Field Guide for Steam Turbines (*Cont.*)

Possible causes	How to check for cause	Recommeded ways to correct
3. Oil being used has wrong characteristics, or it may be contaminated or emulsified.	*n*. Take samples and analyze. Compare with turbine maker's recommendations.	*n*. Drain off oil. Clean system. Refill with correct oil.
COMPLAINT: Scale accumulates on the turbine blades. Sometimes it sets up a vibration; mostly it increases the steam demand and thus affects economy of operations adversely. REFERENCE: Sec. 14-6(e).		
1. Boiler is priming or foaming and throwing over slugs of water with the steam.	*o*. Listen for water. Take calorimeter readings along the steam line. Inspect steam traps for possible failures to act.	*o*. Install line separator. Clean up traps. Consider change of boiler water treatment.
2. Steam demand exceeds the normal peak capacity of the boiler.	*p*. Check steam demand on the boiler meter. Compare with design maximum safe output of boiler.	*p*. Reduce turbine load if practical. Increase boiler capacity if practical.
COMPLAINT: Capacity and efficiency of the turbine appear to be in doubt. Gages indicate it is meeting its rating, but the steam consumption is excessive. REFERENCE: Sec. 14-6(f).		
1. There are various causes.	*q*. Run a complete performance test on the turbine as directed in ASME "Test Code for Steam Turbines" PTC-64 or latest edition.	*q*. Discuss test results with turbine maker.

14-7. GAS TURBINES, CHARACTERISTICS

(a) ***Discussion of Gas Turbine Types.*** Figures 14-7, 14-8, and 14-9, respectively, illustrate schematically three commonly encountered forms of the gas turbine. Whether any one is particularly suited to a given project can be determined only by a detailed analysis. Since this analysis is basically the responsibility of the designer, the trouble-

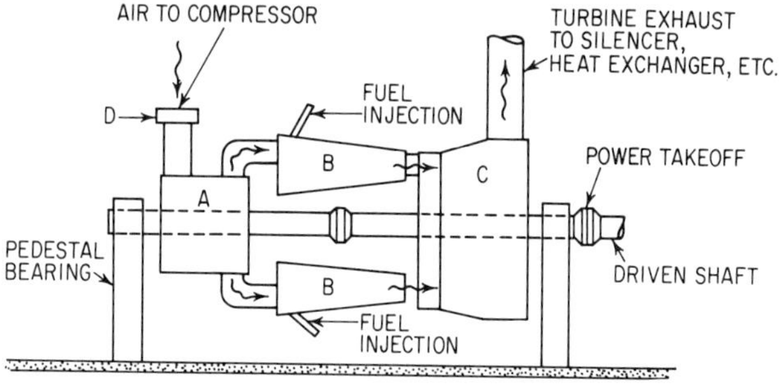

FIG. 14–7. Simple gas turbine. A, air compressor; B, combustor, one or more may be used; C, gas turbine; D, intake air filter or air washer to suit.

shooter is not likely to be equipped for making it. He will most likely need help if the complaint comes within the field of suitable application.

For conventional mechanical troubleshooting, an understanding of the gas turbine, its shapes and peculiarities, is fundamentally desirable. Here are the major parts of the unit and their uses:

The air compressor, which is driven by or from the turbine shaft, provides the air necessary for combustion. As with other air compressors (see Division 13), the gas turbine compressor can have its troubles as these affect an axial design. Also it may be arranged in low-pressure and high-pressure stages with an intercooler between stages, to improve the compression in the high-pressure section. There is no aftercooler, since the heat is desired to speed up combustion in the combustors.

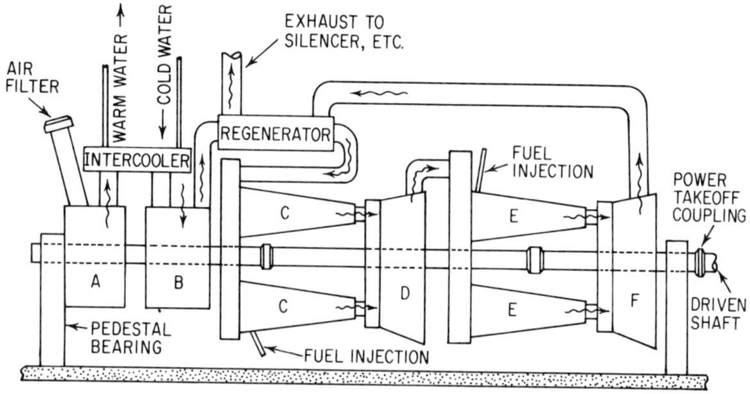

FIG. 14–8. Compound gas turbine, regenerative. *Note:* Each combustor has a fuel injector. For simplicity, not all are shown. A, low-pressure air compressor; B, high-pressure air compressor; C, first-stage combustors; D, first-stage turbine; E, second-stage combustors; F, second-stage turbine.

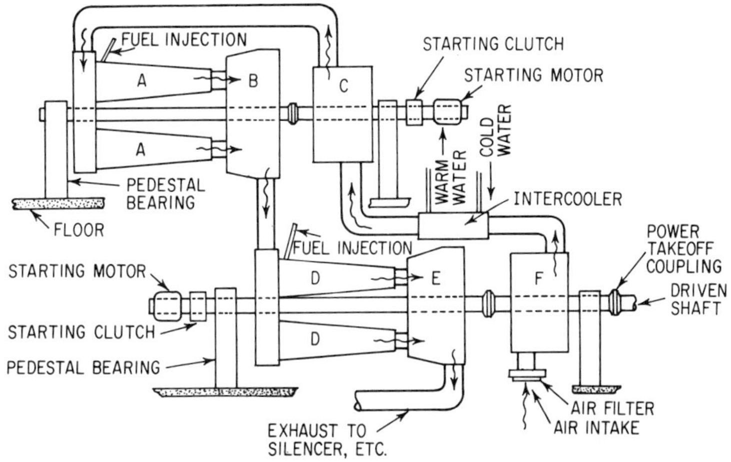

FIG. 14–9. Dual shaft arrangement. A, high-pressure combustors; B, high-pressure turbine; C, high-pressure air compressor; D, low-pressure combustors; E, low-pressure turbine; F, low-pressure air compressor.

The air of combustion, augmented by the products of combustion created in the combustors, then goes to the turbine, passing through the nozzles and vanes similarly to the passage of steam through a steam turbine. A power take-off coupling on the end of the turbine shaft provides the driving force for a generator, pump, blowers, or any other suitable application.

The three basics of the machine are the air compressor, the combustor, and the gas turbine section of the combination. In order to increase the efficiency of the cycle, multicomponent arrangements will be favored in some instances. The possibilities of duplication of the components are shown in the illustrations.

In stationary plants, the fuel for a gas turbine set may swing from piped-in natural or other gas to a suitable fuel oil. Marine installations of course depend on liquid fuels.

The capacity of the gas turbine to deliver heat more than power explains its selection in many industrial circumstances. The heat ratio of 2½ to 1 shaft hp, compared with the 1½ to 1 for a diesel, is well able to furnish process steam by means of a waste-heat boiler or a heat exchanger. In commercial applications, the heat sometimes is delivered, where suitable, to an absorption system of refrigeration. Thus the gas turbine provides power, heat, and cooling. It is the heart of the "total energy" concept in which all services, excepting the delivery of the fuel, originate on the premises. Refer to Sec. 9-3(a) and subsequent sections for a further discussion of the absorption system of refrigeration.

Because the design of a gas turbine rests on delivering the gas to one or more turbine wheels, various arrangements and locations of the components are possible. The troubleshooter, therefore, might face operating units as dissimilar in appearance as those in Figs. 14-10 and 14-11. Both of these are commercial products, accepted, and working their way.

Figure 14-8 shows an arrangement using a regenerator component. This is simply a heat exchanger that transfers heat from the turbine exhaust gases to the compressed air entering the combustor. The results are a lowered waste of heat in the final discharge and a gain in overall efficiency for the machine.

Another method for improving installation economy is to use an airplane type of jet engine to provide hot gas to the turbine. In this design the turbine is free; that is, it has its own shaft and is not mechanically tied to the jet engine shaft. The power take-off coupling is on the turbine shaft; the compressor is on the jet engine shaft.

Jet engines tend to be more critical of their fuel than does the conventional gas-turbine combustor. They operate at temperatures of 1500 F or thereabouts, generally higher than the combustor by from 100 to 300 deg. At the higher temperatures, corrosion shows increased activity,

FIG. 14–10. Packaged gas-turbine power plant developed for aircraft. Bleed air takeoff in center, control panel at right. (*AiResearch Manufacturing Company*.)

FIG. 14–11. Phantom view, gas turbine, showing sound baffles on exhaust. (*Cooper-Bessemer Corp.*)

making highly desirable a fuel oil low particularly in vanadium and sulfur, with sodium also unwelcome.

Tests reported from various sources suggest that a corrosion-removing shutdown might be recommended at the end of each 8,000 hours of operation, unless some distress of operating characteristics occurs sooner. However, at least one manufacturer offers a jet engine turbine combination in which maintenance-free performance is stated as lasting 100,000 hours. In this particular combination, components are so arranged that individual units are readily replaceable. A fouled turbine or jet engine is replaced by another, and the removed unit is returned to the factory for reconditioning. This reduces maintenance shutdowns from days to hours.

As with all air compressors, the gas-turbine compressor is sensitive to contaminated air. To keep foreign dusts and particles out of the system, an effective filter is necessary on the compressor intake. Probably a conventional dry filter will be satisfactory in most installations. If conditions are very bad, an electrostatic type might be considered. What happens to the gas turbine performance following an impairment of the filter is a situation the troubleshooter sooner or later may be obliged to interpret. Further reference to this subject appears in Sec. 14-8 on individual complaints.

14-8. GAS TURBINE TROUBLES

(a) ***Complaint.*** The turbine is noisy. A high-pitched and disturbing whine accompanies its operation.

Causes

1. This is a high-speed machine, some models rotating as high as 45,000 rpm. Unless a silencer is added to the installation, many individuals will be unable to tolerate the sound.

2. With a properly acting silencer, the noise is controlled. If it is not controlled, the silencer is in need of inspection and probably cleaning and adjustment.

Recommended Action

Shut down the machine or bypass the silencer if a bypass is available. Open the silencer and inspect the internals. Look for corroded areas, loosened baffling, or any other abnormality. If defects are extensive, replace entire unit.

(b) ***Complaint.*** Unit is smoking and deposits considerable soot on turbine blades and in heat exchanger.

Causes

1. Injection rate of the fuel feed is too high for the air volume, producing an over-rich mixture.

2. The fuel analysis is not suitable for a gas turbine. Oils with a high percentage of carbon residue and with high viscosity must be avoided in many turbines to avoid excessive maintenance problems.

Recommended Actions

1. Reduce fuel injection rate.

2. Examine analysis of fuel oil. Compare with analyses recommended by turbine builder. Change to the recommended analysis. Do not use residual oils unless specifically approved by builder.

(c) ***Complaint.*** Power delivered at the power take-off coupling has been steadily decreasing.

Causes

1. Air compressor output in cfm is not enough to achieve full power combustion. Dirty air filter is affecting intake.

2. Fuel is carrying impurities into the combustor and promoting corrosion.

3. Back pressure has developed in the heat exchanger, regenerator (if there is one), waste-heat boiler, or silencer.

4. Fuel feed is impoverished by fouling of nozzles or by some obstruction in the oil line.

5. For gas fuel, the pressure of the gas has fallen below the required optimum operating condition.

6. Turbine blades are dirty and corroded, and the condition is intensifying.

Recommended Actions

1. Check all pressures, air, gas, and oil, to the nozzles, as well as ex-

haust gas pressures at the turbine exhaust and thereafter at the final discharge to atmosphere. Clean the air filter.

2. Abnormal pressure conditions, if found, will indicate the areas in which corrections are needed. If the ailing part can be replaced with credit for rehabilitation of the original, consider saving time this way.

3. Check times elapsed since last shutdown and overhaul, in hours. Possibly the machine has been operated too long without regular maintenance.

4. Run a flue-gas analysis to determine the efficiency of combustion. Look for excessive CO as indicating incomplete combustion. Also look for the possibility of too much excess air, indicated by exhaust-gas temperatures running somewhat higher than those of efficient combustion. However, under extreme conditions of excess air, a lowering of exhaust temperatures is possible as the result of this extreme dilution. Correct air-fuel ratio as may be needed to restore good combustion.

5. Since fouling of the nozzles and advanced corrosion in the combustor or in any of the heat exchangers will reduce the output of usable energy, examine each of these components in detail. This may entail considerable dismantling unless the previous noting of pressures and temperatures points to the ailing part. This inspection should include a sighting of the turbine blades for possible accumulations of corrosion and scale.

(d) ***Complaint.*** Gas turbine is furnishing heat for the operation of an absorption cooling system, and the cooling being obtained is insufficient and less than predicted at the time of installation.

Causes

1. The trouble lies in some area of the cooling system and not in the gas turbine.

2. The trouble is resulting from insufficient heat production by the turbine, in the heat exchanger, or in the waste-heat boiler that generates steam for the absorption generator.

Recommended Actions

1. Determine center of trouble, that is, whether it is in the absorption circuits, the gas turbine, or intermediate apparatus.

2. For correction in the absorption circuits, refer to Secs. 9-3 and 9-4, as appropriate.

3. For correction of waste-heat boilers or other forms of steam generator, see Secs. 2-5(a)–(e), 2-6(a)–(e), 4-2(a), and sections of Division 16 as applicable.

4. For correction of the gas-turbine components, see Sec. 14-8(e).

5. For a fast estimation of the problem in each area, check symptoms with the appropriate Troubleshooters' Field Guide.

(e) ***Complaint.*** A drop in thermal efficiency occurs every summer when the intake air temperature to the compressor rises.

Cause

Because of the rise in the ambient temperature, the volumetric efficiency of the compressor is reduced and more power goes into driving the compressor than is the case during the colder weather.

Recommended Actions

1. If cool water is available, consider cooling the inlet air by means of a water washer. The cost of this investment plus the operating costs must be balanced against the power saving in the turbine to find out if such an investment is justified.

2. Some slight increase in turbine speed may be possible, provided that the design of the turbine and compressor is such that an increase is practical and that the result will be satisfactory without development of undesirable symptoms of wear or vibration.

(f) ***Complaint.*** Turbine is difficult to start.

Causes

1. If started by electric motor, voltage may be low. This is for occasional starts as a rule, but if consistent reluctance of the starting motor is noted, low voltage could be the result of heavy additional loads on the electric system occurring simultaneously with the need to start the turbine.

2. For pneumatic starting systems, low air pressure could be a cause.

3. For either method of starting, resistance to turnover of the turbine rotor could be from too-tight bearings, a misalignment of the coupling between the turbine and its compressor, or a misalignment of the power take-off coupling.

4. There is a chance that the starting motor, whether electric or pneumatic, is underpowered for the load.

Recommended Actions

1. Check motor voltage or pneumatic system pressure, as appropriate. Note whether the voltage or pressure equals the design requirements. If the trouble is found here, proceed with the proper steps to correct it.

2. Compare the required starting horsepower with the horsepower available in the electric or air motor.

3. Turn over the turbine slowly and check the alignment of the shaft and couplings, using a dial indicator.

4. Verify the correctness of the clearances in the bearings. If too tight, ease up the fit. Note also the kind of lubricating oil, its viscosity, and the degree of contamination, if any. Change oil if conditions indicate this change to be necessary.

5. If the starting trouble lies in a period of delay that is excessive

before the turbine becomes self-operating, check the fuel-air mixture for being too lean or too rich. Also examine the igniters for fouling by corrosion, burning, or other deterioration that is delaying the ignition. Correct fuel and ignition as found necessary. In troubleshooting this item of delay, compare the time delay with that set by the turbine manufacturer. Generally full ignition may be expected in from 2 to 3 minutes, and the assumption of full load by the turbine will be in about 10 to 12 minutes following the first turnover.

Table 14-4. Troubleshooter's Field Guide for Gas Turbines

Possible causes	How to check for cause	Recommended ways to correct
COMPLAINT: Turbine is intolerably noisy.		
REFERENCE: Sec. 14-8(a).		
1. Turbine is in need of a silencer.	*a.* Compare the current level of sound with the expected level after silencing.	*a.* Install a silencer.
2. Silencer is on the turbine but sound level is still too high. Silencer is dirty or internally maladjusted.	*b.* Open up the silencer. Inspect for deformation, missing baffles, etc.	*b.* Clean the silencer. Repair as found needful.
COMPLAINT: Turbine is smoking.		
REFERENCE: Sec. 14-8(b).		
1. Incorrect fuel-air mixture is reaching the combustor.	*c.* Reduce fuel injection or increase air, and note results.	*c.* Adjust the fuel-air mixture as required.
2. Fuel oil is not suited to the turbine.	*d.* Run an analysis of the fuel oil and compare with oil recommended by builder.	*d.* Replace oil with proper kind.
COMPLAINT: Power output is steadily decreasing.		
REFERENCE: Sec. 14-8(c).		
1. Air compressor is delivering less than its rating.	*e.* Check output of compressor in cfm and in pressure. Inspect air filter.	*e.* Clean air filter and compressor internals.
2. Developed corrosion has impaired turbine blades, compressor	*f.* Open up components and inspect for deterioration.	*f.* Clean if practical. Replace badly damaged parts.

Table 14-4. Troubleshooter's Field Guide for Gas Turbines (*Cont.*)

Possible causes	How to check for cause	Recommended ways to correct
blades, and combustor.		
3. Excessive back pressure on turbine exhaust.	*g.* Take a pressure reading on the exhaust. Compare with design pressure.	*g.* Locate point of trouble in silencer, waste-heat boiler, or heat exchanger. Restore correct condition.
4. Fuel feed is partially blocked or is showing sporadic surges.	*h.* Open fuel lines for inspection if oil pump is normal.	*h.* Clean oil lines if practical or replace if that is easier and cheaper.
5. Gas fuel delivery pressure is too low.	*i.* Compare existing pressure with required pressure. Reset pressure regulator if this is present.	*i.* Go to source of gas fuel and arrange for increase in pressure.
6. Fuel oil is of wrong analysis.	*j.* Take analysis of current fuel oil and compare with specifications by turbine builder.	*j.* Drain fuel oil system and charge with correct oil.
7. Inefficient combustion.	*k.* Run a flue-gas analysis to determine effective level of combustion.	*k.* Adjust air, fuel injection, or both for getting the best combustion.

COMPLAINT: Turbine supplies heat for an absorption refrigeration system. The performance of the absorption cooling system is not up to predictions at time of installation.

REFERENCE: Sec. 14-8(d).

1. Trouble is in the absorption part of the cycle.	*l.* Measure heat output of turbine. If up to rating, go to absorption plant.	*l.* If under rating, see Sec. 14-8(c) & (e). For absorption, see Secs. 9-3 and 9-4.
2. Trouble is in turbine performance.	*m.* Heat output is below rating.	*m.* Do as above.

Table 14-4. Troubleshooter's Field Guide for Gas Turbines (*Cont.*)

Possible causes	How to check for cause	Recommended ways to correct
COMPLAINT: Turbine efficiency drops off each summer.		
REFERENCE: Sec. 14-8(e).		
1. High ambient temperature.	*n*. Note ambient temperature and compare with temperature on which compressor is rated.	*n*. Cool the intake air, if possible, by water sprays. Evaluate a rise in speed for effect and economy.
COMPLAINT: Turbine is difficult to start.		
REFERENCE: Sec. 14-8(f).		
1. Voltage to electric starting motor is low.	*o*. Check the voltage. Learn if low voltage occurs at a certain time only or is continuous.	*o*. Rearrange power demands to allow normal voltage at start of turbine.
2. The air pressure to pneumatic starting system is too low.	*p*. Check air pressure at air motor.	*p*. Limit simultaneous draw of air or arrange for more of it to suit.
3. Turbine shaft bearings are too tight.	*q*. Measure clearances and compare with design clearances. Take temperature of bearings while under load.	*q*. Ease up on the bearing fit. Evaluate lubricating oil to bearings for viscosity and temperature limit.
4. Misalignment of the shaft or couplings.	*r*. Check alignment of shaft and couplings with dial indicator.	*r*. Correct any misalignment found.
5. Delay in starting is from a too rich or too lean fuel-air mixture.	*s*. Pull a sample from a combustor. Check for relative weights of oil and air. Compare with best ratio.	*s*. Adjust the fuel-air mixture.
6. Delay in starting is from deteriorated igniters.	*t*. Inspect each of the igniters. Note any corrosion, sealed or otherwise detracted in performance.	*t*. Clean up igniter or replace.

Table 14-5. Troubleshooter's Field Guide for Gasoline Engines

Possible causes	How to check for cause	Recommended ways to correct
COMPLAINT: Engine is hard to start.		
1. Water or dirt in fuel.	*a*. Draw sample of fuel. Let stand and settle. Strain through muslin. Look for residue.	*a*. Obtain a new source of fuel. Clean the line.
2. Gas-air mixture is too rich or too lean.	*b*. Adjust carburetor setting and note results. Check choke for wrong setting.	*b*. Provide the correct mixture. Adjust choke.
3. Timing is off.	*c*. Listen for a ping or knock. Compare with timing as furnished by engine maker.	*c*. Retime as necessary.
4. Air leak through a gasket.	*d*. Examine all head gaskets and manifold gaskets for leakage.	*d*. Pull up on gaskets or renew.
5. Weak spark.	*e*. Check battery voltage and spark transformer.	*e*. Replace battery or transformer.
COMPLAINT: Engine misses persistently, even when hot.		
1. Bad spark plug.	*f*. Test each plug until bad plug is located.	*f*. Pull plug, clean if dirty but good. Replace if plug is old.
2. Leaky gasket.	*g*. Check gaskets as noted previously.	*g*. Tighten up or replace with new gaskets.
3. Valves are worn or warped.	*h*. Open up and inspect. Look for bent stem.	*h*. Renew valves. Grind seats.
COMPLAINT: Knocking when engine is under load.		
1. Low-octane gasoline.	*i*. Compare octane with recommendation by engine maker.	*i*. Substitute a high-octane gasoline.

Table 14-5. Troubleshooter's Field Guide for Gasoline Englines (*Continued*)

Possible causes	How to check for cause	Recommended ways to correct
2. Cylinders are loaded with carbon.	*j*. Remove the cylinder heads and inspect.	*j*. Scrape and clean out the carbon deposits. Do not use chemicals unless engine maker approves.
3. Connecting rod bearings are loose or worn.	*k*. Drop crankcase and inspect. Check piston pins.	*k*. Renew bearings as necessary. Renew pins if needed.
4. Main crankshaft bearings are worn.	*l*. Knock is deep and solid. Drop crankcase bottom and inspect.	*l*. Replace the bearings.
5. Cylinders are worn and the pistons fit loosely.	*m*. Lift head and measure clearances between pistons and walls of cylinder.	*m*. Fit with new piston rings. Put in an oversize piston. Bore if cylinder is out-of-true.
6. Timing is out, probably too fast.	*n*. Check timing with proper timing as set by engine maker.	*n*. Reset the timing to correct values.
COMPLAINT: Engine runs much too hot.		
1. Water-cooling system is at fault.	*o*. Take temperature of water in and water out of engine. If rise is abnormally low, system cleaning is in order.	*o*. Clean system. Use an acid, if engine maker permits, to break out scale. Flush system under pressure until it runs clear.
2. Lubricating oil is old and contaminated.	*p*. Pull oil sample and analyze.	*p*. Drain and flush lubricating oil system. Fill with clean oil of proper viscosity.
3. Fan belt is slipping.	*q*. Check fan belt for tension.	*q*. Tighten up belt as necessary.

Table 14-5. Troubleshooter's Field Guide for Gasoline Engines (*Continued*)

Possible causes	How to check for cause	Recommended ways to check
4. Engine is air-cooled.	*r.* Measure volume and velocity of cooling air over engine fins.	*r.* Increase if possible. Lower temperature of air by air washer.
5. Cooling tower for engine water is not making its rating.	Refer to Secs. 9-7 and 10-3(f) for guidance in meeting this condition.	
COMPLAINT: Engine performance is not consistent. It sometimes shows a reduction in its power output.		
1. Air leak in the fuel line.	*s.* Examine the entire length of fuel line for leak. Check fuel pump for a leak there.	*s.* Eliminate the air leak.
2. Leaky exhaust valve.	*t.* Check the seating of every valve, both intake and exhaust.	*t.* Grind in the defective seating or replace with new valve as needed.
3. Piston clearances are excessive, due to worn rings.	*u.* Check for excess fuel in the crankcase. Take sample of oil and analyze. Measure clearance between piston and cylinder walls.	*u.* Install new rings. Rebore the cylinder block if deformation indicates this is needed.
4. Leaky cylinder head gasket.	*v.* Examine the tightness of each gasket.	*v.* Tighten up on gasket or renew if gasket is defective and split.
5. Air-intake filter is clogged.	*w.* Examine filter.	*w.* Clean or renew as appropriate.
6. Timing is wrong.	*x.* Compare present timing with timing as set by engine builder.	*x.* Correct the timing.

Table 14-5. Troubleshooter's Field Guide for Gasoline Engines (*Continued*)

Possible causes	How to check for cause	Recommended ways to correct
7. Gas is of nonuniform quality and characteristics.	*y*. Analyze a sample. Compare characteristics with gasoline recommended by engine builder.	*y*. Provide a better grade of gasoline.
8. Dirt in the carburetor.	*z*. Inspect carburetor and check for dirt. Check air filter as possible source of dirt.	*z*. Clean the carburetor. Clean the air filter. Make sure fuel line is clean.

14-9. GASOLINE ENGINES, GENERAL DISCUSSION.

Today, with one or more cars in every family, and a service station on every corner, and repair shops in plentiful supply, the troubleshooting of a gasoline engine is a mystery to very few individuals, at least to the extent of recognizing the existence of something wrong when they hear it.

How to classify the trouble and what to do about correcting it are less familiar. Actually, one or a group of symptoms can originate from more than one source, and the earlier the right source is identified, the more economical becomes the overall cost of the repair job. Downtime is annoying and an inconvenience when an automobile is involved, but downtime can be crucial if the engine's operation is essential to support an industrial process. For example, a shop lighting system depends on an engine-driven generator, or water level controls are maintained by an engine-driven pump, or the compressed air, provided by an engine-driven compressor, fails because of engine trouble and shuts down the plant.

Compressor troubles are examined in Division 13, and pumps in Division 15.

Gasoline engine troubles, because of general familiarity with these engines, are not examined in this section in detail but are summarized in the Troubleshooter's Field Guide, Table 14-5.

DIVISION 15

Pumps

15-1. GENERAL FACTORS AFFECTING PUMP PERFORMANCE

(a) ***Net Positive Suction Head (npsh).*** This is the term given to the absolute pressure available at the suction nozzle or opening of the pump. Npsh is usually expressed in feet *of the liquid being pumped.* If the liquid being pumped is of a different specific gravity than water (and most of them are) and the pressure is expressed in feet of water, the conversion consists of simply dividing the water head by the specific gravity of the liquid being pumped. Thus a water head of say 30 ft becomes a liquid head of approximately 33.4 ft for a liquid having a specific gravity of 0.9.

Calculating the net positive suction head involves these steps:

1. To the pressure existing on the surface of the liquid in its storage reservoir, pond, or tank (this is usually atmospheric) and expressed in feet, add the actual static head as measured from the surface of the water to the center line of the pump. Be careful to convert this static head into feet of liquid, using the specific gravity as explained above.

2. From the total head so determined, we must subtract the vapor pressure of the liquid, this vapor pressure being taken at the temperature of the liquid. Incidentally, the point at which vapor pressure equals atmospheric pressure, or where evaporation begins, is the boiling point of the liquid. The rate of evaporation varies inversely as the atmospheric pressure, meaning the evaporation declines as the pressure increases and evaporates faster if the atmospheric pressure decreases. This means that enough head must be provided to keep the liquid from boiling in the pump suction. This head too is expressed in feet of the liquid.

Another item to subtract is the friction loss in the suction piping, the pump entrance, and any changes in direction that may occur.

If we deduct these values of vapor pressure and friction from the total positive head, the result is our net positive suction head. And it better be big enough; otherwise the pump can suffer from cavitation, noise, lack of capacity, lack of discharge head, or complete failure to pump the liquid.

The nsph formula can be summarized from this equation:

$$\text{Available nsph} = H_a + S - H_{vp} - H_f$$

where H_a = atmospheric pressure or other pressure being exerted on surface of liquid being pumped

S = static head between surface of liquid and center line of pump

H_{vp} = vapor pressure of liquid

H_f = friction loss in suction piping, entrance to pump, etc.

As noted previously, all pressures and heads are to be given in feet of the liquid being pumped.

Vapor pressure corrections are important in the calculated performance of chemical pumps and in the pumping of high-temperature water, notably boiler-feed pumps. In troubleshooting chemical installations, the vapor pressure of the chemicals being pumped should be obtained as a first step by the troubleshooter. For hot water, reference to the standard steam tables of Keenan and Keyes at the saturation pressures and for the water temperature being pumped will make clear the suitability of the available net positive suction head, or its lack, under the conditions.

To illustrate, saturated steam pressure at 180 F stands at 7.51 psi or 15.291 in. of mercury, and water of 180 F will vaporize at this absolute pressure or less. At greater pressures the water will remain water, and providing these higher pressures is the duty of the npsh. Noise in a boiler-feed pump is frequently an indication of a suction temperature that is too high for the available npsh.

Just to overcome vapor pressure of that 180-F water will require about 16.62 ft of positive suction head, this figure determined by the formula

$$\begin{aligned}\text{Inches vacuum} &= 30 \text{ in. Hg} - 15.291 \text{ in. Hg} \\ &= 14.7 \text{ in. vacuum} \times 1.13 = 16.62 \text{ ft}\end{aligned}$$

The conversion of inches vacuum to feet is accomplished by multiplying the inches by the constant of 1.13. Absolute pressure, of course, is the invert of vacuum.

The problem is most critical when the surface of the liquid being pumped is at a level *below* the center line of the pump. In this circumstance the S in the formula is negative, becomes a static lift instead of a head, and is subtracted from H_a.

Since the atmospheric pressure is 14.7 psi, or approximately 34 ft, some npsh may be possible if the water is not too hot or the lift too great. For instance, if the distance from the center line of the pump down to the level of the water being pumped is 5 ft and this level remains reasonably constant, figures in the formula for the 180-F water would be:

$$34 - (5 + 16.62 + 5 \text{ friction}) = 34 - 26.62 = 7.38 \text{ feet npsh}$$

This might or might not be enough, depending on the design of the pump, its necessary discharge capacity, and its discharge head.

(b) *Centrifugal Pump Capacity and Head Laws.* For a quick estimate of basic changes possible in pump performance, here is pump performance related to speed:

1. Pump capacity in gpm varies directly as the rpm, or the variation directly reflects any change in the impeller diameter, expressed as a ratio of the original diameter. Thus, double the rpm, and the gpm likewise is doubled; halve the impeller diameter, and the gpm is halved.

2. The pump discharge head varies as the *square* of the rpm ratio. It also varies as the *square* of any impeller ratio of change. In formula this is

$$\text{Head}_b = \left(\frac{\text{rpm}_b}{\text{rpm}_a}\right)^2 \times H_a \qquad \text{or} \qquad \text{Head}_b = \left(\frac{\text{dia}_b}{\text{dia}_a}\right)^2 \times \text{dia}_a$$

3. Brake horsepower (BHP) varies as the *cube* of the ratio of change in either rpm or impeller diameter.

$$\text{BHP}_b = \left(\frac{\text{rpm}_b}{\text{rpm}_a}\right)^3 \times \text{BHP}_a \qquad \text{or} \qquad \text{BHP}_b = \left(\frac{\text{dia}_b}{\text{dia}_a}\right)^3 \times \text{BHP}$$

In each of the above formulas the subscript *a* equals the original, and subscript *b* equals the value after the change.

The troubleshooter might find in these laws a means of correcting a pump performance, particularly one in which over-pumping, excess head, or excessive power demands are creating unnecessary expenses of operation.

15-2. TYPES AND TROUBLES OF RECIPROCATING PUMPS

(a) *General Notes Concerning This Type of Pump.* Both reciprocating and rotary pumps are classed as positive displacement machinery, since a definite volume of the liquid being pumped is physically ejected from the cylinder by a piston or plunger in the case of the former and a meshing pair of gears or cams by the latter. In each arrangement, either the liquid goes out or the pump stops. This is notably different from the centrifugal pumps, a design that allows the impeller to rotate at full speed with the discharge valve closed and no delivery taking place. In this situation, the work of the centrifugal pump is expended in heating the liquid trapped within it.

The historical workhorse of the past century and the early years of this one is the steam-driven reciprocating pump, either in simple or duplex form. Relatively slow in speed, dependable, and not difficult to maintain, it occurs in both horizontal and vertical arrangements. Figure 15-1 shows a horizontal arrangement's internal parts.

Modifications include the replacement of the simple steam cylinder by a steam engine type of arrangement with a flywheel, usually coupled with duplex water cylinders. This design is obsolete but may still be found in some old industrial plants.

As with centrifugal pumps, the available net positive suction head must exceed the requirements of lift and vapor pressure if the pump is to do its job. The suction pipe should be at least as large as the suction opening on the pump and should be arranged to avoid the formation of any air pockets.

(b) ***Complaint.*** Pump is not delivering its rated capacity in gpm.

Causes

1. The steam pressure may be less than the pump is rated on.

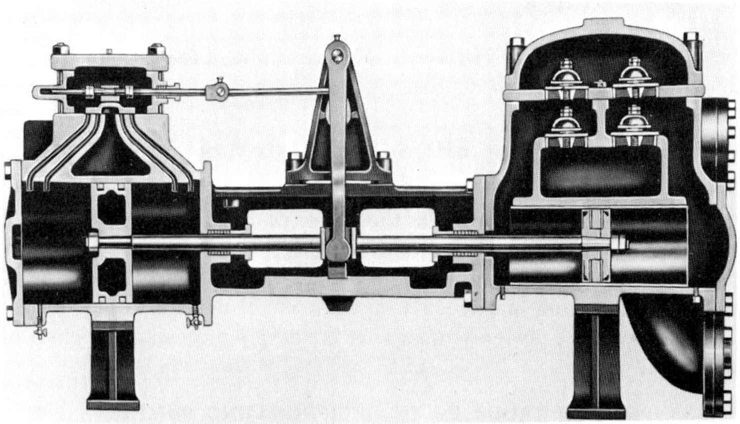

FIG. 15–1. Steam-driven reciprocating pump showing internal arrangement of parts. (*Warren Pumps, Inc.*)

2. Suction troubles include a clogged or too small foot valve or air in the suction line.

3. Suction lift is too great for the pump to realize its full capacity. In other words, there is insufficient net positive suction head.

4. Pump speed is too slow.

5. If pump has been operating for a considerable time, worn valves are leaking off capacity.

Recommended Actions

1. Check speed and steam pressure. Correct as necessary.

2. Measure suction lift in feet. Note temperature of liquid and possibility of vaporization in suction line and pump. Consider a lowering of liquid temperature, if this is found to be critical, or look into ways of establishing a higher npsh. Examine the practicality of decreasing the suction lift.

3. Check for air inleakage through suction or stuffing boxes. Open the air cock on the liquid cylinder and note whether air is present in the effluent. Locate air leak and correct it.

4. Verify size and condition of foot valve, if one is present.

(c) ***Complaint.*** Discharge pressure is low.

Causes

1. Steam pressure is lower tham pump is rated on.

2. High back pressure exists on the exhaust.

3. Worn steam rings keep piston from developing full power.

Recommended Actions

1. Check both throttle steam pressure and back pressure. If differential across the steam cylinder is too small, either step up the steam pressure or reduce the back pressure. *Caution:* Do not increase the throttle pressure above that recommended by the pump manufacturer. Reducing back pressure may involve work on the condenser, if any, or a change of process steam requirements if the exhaust feeds into a process steam header.

2. If the pump has been long years in service and the pressures are found satisfactory, open the valve chest and inspect the valves for conditions of wear that prevent effective performance.

3. Check condition of steam piston rings. Renew if necessary. Reboring of the cylinder may be required if it has worn out-of-true.

(d) ***Complaint.*** Pump vibrates.

Causes

1. The pump is misaligned on its foundation.

2. The foundation itself is not firm and rigid; for example, a pump set on a wooden floor that has not been reinforced and braced to carry it.

3. The packing around the liquid end piston rod is too tight, thus setting up a certain amount of binding and shaking.

4. Pipe pulsations are reflecting back to the pump.

Recommended Actions

1. Examine the position of the pump with respect to its supports. If rigidity of floor is doubtful, set a glass of water on the floor about 12 in. from the pump and note the formation of vibration waves. If these are well defined, look into ways of stiffening the floor.

2. Check the alignment of the steam and water ends with each other. Loosen foundation bolts and correct as necessary.

3. For vibration appearing in the piping connections, install a vibration-absorbing flexible connection between the piping and the pump.

4. Examine packings for binding. Ease up as necessary.

5. If a flywheel is included with the pumping arrangement, verify the balance of the flywheel.

(e) ***Complaint.*** Pump short-strokes.

Causes

1. There may be binding in the packing of the liquid end.

2. There is excessive steam cushioning in the steam end.

3. Steam valves are worn and leaking, particularly the exhaust valves.

4. Valve gear needs adjustment to improve the timing of the valve operating cycle. Evaluate the lost motion compensation method for an inadequate setting.

Recommended Actions

In turn check the possibility of binding, steam cushioning, worn valves, and lost motion arrangement. Correct the fault found by

1. Easing up on the packing.

2. Resetting the valve gear to bring the steam cushion back to normal.

3. Regrinding or renewing worn valves.

4. Adjusting the lost motion between the slide valve and the valve rod. This must be done by lifting the cover of the steam chest.

(f) ***Complaint.*** In order to get full stroke, the piston nut is striking the cylinder head.

Causes

1. Piston nut is of the incorrect size or has been reinforced by a lock nut.

2. Valve-gear adjustment is permitting too long a stroke.

Recommended Actions

1. This situation is dangerous. A ruptured cylinder head is possible. Stop the pump and examine the piston nut or nuts. If more than one nut, or if only one but of wrong size, remove and replace with the correct nut.

2. Adjust the valve gear to bring the stroke within the allowable lim-

its of piston travel. Refer to the pump manufacturer's service manual or to "Plant Operators' Manual," by S. M. Elonka.

(g) ***Complaint.*** Pump refuses to start. This complaint may indicate an emergency condition if the failure of the pump prevents some other essential action.

Causes

1. If pump has been out of service for a lengthy time, either the plunger in the water end or the piston on the steam end might be frozen in the cylinder.

2. Steam leakage through valves in steam chest is preventing full steam pressure from reaching the steam piston.

3. One of the valves has become inoperative. Possibly a disk has become detached from the stem.

Recommended Actions

1. If a frozen piston or plunger is suspected, bar over the pump by hand, noting whether there is excessive resistance. If pump cannot be moved this way, open one or both ends and free up.

2. With steam throttle open, examine exhaust connection for heat and evidence of a stuck and leaky exhaust valve. Remove cover on valve chest and repair or readjust valve as necessary.

3. Check liquid end for a defective valve. Repair or renew as found needed.

Caution: When barring over the pump, be sure that steam throttle valve is closed. Never try this with the valve open. A sudden power movement against the bar under steam pressure could seriously injure persons adjacent to the bar.

(h) ***Complaint.*** The pump becomes vapor-bound.

Causes

1. Liquid temperature in suction is above normal.

2. Suction lift is too high, and npsh is too low.

Recommended Actions

1. If the vapor-binding condition is exceptional, it can be relieved in moments by sluicing with cold water the liquid end of the pump. This condenses the vapor.

2. If the vapor binding is chronic, refer to Sec. 15-2(b) for suggestions.

(i) ***Complaint.*** Occasionally the pump races without increasing the gpm delivery.

Causes

1. There is a stuck or broken liquid valve.

2. The pump is working on inleaking air.

Recommended Actions

1. Stop the pump, open the liquid end, and look for the bad valve.

If found, replace. Since trouble is intermittent, trouble may be in a bent stem or other deformation.

2. Note suction temperature. If it varies from the normal, the variation could be resulting from an influx of air, either up or down, depending on the temperature of the air leaking into the pump or piping. Open air cock for evidence of air in the liquid being pumped.

(j) ***Complaint.*** Pump gets frozen up on overnight shutdowns in winter.

Cause

After shutting down for the night, the pump is not being drained.

Recommended Actions

1. Before leaving the pump at night, make sure that all cylinder drains have been opened on both the steam and liquid ends.

2. If pump must operate intermittently during the night, provide heating for it during its periods of shutdown. This could be accomplished with a gas-fired radiant heater or an electric heater, either of which could be so controlled as to be activated by the pump shutdown and cut out when the pump starts.

3. If there is danger of the pumped liquid freezing, either during the pumping or between pumpings, provide heat for it by coils in the suction tank or electric tracers on the suction piping as appropriate for conditions.

Table 15-1. Troubleshooter's Field Guide for Reciprocating Pumps

Possible causes	How to check for cause	Recommended ways to correct
COMPLAINT: Pump does not deliver its rated capacity in gpm.		
REFERENCE: Sec. 15-2(b).		
1. Steam pressure is too low.	*a.* Take pressure reading at throttle.	*a.* Step up pressure to design value.
2. Back pressure in exhaust is too high.	*b.* Note exhaust pressure. Compare with design back pressure..	*b.* Improve condenser performance. Reduce pressure requirements on process mains.
3. Defective foot valve or valve is too small.	*c.* Inspect valve. Compare size with suction pipe.	*c.* Clean or replace with larger valve.
4. Insufficient npsh.	*d.* Calculate available npsh. Compare with required npsh as asked by pump manufacturer.	*d.* Increase the suction head. Refer to Sec. 15-2(b) recommended actions.

Table 15-1. Troubleshooter's Field Guide for Reciprocating Pumps (*Continued*)

Possible cause	How to check for cause	Recommended ways to correct
5. Pump speed is too slow.	*e*. Compare with designed speed.	*e*. Adjust valve gear to increase speed.
COMPLAINT: Discharge head is too low. REFERENCE: Sec. 15-2(c).		
1. Steam pressure is too low.	*f*. Check throttle pressure.	*f*. See the recommended actions in Sec. 15-2(c).
2. Back pressure is too high.	*g*. Check exhaust pressure.	*g*. See the recommended actions in Sec. 15-2(c).
3. Worn piston rings in steam end are losing power.	*h*. Open up steam end note conditions and clearances.	*h*. Renew rings. Rebore cylinder or install new liner if needed.
COMPLAINT: Pump vibrates. REFERENCE: Sec. 15-2(d).		
1. Pump is misaligned.	*i*. Loosen foundation bolts. Note if pump takes any different position. Examine packing in stuffing boxes for uneven wear.	*i*. Realign if this is the trouble.
2. Foundation is not firm or rigid.	*j*. Examine floor construction for stiffening. Note extent of floor vibration by means of a glass of water.	*j*. Reinforce floor to needed stiffness.
3. Pipe pulsations are reflecting back to pump.	*k*. Test for vibrations on the pump discharge about 2 ft from pump.	*k*. Install vibration-absorbing connections between piping and pump.
4. Flywheel is out of balance.	*l*. Stop pump. Verify balancing weights, if these are used. Try a free turn of the wheel, noting heavy arc.	*l*. Rebalance the wheel, resetting weights as necessary.

Table 15-1. Troubleshooter's Field Guide for Reciprocating Pumps (*Continued*)

Probable causes	How to check for cause	Recommended ways to correct
COMPLAINT: Pump short-strokes.		
REFERENCE: Sec. 15-2(e).		
1. Binding in a stuffing box at the liquid end.	*m.* Inspect packing for uneven wear.	*m.* Repack the box. Evaluate for overpacking or misalignment. Ease packing, or realign as found necessary.
2. Too much steam cushion in steam cylinder.	*n.* Shut down pump and bar over to determine where piston stops, relative to cylinder head.	*n.* Adjust valves to suit the proper stroke.
COMPLAINT: Stroke is too long, with piston nut striking cylinder head.		
REFERENCE: Sec. 15-2(f).		
1. Valve gear setting is incorrect.	*o.* Bar over pump and determine what correct stroke should be.	*o.* Adjust gears to manufacturer's manual.
2. Piston nuts have been changed and incorrect sizes used.	*p.* Inspect the piston. Take nut dimensions. Compare with manufacturer's parts list.	*p.* Correct nuts as found necessary. Adjust stroke to suit correct conditions.
COMPLAINT: Pump refuses to start.		
REFERENCE: Sec. 15-2(g).		
1. Moving internals have become "frozen" as result long nonoperating period.	*q.* Bar over the pump to note excessive resistance, as indicator of frozen condition.	*q.* Open pump and clean. Open and clean steam end, if liquid end cleaning does not release the piston.
2. A valve has become inoperative.	*r.* Examine liquid end for defective valve.	*r.* Install new valve.
3. Steam valve is stuck.	*s.* Lift valve chest cover and inspect valves.	*s.* Free up valve, repair or replace as necessary.

Table 15-1. Troubleshooter's Field Guide for Reciprocating Pumps (*Continued*)

Probable causes	How to check for cause	Recommended ways to correct
COMPLAINT: Pump is vapor-bound.		
REFERENCE: Sec. 15-2(h).		
1. Liquid temperature is higher than usually handled by the pump.	*t*. Verify high temperature by comparing reading with normal.	*t*. Sluice liquid end of pump with cold water to condense vapor. Correct excess temperature cause.
2. Npsh is inadequate.	*u*. Vapor binding is chronic if this is case. Calculate available npsh. Compare with pump requirements.	*u*. Increase npsh if possible. Lower temperature of liquid. Reduce lift.
COMPLAINT: Pump races without increasing gpm output.		
REFERENCE: Sec. 15-2(i).		
1. Air is leaking into suction.	*v*. Test for air. Open cock on liquid end. Note air in effluent. Check packing for leaks.	*v*. Correct air leaks as found.
2. A liquid valve is stuck or broken.	*w*. Remove valve chest cover and inspect valves.	*w*. Replace bad valve.
COMPLAINT: Pump becomes frozen in winter when shut down overnight.		
REFERENCE: Sec. 15-2(j).		
1. Pump cylinders are not drained after shutdown.	*x*. Check temperature of liquid. This might need some heating.	*x*. Provide spot heating for the pump and for the liquid being pumped.

15-3. TYPES AND TROUBLES OF ROTARY PUMPS

(a) ***Discussion.*** Rotary pumps generally are found installed in systems of medium-to-small capacity and in ones calling for a high discharge pressure. The various designs employ gears, cams, and sliding blades or vanes that get behind the liquid being pumped and force it out through the discharge nozzle. Figure 15-2 is a sketch of a gear arrangement.

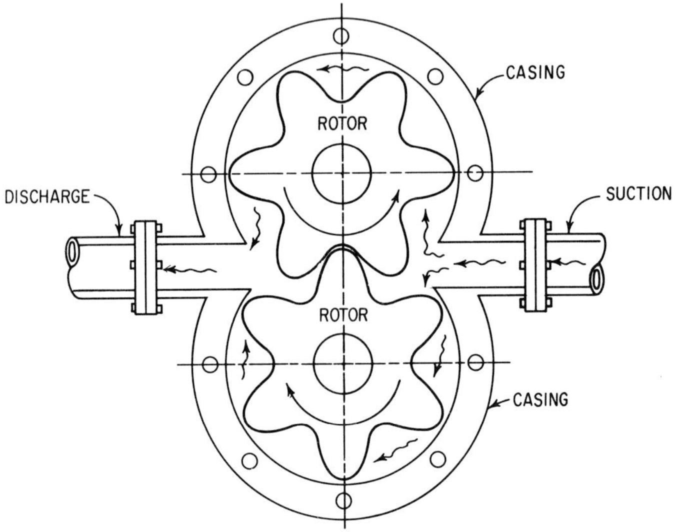

FIG. 15–2. Rotary gear pump.

Rotary pumps may be classed also as "power pumps," meaning that their operation requires some form of driver. Electric motors are common choices, but steam engines and internal-combustion engines will be encountered from time to time. The internal-combustion engine is a favored driver for field construction projects where the availability of steam or electricity may be limited.

Depending on conditions, the pump may be direct-connected to its driver or driven through a belt or gear train. Figure 15-3 shows a belt-driven rotary pump.

Because of the possible accumulation of air in a rotary pump, good arrangement calls for at least a short vertical discharge pipe leaving the pump before the flow enters a horizontal run. Air pockets are countered in this manner. In some instances, the mounting of an air vent valve, either automatic or manually operated, will be useful if placed at the top of the vertical pipe.

Troubleshooters facing rotary pump malfunctions will remember that pump misbehavior actually can be a reflection of driver difficulties. If their shooting stops with the pump, they well may be stopping too soon.

(b) ***Complaint.*** During cold weather, the circuit breaker on the motor service frequently kicks out. Material being pumped is an oil.

Cause

Cold weather increases the viscosity of the oil and consequently increases the load on the driving motor.

Recommended Actions

1. Examine the possibilities of heating the oil by means of coils in the oil tank or by a steam or electric tracer on the suction piping.

2. Protect the pump casing against potential rupture from overpressure by means of a pressure relief valve. This valve may be mounted in a bypass from the pump discharge to the suction side of the pump.

(c) ***Complaint.*** The pump has stopped and refuses to start.

Causes

1. Pump shaft may have developed a bend which is jamming the gears.

2. Stuffing boxes are too tight, with a resultant binding.

3. Shifting of the foundations has produced a misalignment.

4. If pump is fitted with an excess-pressure governor, the governor may be taking charge.

5. Electrical trouble such as low voltage may be present.

Recommended Actions

1. Verify the voltage at the motor terminals. Measure the resistance across the motor for indications of shorting or other motor defect. Replace motor if found defective.

2. Inspect stuffing boxes for binding. Slack off on packing.

3. Check alignment by means of a straight edge laid across the tops of the couplings while the couplings are rotated one-quarter turn at a time.

FIG. 15–3. Belt-driven form of rotary pump. (*The Kraissl Company.*)

If wedges have been used in the original connection, examine them for possible resetting. Do this preferably at operating temperature.

4. Inspect holding-down bolts and slack off or pull up as may be found necessary to correct alignment.

5. With pump disconnected from its driver, rotate the pump manually and note whether there is any excessive degree of resistance. If there is, the pump must be disassembled, the rotors removed, and both shaft and rotors evaluated for trueness.

(d) ***Complaint.*** Pump operates, but the draw of electric power is excessive.

Causes

1. Any of the first three troubles named in Sec. 15-3(c) can impose an abnormal load on the driver.

2. Pump is running too fast.

3. Viscosity of the liquid being pumped is greater than was expected.

Recommended Actions

1. Refer to recommended actions in Sec. 15-3(c).

2. Reduce the speed by changing the ratio between the driver and the pump. If driver is an electric motor, consider replacement by one of lower speed.

3. Compare viscosity of liquid with that on which the selection of the pump was made originally.

(e) ***Complaint.*** Pump delivers either no liquid or a mere dribble, compared with its rating.

Causes

1. Pump is losing its prime by reason of air leaks on the suction.

2. Lift is so high as to be on the critical point, and therefore it is subject to very small changes of temperature or condition of the pumped liquid.

3. Pump is running too slow.

4. Suction pipe is not sufficiently submerged. Two or three feet should be provided when the level in the suction well is at its lowest.

5. Foot valve or strainer is clogged.

Recommended Actions

1. Take vacuum reading on suction to determine if air is leaking into the pipe. If this is the case, inspect all joints and the pipe submergence to identify the point of leakage and correct it.

2. If lift is more than 12 ft, or less for liquids of high viscosity, calculate the net positive suction head and compare this with the needs of the pump. If npsh is insufficient, consider reducing the lift and find a way to increase the static pressure head.

3. Count the rpm of the pump and compare with the design speed. Increase the speed if this shows it to be too low.

4. Examine foot valve, if any, for obstructions to liquid entry. If no

valve is being used, measure the submergence of the suction pipe. Free up the foot valve and strainer or increase the submergence, or both, to relieve the conditions found.

(f) ***Complaint.*** Pump operation is extremely noisy.

Causes

1. Pump is operating on a reduced flow. Liquid is slugging.
2. Liquid vapors are forming bubbles that collapse in the pump, this collapse being accompanied by crackling sounds and minor explosions.
3. Air is leaking into the suction and producing crackling.

Recommended Actions

1. Investigate the reason pump is operating on reduced flow; for example, a partially closed valve, an obstruction in the suction or discharge line, a dirty strainer or clogged foot valve, or a restricted flow into the tank or basin from which the pump is taking its suction. When the cause is identified, correct it. Note the size of the suction pipe. If it is smaller than the pump inlet nozzle, it is too small and should be replaced by a pipe of the proper size.
2. Vaporization condition can be determined by taking the temperature of the liquid entering the pump and also by noting the existing vacuum in the suction. Relief to the vaporizing may require cooling of the liquid, the reducing of the suction lift, or the imposing of an increased static head on the suction side of the pump.
3. Examine for air and, if inleakage is found, reseal the joints or other openings where it is occurring.

Table 15-2. Troubleshooter's Field Guide for Rotary Pumps

Possible causes	How to check for cause	Recommended ways to correct
COMPLAINT: In the winter season, the pump motor circuit breaker kicks out frequently.		
REFERENCE: Sec. 15-3(b).		
1. Increased viscosity of liquid being pumped from cold weather increases load beyond capacity of breaker.	*a.* Sample the liquid and check viscosity with design liquid.	*a.* Heat liquid by coils in tank or by steam or electric tracer on suction pipe.
COMPLAINT: Pump has stopped and refuses to start.		
REFERENCE: Sec. 15-3(c).		
1. Bend in pump shaft is causing binding in rotors.	*b.* Turn shaft manually and with dial indicator check for a bow in shaft.	*b.* Remove rotors and either true up shaft or replace.

Table 15-2. Troubleshooter's Field Guide for Rotary Pumps (*Cont.*)

Possible cause	How to check for cause	Recommended ways to correct
2. Stuffing box packing is too tight.	*c.* Slack off on packing and check ease of rotation, manually.	*c.* Leave packing in slackened state.
3. Misalignment of pump and driver from shift in foundation.	*d.* Test alignment by means of straight edge over top of couplings, while rotating shaft.	*d.* Correct alignment by trueing up foundation bolts or by shims in couplings.
4. Excess pressure governor is taking charge.	*e.* Note pressure at which governor is set. Compare with pump discharge needed.	*e.* Reset the governor or reduce the operating pump pressure.
5. Low voltage.	*f.* Check voltage.	*f.* Refer to source of electric service. Reduce load other than for this pump.
COMPLAINT: Pump uses an excessive amount of electric power. REFERENCE: Sec. 15-3(d).		
1. Refer to the first three causes in Sec. 15-3(c) and follow recommended actions.		
2. Viscosity of liquid being pumped is greater than pump is designed to handle.	*g.* Pull sample of liquid. Compare viscosity with that on which pump is rated.	*g.* Change viscosity or rerate the pump. If motor is overloaded, install larger motor.
3. Pump is operating too fast.	*h.* Compare present speed with designed full-load speed.	*h.* Adjust speed ratio between driver and pump. This may involve gears or a belt drive. Can job accept smaller capacity?

Table 15-2. Troubleshooter's Field Guide for Rotary Pumps (*Cont.*)

Possible cause	How to check for cause	Recommended ways to correct
COMPLAINT: Pump discharge varies from a dribble to none. REFERENCE: Sec. 15-3(e).		
1. Pump is losing prime due to air leak.	*i.* Take vacuum reading on suction. Lower than normal suggests air leak.	*i.* Check all joints, and make tight. Check submergence. Increase if air is entering through foot valve.
2. Suction npsh is too small for temperature of liquid.	*j.* Calculate the npsh and compare with what is available and what is needed by pump.	*j.* Increase static head if possible. Reduce suction lift. Reduce temperature of liquid.
3. Pump is running below design speed.	*k.* Take rpm and compare with design rpm. Look for binding in the stuffing boxes.	*k.* Increase speed by changes to ratio of driver to pump. Slack off on packing.
4. Foot valve is clogged, strainer is dirty, or both.	*l.* Inspect both foot valve and strainer.	*l.* Clean both foot valve and strainer.
COMPLAINT: Pump noise is annoying. REFERENCE: Sec. 15-3(f).		
1. Pump is running less than capacity and liquid is slugging in pump.	*m.* Inspect system for a partially closed valve, a partially obstructed suction or discharge, or a fouled foot valve or strainer.	*m.* Remove obstruction. Clean system as found necessary.
2. Liquid is vaporizing in suction or in pump.	*n.* Listen to pump for internal crackles or minor explosions. This is evidence of bubble collapse and vaporization.	*n.* Correct suction conditions by an increase of npsh, a lowering of suction temperature.

Table 15-2. Troubleshooter's Field Guide for Rotary Pumps (*Cont.*)

Possible cause	How to check for cause	Recommended ways to correct
3. There is an air leak on the suction.	*a*. Take vacuum reading on suction. An abnormally low vacuum is evidence of air leakage.	*a*. Inspect suction system joints, valves and submergence. Correct leaks where found.

15-4. TROUBLES OF CENTRIFUGAL PUMPS

(a) ***Discussion.*** The impeller of a centrifugal pump imparts sufficient rotative speed to the liquid that enters it to discharge this liquid in whatever volume and at whatever pressure may be desired.

The design of the impeller varies in order to satisfy various conditions. Basically, it may accept the liquid through an annular opening at its center on one side only. This design is termed "single suction." When the opening is on both sides of the impeller, we call it a "double suction."

Both single-suction and double-suction pumps are available as single-stage or multistage arrangements, each impeller constituting a stage. Since the stages are connected in series, with the discharge of the first stage leading into the suction opening of the second stage and so on, multistage pumps are highly useful in stepping up the ultimate discharge pressure for a constant volume of liquid. Each impeller pushes up the liquid to a pressure higher than that at which it was received.

In some designs, one or more stages of a multistage pump are arranged back-to-back, in this way balancing the axial thrust that is given to the shaft by the jet-like discharge of each impeller. Without this balancing, it is often necessary to equip the pump with a thrust bearing on the rotor shaft.

To suit the conditions of operation, centrifugal pumps come in horizontal, vertical, and deep-well, or submerged, forms. They may be mated with various forms of drivers, such as electric motors, steam turbines, gas turbines, diesel engines, and steam engines. The transmission of power to the pump may be direct or by reduction gear or belt.

They also take the in-the-line design of the "canned pump" illustrated by Fig. 15-4. The canned pumps are encountered most often in chemical activities, including such services as the circulation of sodium in certain nuclear reactor systems. Incidentally, the troubleshooter faces no difficulty in locating a sodium leak; sodium ignites spontaneously on contact with the air.

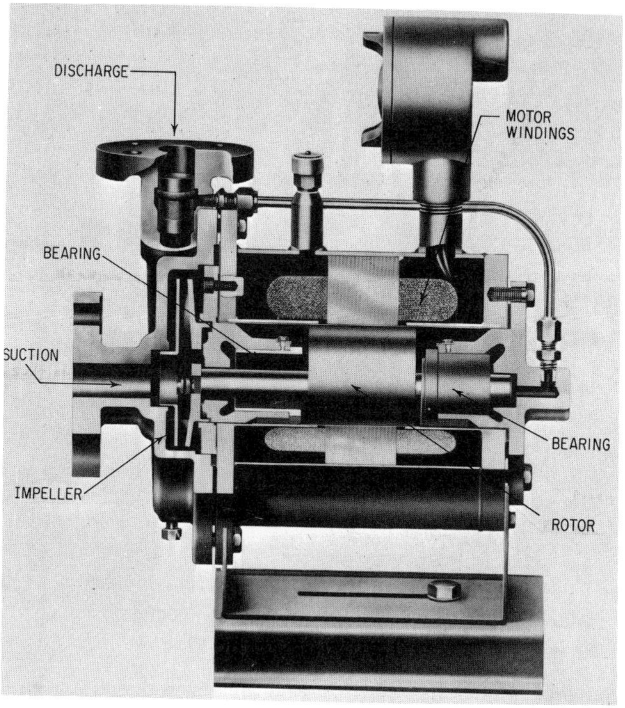

FIG. 15–4. Internal arrangement of a "canned" pump. (*Crane Company*.)

Impellers have their particular styles, too. The closed impeller, usually observed on clean water or clean liquid systems, is built with its vanes shrouded on both sides. This construction creates a succession of enclosed liquid corridors extending from the center of the impeller to the periphery, and the liquid being pumped is ejected by the impeller through these passages. The result is a highly efficient handling of the liquid.

Open impellers omit the shroud plate on one or both sides, depending on how much solid material must pass through the pump. This design is usually preferred for the handling of sewage, some sand, and slurries. In order to withstand this sort of service for long, low-maintenance periods, the pump is rugged and gives up some efficiency as compared with the enclosed impeller.

Large pumps sometimes introduce priming problems. A steam jet ejector, mounted on the pump casing, can be used to evacuate the air from the pump and its suction and allow the liquid to enter. In other

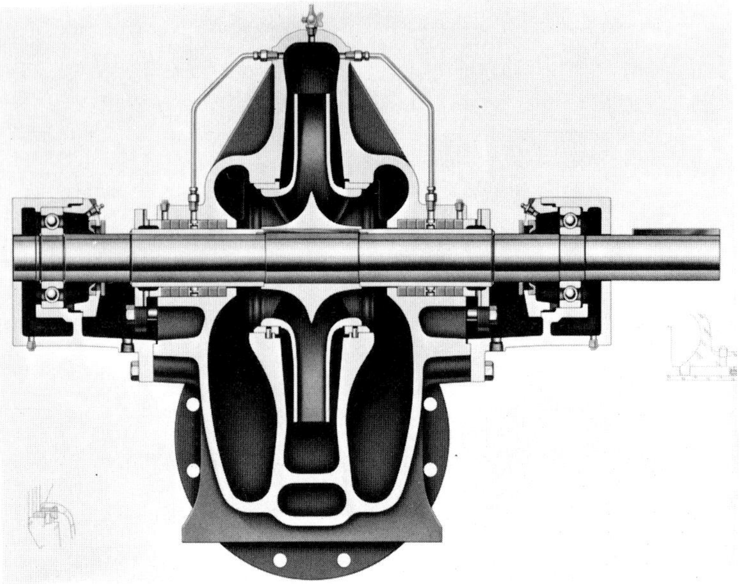

FIG. 15–5. Single-stage, double-suction centrifugal pump. (*Allis-Chalmers.*)

arrangements, a foot valve allows liquid to be introduced into the pump and then build up a liquid column right down to the foot valve. Smaller pumps generally are self-priming.

(b) *Complaint.* Pump is delivering only about one-third of its rated capacity.

Causes

1. Rotor is turning backward.
2. Suction piping contains an air leak.
3. Pump is running below its designed speed.
4. Required discharge head is higher than the pump can provide, because of an undersize discharge pipe or a partially open valve.
5. There is insufficient npsh.
6. Worn wearing rings are permitting churning and leakage.
7. Impeller is partially plugged or defective.
8. Suction pipe is not sufficiently submerged.

Recommended Actions

1. Compare the direction of rotation on the driving motor or turbine with the designed rotation of the pump, as marked by an arrow on the pump casing. Disconnect the pump and establish the correct direction of rotation. Since procedures differ, and in some cases the rotation

cannot be changed, do not disassemble the rotor or make any modification except under the exact directions of the pump builder.

2. Check the suction pipe for an air leak by means of a vacuum gage. If presence of a leak is indicated, test all joints and submergence until found and then correct.

3. Examine size and configuration of the discharge piping. Make sure that all valves are *wide* open. Consider reducing friction by replacing the discharge pipe with a pipe at least one full size larger.

4. Verify the established speed of the driver and compare this with its current speed. For motor drives, check against low voltage. For turbine drivers, make sure the low speed is or is not the result of a governor malfunction. Take steps to correct the low voltage or to reset the governor of a turbine.

5. Calculate the npsh and note temperature of liquid. If together they are the cause of the low delivery, consider modifying the liquid temperatures, increasing the static head, or both. The latter expedient could involve relocating the pump to a lower level with respect to the suction lift.

6. Vibration and a drop in efficiency, singly or together, may indicate a condition of worn wearing rings on the impeller or worn bearings, or both. Inspect all bearings and rebabbitt or replace with new roller or ball bearings as the case may call for. For horizontally split casing, lift the top half and inspect the impeller wearing rings. Replace if necessary.

7. Lift top half of casing and inspect impeller for clogging or defects or foreign material passing through the pump. Remove impeller, clean, repair, and balance before reinstalling it in the pump.

8. Measure submergence of suction and evaluate the foot valve for size. Make sure it holds the prime, repairing it if repairs are indicated.

(c) ***Complaint.*** Discharge pressure is lower than wanted.

Causes

1. Pump speed is less than design calls for.
2. Npsh is lower than the installation needs.
3. The impeller is too small for the required operating conditions.

Recommended Actions

1. Follow instructions given in recommended action no. 4 of Sec. 15-4(b).

2. Refer to recommended action no. 5 of Sec. 15-4(b).

3. If a too small impeller is suspected, communicate with pump builder. The head varies as the square of the ratio of the impeller diameters. Thus if the diameter of the existing impeller is increased by 50 percent, the discharge head theoretically will be increased $2\frac{1}{4}$ times.

(**d**) ***Complaint.*** Pump vibrates excessively.

Cause

The pump rotor shaft is bent as a result of thermal distortion, a misalignment, or an impeller plug that throws the impeller out of balance.

Recommended Actions

1. By means of a dial indicator and a manual rotation of the shaft, determine whether a bend or bow actually exists. Deflection of indicator should not exceed 0.0002 in. or as stipulated by pump builder.

2. To correct a thermal distortion, the rotor must be removed from the pump and heated for straightening. Do not attempt this except under specific directions of the pump builder.

3. Check the possibility of misalignment by rotating the pump manually, while aligning the coupling tops with a straight edge. Correct by resetting the foundation bolts, shimming, etc., to establish a true alignment.

4. If the results of recommended actions 1 through 3 above turn out negative, a plugged impeller is reasonably certain. Lift top half of casing and pull rotor. Inspect, clean, repair, and balance before reinsertion in the pump.

(**e**) ***Complaint.*** After delivering its full-rated capacity for some time, the gpm falls off rapidly although the pump speed continues as before.

Causes

1. There is probably an obstruction in the suction or discharge piping. This obstruction has taken its position since the pump went on the line.

2. A horizontal run of suction piping might be creating an air pocket that in time becomes large enough to interdict the effective operation of the pump.

3. Air leaks are entering pump through stuffing box.

Recommended Actions

1. Open pipe joints at pump and probe for stoppages.

2. If horizontal suction pipe is found, check level of the horizontal run with bubble level. Correct pitch of pipe to make sure any air in the pipe will travel upward with the liquid.

3. Generally a trickle of liquid is desirable through a stuffing box to control temperatures and reduce frictional wear. If no fluid is leaking from the stuffing box, slack off on the packing to produce a limited flow. New packing may be required if the stuffing box remains dry persistently.

(**f**) ***Complaint.*** Pump operation is excessively noisy.

Causes

1. Cavitation in the impeller is accompanied by the formation of

bubbles which collapse with minor explosions in the pressure stages.

2. The pump is churning, meaning that the volume of liquid is much too small to suit the operating characteristics of the pump.

3. High fluid velocities in undersized pipelines create sounds that may seem to be originating in the pump but are not.

4. Noise could be in the motor rather than the pump.

Recommended Actions

1. Throttle the discharge valve on the pump down to from half to three-fourths flow, and note whether noise stops. A stoppage from throttling is definite evidence of a cavitation condition and an inadequate npsh. Correct the condition, if possible, by lowering the pump with respect to its suction or imposing additional static head on the suction. If neither expedient is feasible, a reduction of the liquid temperature might be considered.

2. Investigate the reason that liquid flow handled by the pump is less than the pump capacity. Consider replacing the pump by one of proper size or installing a smaller impeller.

3. Calculate the velocity of the liquid in the discharge piping. Because of the noise it generates, a high velocity may be unsuitable in quiet environments but acceptable in a commercial or industrial area. Where quiet is essential, the velocity should not exceed 4 or 5 ft per second.

4. Listen to the motor, as well as to the pump, for noises. Motor fan noise could be the culprit, and sometimes the bearings are to blame. Disconnect the motor and run it separately, if necessary, to establish its guilt or innocence.

(g) ***Complaint.*** Every time the pump is started, the start is followed by a widespread water hammer throughout the system. This is annoying and at times so violent as to frighten persons in the area.

Cause

This is the result of initiating a sudden and powerful movement of the liquid in the discharge piping.

Recommended Action

Start the pump against a closed discharge valve and then open the valve slowly until a full flow is established. This will eliminate the hammer. Stopping the pump, reverse the procedure, the discharge valve being closed slowly before the pump is stopped.

(h) ***Complaint.*** Replacement of packing is much too frequent, causing undesirable expense and inconvenience.

Causes

1. The type of packing is unsuited to the type of stuffing box in use.

2. Sealing cage is in wrong position or is without the sealing liquid because of a stoppage of the liquid line to the stuffing box.

3. The rotor shaft is bent or bowed and is placing uneven stress on the stuffing box.

4. Shaft is off-center because of a misalignment.

5. Packing is too tight.

6. Rotor is out of balance and is vibrating.

Recommended Actions

Packing failures can point to serious conditions elsewhere in the pump, such as a bowed shaft, an out-of-balance rotor, or a considerable misalignment. Look for the source of the trouble in this order:

1. Verify the suitability of the packing in use. Change it if its specifications differ from those of the pump builder. Slack off.

2. Verify the position of the sealing cage and the flow of sealing liquid to it. Correct a faulty condition as necessary.

3. With a straight edge laid over the top of the couplings, rotate the pump shaft by hand and note any indication of a misalignment or bowed shaft. Correct a misalignment as suggested in Sec. 15-4(d). Do the same for a bowed rotor.

4. Disconnect the pump from its driver and turn it over slowly by hand to determine a condition of imbalance. Lift rotor and correct the balance as recommended by pump builder. In some instances, the lack of balance will be found to be resulting from foreign material wedged into the impeller. If this can be removed in the field and the impeller is found to be undamaged, lifting the impeller from its bearings may not be required.

(i) ***Complaint.*** Although handling unheated liquids, the pump always runs hot and at times seizes and stops.

Causes

1. Pump is operating way down on the capacity curve, and consequently it churns the liquid until a condition of overheating is reached. This can result from an insufficient npsh.

2. Rotor is rubbing against a diaphragm, diffuser ring (if pump uses them), or the casing at the wearing rings.

3. The rotor shaft is misaligned.

4. Excessive thrust has developed as the result of a failure in the balancing device (if pump has one), or the hydraulic balance of the impellers has been thrown out by the wedging of foreign material in one of the vanes.

Recommended Actions

1. Compare pump discharge in gpm with the design curve and the two delivery heads. If gpm is on the unstable and inefficient part of the curve, investigate ways to increase the volume of liquid to be handled. Also consider the replacement of the pump with one of more suitable capacity, if the trouble is chronic.

2. Turn over the pump rotor manually and slowly. Listen and feel for any evidence of contact friction. Lift the casing and inspect the rotor. Look for marks of rubbing on the rotor. Paint suspected areas with Prussian blue; then rotate the shaft and impellers. Rubbing areas will then show up clearly. Reposition the shaft to the correct axial clearances.

3. Misalignment of the rotor shaft can be found by laying a straight edge over the tops of the couplings while the rotor is turned through its full 360 deg. Correct alignment as may be found necessary by retensioning the foundation bolts or shimming up base plates.

4. Excessive thrust can be found by measuring the axial clearance with feelers and by examining the rubbing surfaces. Correct this by reconditioning the thrust bearing, replacing it, or removing foreign material from the balancing impeller.

5. If pump must operate over a wide range of capacities, from the lower curve to the top of its head-capacity characteristic, install a circulating pipe from the pump discharge back to the pump inlet. By means of this pipe, the pump can operate with full impellers at all times, even at periods of partial load.

Table 15-3. Troubleshooter's Field Guide for Centrifugal Pumps

Possible causes	How to check for cause	Recommended ways to correct
COMPLAINT: Pump delivery is a small fraction of its rated gpm.		
REFERENCE: Sec. 15-4(b).		
1. Rotor is turning backward.	*a.* Compare the arrow-marked direction of pump rotation with motor rotation.	*a.* Follow the pump builder's instructions. Some pumps cannot be reversed.
2. Air leak in suction.	*b.* Check with vacuum gage on suction. Inspect the pipe and stuffing boxes.	*b.* Seal up leaky joints. Tighten up on packing.
3. Pump is running below its design speed.	*c.* For motor drives, use tachometer and check for low voltage. For steam turbines, check throttle pressure and exhaust pressure. Also verify governor function.	*c.* Act to correct low voltage. Step up steam pressure or decrease exhaust pressure. Reset governor if faulty.

Table 15-3. Troubleshooter's Field Guide for Centrifugal Pumps (*Continued*)

Possible cause	How to check for cause	Recommended ways to correct
4. The available npsh is too low.	*d.* Measure vacuum with gage. Compare with npsh needed.	*d.* Increase static head or lower suction lift.
5. Insufficient submergence of suction pipe.	*e.* Measure submergence in feet. It must be at least 4 times inside diameter of pipe and not less than 2 ft for small pipes.	*e.* Increase submergence as necessary.
6. Impeller may be partially clogged or defective.	*f.* If discharge restriction is recent, suspect a clogged impeller. If delivery under rating from start, look for a defective impeller. Note vibration.	*f.* Lift top half of casing and check impeller. Remove and clean or repair as needed.
7. Wearing rings are worn and need replacement.	*g.* Note head developed with the existing suction. Compare with design values, shown on head-capacity curve.	*g.* Lift impeller and replace rings.

COMPLAINT: Discharge head is below its rating on the head-capacity curve.
REFERENCE: Sec. 15-4(c).

1. Pump speed is lower than design speed.	*h.* Refer to Sec. 15-4(b).	
2. Net positive suction head is too low to meet pump requirements.	*i.* Take vacuum reading on suction. Note temperature of liquid. Calculate npsh and compare with required.	*i.* Add static head. Lower the pump. Reduce the temperature of pumped liquid.
3. Impeller is too small.	*j.* Take rpm by tachometer. Take capacity in gpm at measured speed. Check	*j.* Communicate with pump builder. Follow his instructions.

Table 15-3. Troubleshooter's Field Guide for Centrifugal Pumps (*Continued*)

Possible cause	How to check for cause	Recommended ways to correct
	against head-capacity curve. If capacity up and head down, suspect an undersize impeller.	
COMPLAINT: Vibration of pump is excessive. REFERENCE: Sec. 15-4(d).		
1. Rotor shaft is bent from thermal distortion.	*k*. Set dial indicator on shaft. Rotate shaft slowly. If deflection exceeds 0.0002 in. or as stipulated by builder, shaft is distorted.	*k*. Remove impeller rotor. Follow builder's instructions in correcting bend.
2. Distortion is from pump misalignment.	*l*. Check driver-pump alignment by straight edge over coupling while slowly rotating shaft.	*l*. Correct by resetting foundation bolts, shims, or bedplate.
3. Impeller is out of balance.	*m*. If no bend or alignment distortion is found, lift pump casing and turn impeller for inspection of vanes.	*m*. Remove materials causing out-of-balance. Repair and true up as necessary.
COMPLAINT: Pump tends to lose capacity after operating at rating for several hours. REFERENCE: Sec. 15-4(e).		
1. Air leaks through stuffing box.	*n*. Note whether liquid is trickling out through packing. If dry, leak is likely.	*n*. Repack to produce trickle of liquid.
2. An obstruction may be present in either the suction or the discharge piping.	*o*. Look for a partly closed valve. Open a pipe joint and probe for foreign objects or materials.	*o*. Make sure all valves are wide open. Remove obstruction.

Table 15-3. Troubleshooter's Field Guide for Centrifugal Pumps (*Continued*)

Possible cause	How to check for cause	Recommended ways to correct
3. An air pocket is in the suction line.	*p*. Look for a horizontal run of suction piping. With a level, see if horizontal run pitches up or down.	*p*. Repitch the horizontal run toward pump inlet nozzle.
COMPLAINT: Pump is noisy in operation. REFERENCE: Sec. 15-4(f).		
1. Cavitation.	*q*. Test-throttle the discharge valve. If noise stops, cavitation most likely is there. Check npsh.	*q*. Increase static head. Lower pump; reduce liquid temperature.
2. Pump is operating low on its capacity curve.	*r*. Investigate reason pump liquid feed is less than pump is designed for.	*r*. Step up volume through pump. Consider smaller pump or a smaller impeller.
3. Noise is from high velocity in discharge piping.	*s*. Calculate gpm velocity through the existing size piping.	*s*. Increase pipe size.
4. Noise is actually in driver.	*t*. Listen with stethoscope to motor or turbine drive.	*t*. Eliminate the condition. Install flexible pipe connections in pump discharge.
COMPLAINT: Whenever pump is started, the start is followed by heavy water hammer in the discharge piping. REFERENCE: Sec. 15-4(g).		
1. Start is too rapid.	*u*. Start pump against a closed or nearly closed discharge valve. Then open valve slowly. If no hammer, this is the reason.	*u*. Establish a policy of slow starting and slow stopping.

Table 15-3. Troubleshooter's Field Guide for Centrifugal Pumps (*Continued*)

Possible cause	How to check for cause	Recommended ways to correct
COMPLAINT: Pump packing wears out with abnormal rapidity. REFERENCE: Sec. 15-4(h).		
1. Type of packing or material is unsuited to this duty in this particular pump.	*v*. Compare characteristics of packing with the builder's selection of packing.	*v*. Replace with proper packing.
2. Lubricating trickle of liquid to stuffing box is stopped or restricted.	*w*. Examine packing for degree of dryness.	*w*. Free up fluid flow to stuffing box.
3. Rotor shaft is bowed.	*x*. Check with dial indicator as given in Sec. 15-4(d). Note whether there is excessive vibration.	*x*. True up the shaft. Follow instructions of pump builder.
4. Shaft is off-center from a misalignment.	*y*. Test for alignment with straight edge on couplings.	*y*. Eliminate the misalignment by resetting foundation bolts, shims, etc.
5. Packing is too tight.	*z*. Note temperature of stuffing box. Is liquid trickle present?	
6. Rotor is out of balance.	*aa*. Turn slowly by hand. Tendency to oscillate indicates an unbalanced rotor.	*aa*. Lift pump casing. Locate point of unbalance and correct; it may need shop repairs.
COMPLAINT: Pump always runs warm although liquid pumped is cool. At times pump seizes, and further operation must await completion of a period of cooling. REFERENCE: Sec. 15-4(i).		
1. Pump is operating low on capacity	*bb*. Check pump discharge gpm versus	*bb*. Install a smaller pump or a smaller

Table 15-3. Troubleshooter's Field Guide for Centrifugal Pumps (*Continued*)

Possible cause	How to check for cause	Recommended ways to correct
curve, and so it churns and overheats.	the design gpm at normal speed.	impeller. Talk with pump builder. Put in a circulating pipe.
2. Pump is misaligned.	*cc.* Check with a straight edge laid on the coupling while slowly rotated.	*cc.* Correct misalignment as suggested in Sec. 15-4(h).
3. Rotor is in rubbing contact with other pump internals.	*dd.* Listen with stethoscope for rubbing sounds with rotor rotating slowly.	*dd.* Lift upper half of casing. Inspect clearances of diaphragms, wearing rings, etc. Reset rotor to correct clearances.
4. Heat is generated from an excessive thrust.	*ee.* Inspect thrust adjustments. Compare with design clearance at thrust bearing. Check the balancing drum or other device. Note whether any impeller is clogged or damaged.	*ee.* Reestablish proper clearances by the marked shaft adjustments. Repair thrust device if found defective.
COMPLAINT: Pump bearings have an uneconomical, short life.		
REFERENCE: Sec. 15-4(j).		
1. Bearing lubrication is insufficient.	*ff.* Record the bearing temperatures. Note rise in lubricating oil temperature across the bearing.	*ff.* Increase the oil flow. Clean out the oil reservoir and piping to bearing.
2. Oil cooling system is not cooling effectively.	*gg.* If water-cooled, note temperature of water in and water out. Compare with design values.	*gg.* Clean water system. Circulate a suitable chemical cleaning compound, as 50% muriatic solution. Increase flow or find another, cooler source.

Table 15-3. Troubleshooter's Field Guide for Centrifugal Pumps (*Continued*)

Possible cause	How to check for cause	Recommended ways to correct
3. Foreign materials are in bearings.	*hh.* Inspect for such materials. Locate point and manner of entry.	*hh.* Clean out the bearings. Clean lubricating oil strainer. Replace contaminated oil.
4. Misalignment of the shaft puts uneven stress on bearings.	*ii.* Check for alignment as described in Sec. 15-4(h).	*ii.* Correct as per Sec. 15-4(h).
5. Shaft is bowed or bent.	*jj.* Check as described in Sec. 15-4(h).	*jj.* Correct as per Sec. 15-4(h).

(j) ***Complaint.*** Pump bearings do not last; they must be replaced frequently.

Causes

1. Lubrication is faulty.
2. If bearing has a cooling system, this system could be defective in design or operation.
3. Dirt or other foreign material is getting into bearings.
4. Bearings are rusting because of a leakage into the bearing of cooling water.
5. Shaft is bowed or out of alignment.

Recommended Actions

1. Take sample of lubricating oil and analyze for viscosity and other characteristics. Compare characteristics with those recommended by pump manufacturer. Change oil if this appears advisable.
2. Prove out the performance of the bearing cooling system. Make sure the water supply piping and the drainage are normal and clear of obstructions. Replace piping as necessary.
3. Presence of dirt will show up in the analysis of the lubricating oil sample if this sample is taken from the pump bearings. A sample from the oil storage tank probably would not be significant unless the tank itself were badly contaminated.
4. Inspect bearings for rusting. If observed, examine all parts to locate point of cooling water entry to bearing. Correct this.
5. Check for shaft bowing or misalignment as suggested in Sec. 15-4(i). If found, correct as necessary.

15-5. TROUBLES WITH SHALLOW-WELL PUMPS

(a) ***Discussion.*** Shallow-well pumps may be of the positive displacement, centrifugal, or jet type. Whatever the form, the application is limited by the ability of the pump to handle a suction lift, meaning that when operating it creates sufficient vacuum in the suction pipe to allow the atmospheric pressure on the water in the well to push up this water into the pump. In the trade, a rule of thumb says to use a deep-well technique whenever the level of the water in the well is more than 25 ft below the ground. Since in most cases starting the pump lowers the level, depending on how fast the natural inflow to the well can make up for the draw, the 25-ft rule should be regarded critically; perhaps 20 ft would be safer. In any event, a trial pumping to determine the well flow prior to selecting the pump capacity should be considered as essential. If the pump draws out the water faster than the water table can put it in, the whole water system is in trouble. This writer has witnessed a reportedly inexhaustible spring pumped dry in 9 minutes at a rate of 450 gpm.

Should the troubleshooter be without a pump of sufficient capacity to test-pump a well, it is sometimes possible to borrow the services of the nearest fire station pumper. These trucks are set up for pumping out wells, springs, and rivers.

For troubleshooting reciprocating-type pumps, refer to the applicable situations in Sec. 15-2; for rotary pumps, Sec. 15-3; and for centrifugal pumps, Sec. 15-4.

(b) ***Jet Pumps for Shallow Wells.*** The operating principle of the jet pump for shallow wells is this: part of the discharge under pressure of a centrifugal pump is diverted into a diffuser tube, which is a form of venturi, and thereby creates a vacuum in the throat of this tube by the simple process of changing the discharge pressure into velocity. This vacuum causes the water to rise in the suction pipe from the water in the well, and, together, the lifted water and the diverted water from the pump discharge enter the eye of the impeller and are thus pressurized to repeat the cycle. The total discharge of this jet pump is therefore greater than the useful discharge, since a part of it must be diverted to maintain the supply to the pump.

The diffuser and the nozzle that injects water under pressure into the diffuser are housed in the pump casing for shallow wells, the whole thing being constructed as a unit. In this way it differs from jet pumps used for deep wells, the latter operating with the nozzle and diffuser submerged at all times in the water of the well. The shallow-well jet needs only one pipe, the suction pipe, to dip down into the well; the deep-well installation needs a tail pipe on the diffuser discharge mouth as well as the driving pipe through which the diverted water descends to the submerged nozzle.

The troubles with a shallow-well jet pump, other than mechanical or electrical as examined in Sec. 15-4, involve malperformance of the water system as a whole and will be treated in Sec. 15-7 and Table 15-4.

15-6. TROUBLES WITH DEEP-WELL PUMPS

(a) ***Discussion.*** Deep-well pumps are encountered in the forms of *jet pumps* in which the motor and pump are above ground with the nozzle-diffuser assembly submerged in the water of the well and *submersible pumps* with the whole unit, impellers, motor, and bearings, permanently located below the water level in the well.

The former is illustrated by the sketch in Fig. 15-6; the latter is shown in the photograph in Fig. 15-7. Water enters the illustrated submersible pump through the screened openings in the middle of the unit, a point between the impellers and the driving motor.

The submersible pump is more efficient than the jet pump, the entire discharge of the submersible being available for useful applications. With the same horsepower motors and the same discharge pressures, the submersible will deliver upward of 50 percent more water than the jet.

Construction of the submersible usually incorporates multiple im-

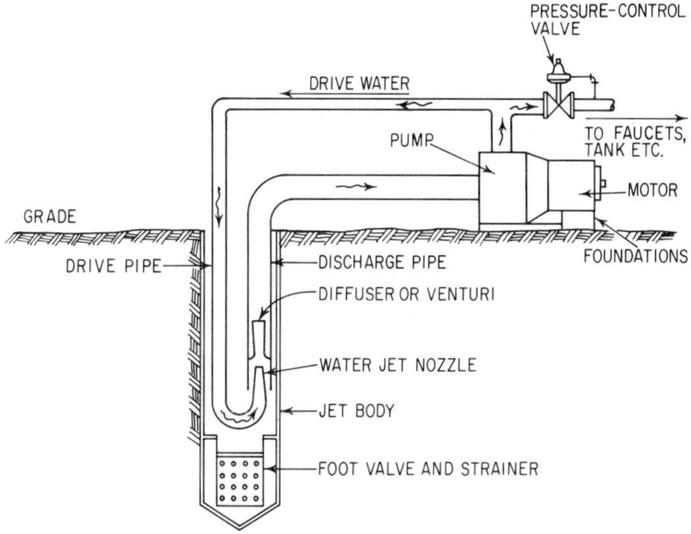

Fig. 15–6. Deep-well jet pump (no scale).

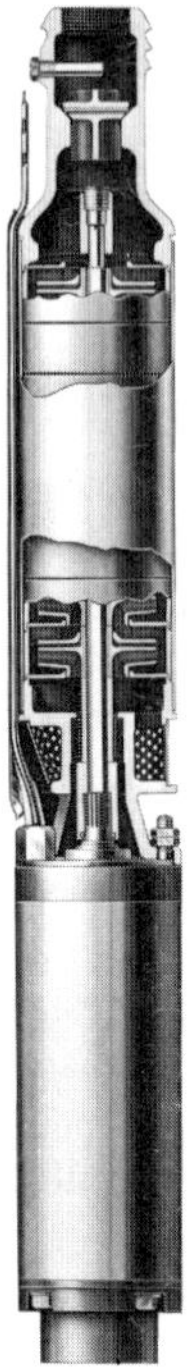

FIG. 15-7. Submersible form of well pump with motor in bottom compartment and inlet strainer between motor and impeller sections above. (*Goulds Pumps, Inc.*)

pellers, or stages, with flow-directing diaphragms (or diffusers) between impellers. The discharge pipe at ground level is check-valved to prevent a back flow from the water system when the pump stops.

The troubleshooter is advised to check the materials of construction as published by the manufacturer. This can be important, for example, if the impellers are of plastic and the troubleshooter is prepared only for metal replacements.

Standard pump sizes are designed to fit 4- and 6-in. wells, with discharges from 35 to 350 gpm. Well depths to receive them run from 300 to 1,000 ft. Motor horsepowers go up to 5, depending on output and discharge head.

(b) *Vertical Pumps.* The two best known of the vertical-pump family are the *turbine* design, suitable for either shallow- or deep-well operation, and the *sump pump* design.

In the turbine pump, each stage consists of a suction bowl and an impeller. Any reasonable number of stages is possible, but all the impellers are attached to a common shaft, and any kind of bend or sway in this shaft can have most undesirable consequences. Frequent practice is to guide the shaft within bearings spaced at 10-ft intervals.

The intake is at the very bottom of the pump through a strainer. Driving motor and all controls are mounted above ground and clear of the water.

Turbine pumps are larger than the submersibles, the off-the-shelf units starting at 4 in. and going on up to suit the requirements of the job.

Lubrication of the shaft bearings, either water or oil, can offer a problem, the latter being carried down to the bearings by a pipe designed for this purpose.

Turbine pumps may be used at depths beyond 100 ft; consequently, a straight and true drive shaft is essential.

Sump pumps commonly operate on sump pits of shallow depth, with the pump impeller submerged at the bottom of a short shaft while the motor is mounted above the pit. The pump is started and stopped by a float switch. Sump pumps may be centrifugal, turbine, or rotary in design.

(c) *Water System Components.* In addition to the pump, the usual water system includes a pressure tank for storing the water between pump-operating periods, plus a pressure switch that starts and stops the pump at designated pressures. Pressures commonly range from 20 to 60 psig.

The tank should be sized to balance the capacity of the pump. When the draw pulls down the level in the tank and so reduces the pressure leaving the tank that the pressure switch starts the pump, the pump should be able to meet the continuing demand. A sign of a too small pump is the loss of water to one or more of the faucets that happen to be in simultaneous operation. Alternatively, the flow of water may not stop entirely but decrease to a trickle, depending on the relationship between the pump output and the simultaneous demand.

In critical installations, with the pump output and the demand teetering between enough and not enough, the situation may be helped by elevating the shutoff pressure of the pump. For example, if the original setting is 40 psig, by changing this to 60 psig, the pump stores more water in the tank than at the 40-psig limit, and thus the tank and pump together may be able to meet higher demands for a short time.

Air cushioning in the pressure tank is gradually lost by absorption of air by the contacting water, and if not replaced, this results in increased operation of the pump and a drop of pressure to the faucets. Replenishment of the air is accomplished automatically by various devices such as an aspirator for jet pumps, both shallow and deep well, and a combination check and snifter valve for submersibles. Figure 15-8 shows a Brady aspirator.

(d) ***Plunger Pumps.*** Figure 15-9 illustrates the principles of this design, which is sometimes called a suction rod pump. In operation, the withdrawal of the plunger from the suction chamber fills the cham-

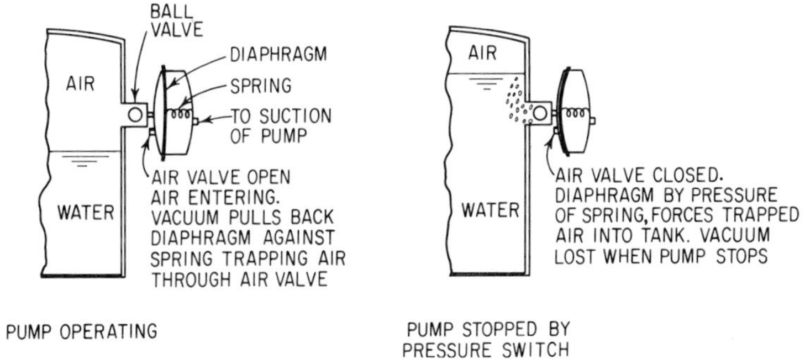

FIG. 15–8. Brady aspirator (no scale). *Note:* Brady Air Controls, Inc. recently has developed a tank for use on water systems with a factory-applied mineral-oil blanket to separate the water from the air. This oil layer prevents absorption of the air by the water.

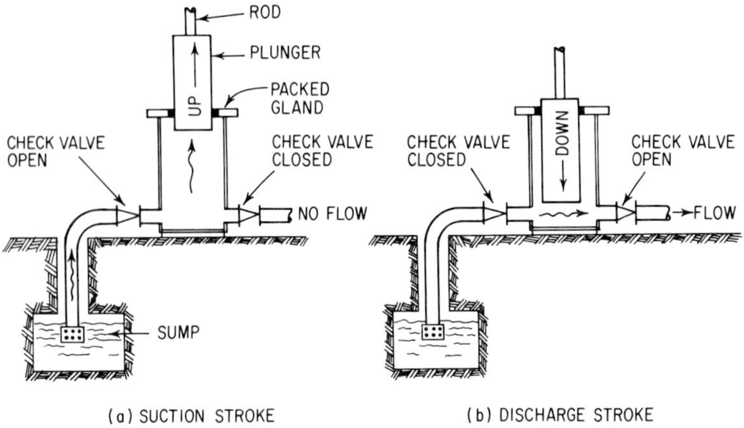

FIG. 15–9. Single-action plunger pump (no scale). *Note:* Pump is available in horizontal form also.

ber with water through the bottom admission valve. The return stroke of the plunger into the chamber creates enough pressure to close the admission valve and open the discharge valve, the water leaving the pump through the latter and being thus forced into the piping system.

This pump depends on the development of a partial vacuum in the suction pipe to establish the flow into the pump; consequently it is sensitive to the need for a net positive suction head in order to perform satisfactorily. Since this type of pump is often used for pumping oil, the formation of oil vapors in the suction, particularly if the oil is warm, can be a source of trouble.

Other than this, the pump is subject to the usual mechanical or electrical troubles of the driving head and the occasional fouling of the suction strainer. Failure of the valves to seat properly, due to corrosion or abrasion from foreign materials, also results in a reduced or stopped discharge.

(e) ***Complaint.*** Delivery of pump is diminished, interrupted, or stopped.

Recommended Actions

1. After checking driving machinery as to voltage, steam pressure, and such, take a sample of discharge for presence of excessive mud, debris, or other valve-fouling materials.

2. Check level of water or oil in the well. If so low as to affect suction, stop the pump, wait for the water to rise to its normal level, and then restart the pump with a partly closed discharge line valve. This throttles the discharge down to the inflow of the well.

3. Dismantle pump and examine the admission and discharge valves and their seats. Check tightness of their closing. Replace or regrind the valves and seats as found necessary. If springs are in poor condition, replace them with springs of proper tension.

4. For evidence of excessive corrosion, consider replacement of the fluid-contacting parts with materials more corrosion-resistant. Confer with the manufacturer of the pump regarding this.

(f) ***Air Lift Pumping.*** Figure 15-10 shows an air lift installation in a deep well. If an air compressor must be provided at the well, it could represent a relatively expensive installation. If the compressed air is available from an already existing system, investment costs could be relatively lower than for some other form of pumping.

Pipe corrosion and fluctuations or interruptions in the pressure of the compressed air are the most likely troubles with this type of pumping.

(g) ***Complaint.*** Delivery of water is curtailed or stopped.

Causes

1. Air pressure has dropped below needed level.
2. Debris, mud, or other foreign material has choked the delivery pipe.

Recommended Actions

1. Verify air pressure.
2. Check level of water in well. The trouble could be a falling off of the well inflow from the surrounding earth. If so, wait for well to revive and then limit discharge to the measured inflow.
3. Pull the air pipe and clean out the delivery pipe.

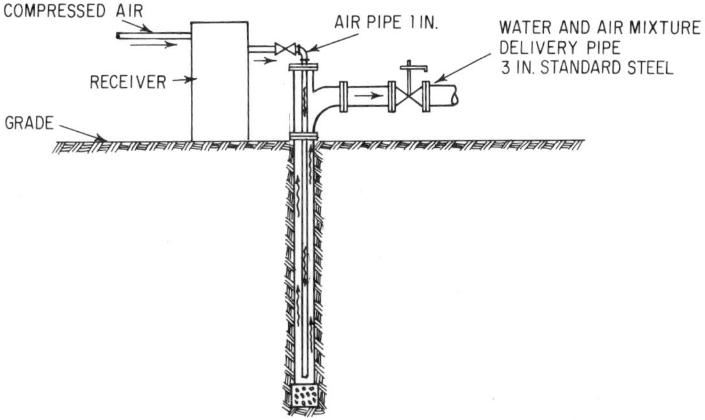

FIG. 15–10. Simple air lift.

15-7. COMMON TROUBLES WITH WELLS AND WATER SYSTEMS

(a) *Shallow-well Pumps and Systems.* A common complaint concerns the system pressure. For example, with a lawn sprinkler in action, water to second-story faucets or even first-floor in some instances becomes a mere trickle, coughs air and water in a mixture, or discharges nothing.

Reference to a pump head-capacity curve for this particular installation is likely to show that the demands in gpm of the sprinkler are so great that the head for such a large discharge is way down on the curve. There simply is not enough head available under the sprinkling conditions. Or in plain words, the pump is too small for the simultaneous demands made on it.

The trouble may be relieved for short periods by changing the shutoff pressure of the pump from 40 psig (which is a common setting) to 60 psig. This change puts more water in the storage tank with enough head to back it up for short intervals.

Permanent cure means either the replacement of the pump with a larger one (and this might create new trouble by pumping the well dry too fast) or a rescheduling of water use by the people in the building. If the occupants refrain from shower bathing while the lawn sprinkler is in operation, or vice versa, the overall performance of the system might be considered satisfactory. In any event, the ultimate capacity of the system is limited by the flow rate of the well or spring that is supplying the water.

Another trouble is the development of a waterlogged condition of the pressure tank. Air compressed in the tank by the pump's input of water is for the purpose of providing adequate circulating pressure on the water in the system.

Disappearance of this air by absorption in the water, or any significant reduction in its volume, removes or diminishes the circulating force and produces a state of unsatisfactory flow from the faucets.

Most water systems are fitted with automatic air replacement devices; therefore, a loss of air pressure points to a possible malperformance of the device in service.

The troubleshooter, finding no air in the tank, will note first the kind of air replacer on the system and check its operating condition. The Brady aspirator in Fig. 15-8 requires the development of a vacuum, usually provided by a connection to the pump suction, in order to trap air from the surrounding atmosphere and inject it into the tank. About 3 in. of vacuum is the minimum needed, and by putting a vacuum gage on the pump suction, we can observe whether the pump is pulling that much. If not, there is probably a leakage in the pump suction line; if there is 3 in. or more, the aspirator is the area of the trouble. Or possibly, the tube connecting the aspirator to the pump suction is leaking so that adequate vacuum at the pump is being lost in transit.

In summary, any inadequacy of vacuum at the air intake valve calls for a check of the suction piping for leaks by the troubleshooter. He should be prepared to make this check when answering a service call.

(b) *Deep-well Pumps and Systems.* Turbine pumps reported to have lost their rated volume of discharge may be suffering from a clogged strainer, mud or debris accumulated in the bowls or impellers, or a warped shaft. These are below-ground troubles; above ground, the driving motor, controls, or line valves may be failing to perform as designed. The troubleshooter should therefore consider the following possibilities:

1. Electrical troubles such as low voltage, short-circuiting may exist. Refer to Division 16 and the appropriate type of motor for an evaluation of the symptoms.
2. Verify correct operation of all controls.
3. Listen to the discharge pipe with a stethoscope for sounds of rubbing or striking down in the well. Such sounds indicate a warped or bent shaft, a bad bearing, or a combination of these. Pump must be stopped and shaft pulled for correction. In some instances, the noise may be so definite as to be audible without the stethoscope.
4. Determine normal well water level and compare with observed

Table 15-4. Troubleshooter's Field Guide for Well Pumps and Water Systems

Possible causes	How to check for cause	Recommended ways to correct
COMPLAINT: There is insufficient pressure on the system. More than two faucets cannot be turned on at the same time. REFERENCE: Sec. 15-7(a).		
1. Pump is too small.	*a.* Calculate maximum demand in gpm, and compare with capacity-head curve of the pump.	*a.* Install a larger pump. For temporary relief, increase tank pressure.
2. The pressure tank is waterlogged.	*b.* Test tank for air.	*b.* Check air intake device and repair if necessary.

Table 15-4. Troubleshooter's Field Guide for Well Pumps and Water Systems (*Continued*)

Possible cause	How to check for cause	Recommended ways to correct
3. Poor suction vacuum.	*c.* Verify vacuum at suction inlet to pump. If low, examine suction for leaks.	*c.* Correct leaks in pump suction or vacuum connection to tank.
COMPLAINT: Deep-well pumps are failing to meet their rated capacity. REFERENCE: Sec. 15-7(b).		
1. Strainer is clogged.	*d.* Check pump rpm.	*d.* Withdraw pump and clean or replace strainer. If practical, try vacuum cleaning well with extension hose. Drag out floating debris with hooks on wire.
2. The pump shaft is warped or bent. Bad bearing.	*e.* Check rpm for slowdown from rubbing. Listen for abnormal sounds with stethoscope.	*e.* Pull pump and correct defect.
3. Well is overpumped.	*f.* Stop pump. Wait for 1 hour, start, and check discharge.	*f.* Throttle discharge to well capacity.
COMPLAINT: Submersible deep-well pump is noisy and fails to meet rated capacity. REFERENCE: Sec. 15-7(c).		
1. Foreign material, sand, etc., going through impellers has abraded them.	*g.* Check pump rpm and compare with design speed. Check pressure and compare with design pressure at this speed. Listen for unusual sounds and examine a sample of discharge for foreign materials.	*g.* Pull pump, if evidence indicates leaky impeller. Install new parts or complete rotor as found necessary.

Table 15-4. Troubleshooter's Field Guide for Well Pumps and Water Systems (*Continued*)

Possible cause	How to check for cause	Recommended ways to correct
COMPLAINT: Delivery from plunger type of pump has become erratic. REFERENCE: Sec. 15-6(d).		
1. Water level in well has fallen below normal.	*h*. Stop pump. Wait at least an hour, then start and measure head and gpm at discharge. Gage well for level check if possible.	*h*. Throttle pump to balance rate of inflow to well.
2. Valves have become scored and leak.	*i*. Examine valves.	*i*. Replace damaged valves or regrind, if this will remove trouble.
COMPLAINT: Delivery from an air lift is below normal. REFERENCE: Sec. 15-6(e).		
1. Air pressure is too low.	*j*. Check air pressure, against rated pressure.	*j*. Increase air pressure by whatever means are available.
2. Mud and debris are in nozzle or delivery pipe.	*k*. Probe delivery pipe with wire. Sample delivery water for mud and silt.	*k*. Fish up obstructions manually if possible. Otherwise pull pipe and clean.
3. Water level in well is too low.	*l*. Gage water through delivery pipe. Let well rest for an hour, then measure delivery in gpm.	*l*. Throttle discharge to equal inflow of well.

level. Overpumping, apparent in a subnormal water level that is steadily falling, requires a throttling of the pump discharge to bring it within the flow rate of the well.

(c) ***Submersible Pumps.*** Since these are entirely inside the well, they must be pulled above ground for verification of any physical defects as evidenced by abnormal noise or vibration. These are centrifugal pumps; consequently, debris and abrasive materials going through the impellers and diffusers can damage them and thus reduce the pump volume of discharge.

Since all the probable troubles will be found down in the well, the troubleshooter is obliged almost invariably to pull the pump. Otherwise repairs are not practical.

The general troubles of *plunger pumps* and *air lift pumping* are discussed in Sec. 15-6(d) and (e).

DIVISION 16

Electric Driving Motors

16-1. TYPES AND TROUBLES OF ELECTRIC MOTORS DRIVING MECHANICAL EQUIPMENT

(a) *The Troubleshooting Approach.* For the purposes of the mechanical troubleshooter, the electrical factor in a complaint involves primarily the determination of whether the trouble lies in the driven mechanical machine, the driving motor, or both. An example of the "both" category is observed in cases of excessive vibration. If the vibration is intolerable in the mechanical machine, probabilities are good that the motor is suffering damage also. Such vibrations in a squirrel-cage induction motor are known to break or loosen the bars in the rotor or to separate bars from the end rings.

Testing and the correction of motor troubles are the work and responsibility of the electrical expert and his shop. However, for the mechanical fieldman who wants to go into electrical troubles, a good guidebook is Philip T. Green's "Electrical Testing and Troubleshooting."

For the electrical troubleshooter, Green recommends the following equipment for field testing:

four No. 6 dry cells
test panel with a bell set
dc volt-ohmmeter
0-150-300-600 ac voltmeter
pocket megger
magnetic compass
wiggen voltage tester
VR-75-30 regulator tube
phasc sequence indicator
test leads

This entire packet of testing items can be stowed and carried in a box only 7 × 11 × 22½ in. for a total weight of 35 lb.

Electrical troubles may be suspected in the presence of any or several of the following conditions:

1. The machine refuses to start.
2. Noise and vibration are excessive.
3. Motor is running excessively hot and may be smoking.
4. Closing the starting switch kicks out the breaker or blows a fuse.
5. Motor does not come up to speed.
6. On dc motors, there is sparking at the brushes.

The simplest field check of the motor, after determining that normal voltage and current values are present, is to disconnect it from the driven machine and start it. If the trouble symptoms persist, the electrical troubleshooter should take over.

Causes of electrical trouble vary somewhat with the type of motor involved; that is, whether it is a squirrel-cage induction motor, a wound rotor form, a dc unit, or a synchronous motor.

As general information for the mechanical troubleshooter, each of these types will be examined and its more obvious troubles pointed out in the appropriate field guides, Tables 16-2, 16-3, and 16-4.

(b) *Motor Lubrication.* The mechanical troubleshooter, observing evidence of inadequate lubrication of a driving motor, should not attempt to regrease or reoil the ailing motor unless he knows how. There are sleeve bearings and ball bearings, some with sealed oiling facilities, with or without pressure fittings. Each has its own technique.

Sealed bearings normally are good for 2 years of operation before a regreasing is needed. This is a common arrangement with fractional horsepower motors, and the motor must be opened and repacked. Single-phase and fractional horsepower motors are sometimes fitted with grease gun fittings and relief valves. These designs must be greased only when stopped, and the proper operation of the relief valve in order to admit the new grease should be verified.

Table 16-1 shows the relationship between motor sizes and times between greasings, as recommended by General Electric Company.

A good grease should not increase in hardness significantly either when standing in containers or when in the motor bearings. Furthermore, it should not separate into oil and soap nor corrode metal surfaces and should operate satisfactorily at temperatures as low as −25 F and as high as 175 F.

Oil for the lubrication of gear motor units that are splash-oiled or use an integrally built lubricating oil pump, call for petroleum oils of pre-

Table 16-1. Motor Greasing

	Motor Horsepower		
Level of service	Up to 7½	10 to 40	50 to 150
Occasional	7 years	5 years	3 years
Average	5 years	3 years	1 year
Hard	3 years	1 year	½ year
Brutal	½ year	¼ year	¼ year

mium quality that, in addition, are noncorrosive, defoam readily, and will operate at whatever low temperatures will be experienced.

If the motor is using waste-packed sleeve bearings, it should be re oiled every 1,000 hours of operation. Dates of recent oilings should be obtained by the troubleshooter.

Caution: Never attempt to introduce oil into the bearings while the motor is running. Neglect of this caution can result in oil getting into

Table 16-2. Troubleshooter's Field Guide for Induction Motors

Probable causes	What to do
COMPLAINT: Squirrel-cage type of induction motor will not start.	
1. Overload has tripped out the circuit breaker or blown fuses.	*a.* Refer to the appropriate mechanical chart to discover causes of overload. Reduce load to a level that can be handled.
2. Low voltage.	*b.* Check nameplate values for power supply. Look for ways to bring up the voltage to the required level.
3. Open circuit or short circuit in motor windings.	*c.* If verified by field test, you will probably have to send motor back to shop.
4. Grease is stiff in bearings from cold or other cause.	*d.* Clean lubricating system. Regrease with grease of proper specification.
5. Loose terminal-lead connection.	*e.* Check this, and if found, tighten up the connections.

Table 16-2. Troubleshooter's Field Guide for Induction Motors (*Continued*)

Probable causes	What to do
COMPLAINT: Motor is exceedingly noisy.	
1. Motor is running single-phase.	*f.* Stop the motor. If running single-phase it will not restart. Check for an open circuit.
2. Shaft running out-of-true in a sleeve bearing.	*g.* Check the alignment of the shaft. Also look for excessive end play and correct.
3. Vibrating as result of a rotor being out of balance, or vibration is from driven machine.	*h.* Disconnect the motor from the driven machine. If noisy or vibrating when running alone, take to shop and repair or rebalance rotor. Look for broken or loose bars. Measure extent of vibration in driven machine. Act on this as directed in appropriate sections.
4. Air gap is not uniform.	*i.* See that rotor is properly centered. Put in new bearings if necessary to get correct centering.
5. Motor is driving through a loose coupling.	*j.* Examine coupling. Check alignment of coupling by means of feelers with the coupling slowly rotating. Tighten bolts as necessary.
COMPLAINT: Motor is running hot. At times smoke is observed.	
1. Motor is operating under a severe overload.	*k.* Refer to applicable sections of the mechanical divisions that deal with overloads of the driven machine. If overload is found present, reduce as necessary.
2. Motor is single-phasing.	*l.* Verify all circuits. Proof of single-phasing is in refusal of motor to start if stopped.
3. Motor location is such as to restrict necessary ventilation. Alternatively, the air passages through	*m.* Note ambient temperatures. Compare this with the base on which the motor's temperature

Table 16-2. Troubleshooter's Field Guide for Induction Motors (*Continued*)

Probable causes	What to do
the motor are clogged with lint or other foreign materials.	rise is stated. Improve the area ventilation. If area is satisfactory, clean all air passages through the motor.
4. Motor is operating on unsuitable voltages and frequency.	*n*. Compare power service with the conditions specified on the name-plate.
5. Bearings are too tight.	*o*. Take bearing temperatures. Check on the rotor's resistance to rotation. Ease up on the bearings. Verify the correctness of the bearing lubrication.
6. If motor is driving through a belt, the belt is too tight.	*p*. Check on belt tension. Slack off if found too tight.
7. Motor is not set level, thus producing an end thrust.	*q*. Verify correctness of motor mounting. Check with bubble level and correct if needed. Do not reset the motor unless the driven machine is aligned and reset at same time.
8. Grease is different than specified by motor manufacturer.	*r*. Clean out lubricating system and regrease with proper materials.
COMPLAINT: Wound rotor type of motor runs under design speed with load removed.	
1. There is an open circuit in the rotor circuit, including cable to the control.	*s*. Check each circuit. Repair open circuit when located.

Table 16-3. Troubleshooter's Field Guide for Direct-current Motors (*Continued*)

Probable causes	What to do
COMPLAINT: Motor will not start.	
1. There is an open circuit in the control system.	*a*. Check the starting resister, and look for an open switch or a burned-out fuse. Close the circuit as found necessary.

Table 16-3. Troubleshooter's Field Guide for Direct-current Motors (*Continued*)

Probable causes	What to do
2. The terminal voltage is low.	*b.* Verify voltage service in comparison with the nameplate requirements.
3. Motor is overloaded.	*c.* Compare design load and actual load on the driven machine. Reduce the load. If this is not practical, replace with a larger motor.
4. There is excessive internal friction in the motor.	*d.* Verify this by disconnecting the motor from the driven equipment. Note resistance to rotation and rise in temperature. Check the bearings for faulty lubrication or a fit that is too tight. Correct as necessary. If still resistant, strip down the motor and check clearances and fits, part by part. If trouble is in a bent or sprung shaft, straighten or replace as found requisite.
COMPLAINT: Motor starts, but immediately the overload relays cut out and it stops.	
1. Motor field is weak or absent.	*e.* For adjustable-speed motors, look into the rheostat for defective operation or incorrect setting. Adjust or repair as needed. Examine field coils for an open winding. Search for loose or broken wiring connections.
2. Load is in excess of the motor starting torque.	*f.* Replace the motor with one that is designed to meet the starting and running conditions of the connected load.
COMPLAINT: Motor does not come up to speed when under load.	
1. Motor is overloaded.	*g.* Compare actual load with the design load as indicated on the motor nameplate. See applicable mechanical sections for determination of overloads.

Table 16-3. Troubleshooter's Field Guide for Direct-current Motors (*Continued*)

Probable causes	What to do
2. Line voltage is lower than motor is designed for.	*h.* Check running voltage against nameplate voltage. If too low, look for excessive resistance in area demands, for whatever cause, and relieve.
3. Brushes are not in the optimum neutral position, especially if motor has been in service for a considerable time. Will be ahead of neutral.	*i.* Reset the brushes on neutral.
COMPLAINT: Motor is running at an overspeed rate although under full load.	
1. Brushes are not in neutral but are behind neutral.	*j.* Reset the brushes correctly to neutral.
2. Line voltage is higher than motor design.	*k.* Check voltage at terminals and compare with nameplate voltage.
3. Field is weak.	*l.* Look for grounds and correct if found.
COMPLAINT: Motor is sparking heavily at brushes.	
1. Commutator surfaces are in poor condition.	*m.* Clean up and grind the commutator and the brushes. Reset the brushes properly. DO NOT USE EMERY CLOTH. The dust is metallic and will cause a short.
2. Armature is out of balance.	*n.* This shows up in excessive vibration. Disconnect motor and make sure the vibration is in the motor and not coming from the driven machine. Take armature to shop and balance.
3. Short circuit in the motor.	*o.* Test for shorts in the armature. Clean out loose metal particles that may have accumulated between segments of the commutator. Temporary repairs can be made by cutting out the commutator bar in which the short is found.

Table 16-3. Troubleshooter's Field Guide for Direct-current Motors (*Continued*)

Probable causes	What to do
4. Brushes are poorly fitted.	*p*. Clean the brushes and holders and set properly.
5. Paint, chemicals, or atmospheric deposits on the commutator.	*q*. Clean thoroughly. Consider replacing the motor with one designed to operate totally enclosed in a contaminating atmosphere.

Table 16-4. Troubleshooter's Field Guide for Synchronous Motors

Probable causes	What to do
COMPLAINT: The motor will not start.	
1. There is an open circuit in one phase.	*a*. Test each phase to locate fault. Repair as found necessary.
2. Frictional resistance is too great for the motor to overcome on starting.	*b*. Refer to the mechanical sections applicable to the driven machine (pump, compressor) for suggestions to reduce load. Check bearing friction of motor and effectiveness of bearing lubrication. Look for faulty alignment relative to driven machine and correct if found.
3. Voltage is low.	*c*. Compare available voltage with nameplate requirements.
4. Connections are contrary to designed direction of rotation.	*d*. Reverse any two of the main connections.
COMPLAINT: Motor trips the breaker when put on the line.	
1. The exciter voltage is too low.	*e*. Provide increased excitation.
2. There is an open circuit or short circuit in the field.	*f*. Test out to locate open circuit or short circuit and correct. Use a polarity indicator checking for the short circuit.

Table 16-4. Troubleshooter's Field Guide for Synchronous Motors (*Continued*)

Probable causes	What to do
3. Load fluctuations are excessive.	*g*. Refer to mechanical sections applicable to load control. If motor hunts, try an increased or decreased field current. This might work.
COMPLAINT: Motor overheats in particular areas.	
1. Stator overheating can result from an off-center rotor.	*h*. Check position of rotor. Shim bearings as necessary to center it.
2. Alternatively, it can follow an open phase.	*i*. Search out the faulty phase and correct.
3. Overheating is concentrated in one or more coils.	*j*. This is from a short circuit. A temporary repair can be made by cutting out the defective coil.
4. When overheating is in the field coil.	*k*. Replace or repair the coil as necessary, since the condition can result from a short circuit in the coil. Also, overheating can follow development of an excessive field current. Correct by reducing this current to the nameplate value called for.
COMPLAINT: Motor overheats all over.	
1. Operating under an excessive overload.	*l*. Refer to mechanical sections applicable to load control. If load cannot be reduced, replace motor with a larger one. Evaluate frictional level in bearings and the effectiveness of the lubrication.
2. Excitation is not adjusted to nameplate requirements, or there is no excitation.	*m*. Adjust excitation. Verify the competency of the exciter to perform as intended. If shop repairs are needed, turn it in.
3. The environment of the motor is higher in temperature than the motor is designed to accept.	*n*. Provide area ventilation sufficient to establish the design ambient.

Table 16-4. Troubleshooter's Field Guide for Synchronous Motors (*Continued*)

Probable causes	What to do
4. Ventilating passages through motor are cluttered with lint, dust, or other obstructing materials.	*a*. Clean the motor as required.

the motor windings, possibly being thrown free into the area and setting up a fire and accident hazard.

(c) ***Precautions for the Mechanical Troubleshooter.*** When electrical trouble is suspected in a motor-driven machine and a competent electrician is not available, the mechanical troubleshooter is cautioned against undertaking any adjustments or repairs of electrical equipment with which he is not intimately familiar. Grounds and short circuits can be particularly dangerous to the uninitiated inspector and checker.

Many of the suggested actions listed on the Field Guides can be performed safely only with the electrical equipment shut down. After establishing this condition, the troubleshooter should tag every switch and control against their being reenergized by any person but himself. In this way, closing of a switch by persons unfamiliar with the temporary need to keep it open is reasonably well avoided.

Motor manufacturers are particularly sensitive to the kind of lubricants to be used in their products. If the recommended oil or grease is not available at the moment, the troubleshooter is advised to get it. An almost right specification can be just enough wrong to void a guarantee or to precipitate further trouble. Stay off the end of a limb and either do it properly or don't do it at all.

DIVISION 17

Incinerators and Fireplaces

17-1. INCINERATOR TERMINOLOGY

Air, Excess: Air passing through the furnace in greater quantities than is theoretically necessary to burn the fuel.

Air, Overfire: Air entering the combustion space above the fuel bed or above the grate is termed overfire air. This is the usual route for introduction of the secondary air.

Air, Primary: That volume of air for combustion which is introduced through the fuel bed, or immediately adjacent thereto, and intended to support the burning of the fuel by intimate contact.

Air, Secondary: That volume of air which is introduced above or beyond the fuel bed, with the purpose of furnishing oxygen to, and completing the combustion of, those combustible materials which have escaped burning by the primary air and are being carried through the furnace as atmosphere-polluting particles.

Air, Underfire: Air that is introduced beneath the grate or fuel bed. Primary air is customarily brought in as underfire air.

Baffle: A refractory obstruction used to change the direction of movement of the products of combustion in their journey through the incinerator. A bridge wall between chambers of the incinerator is a form of baffle, forcing the gases to rise up and pass over it.

Capacity: The weight of material being burned, expressed in pounds per hour, when handling the type of waste specified. For descriptions of the waste classifications, refer to Sec. 17-2.

Dampers, Barometric: A balanced plate or disk floating between the stack draft and the ambient atmosphere, which by admitting or rejecting ambient air to the breeching, maintains a steady negative draft over the fire in the incinerator. Barometric dampers are commonly used in heating boilers and furnaces.

Dampers, Butterfly: A plate or disk design that is able to rotate on its axis and controls the draft by this rotating action.

Dampers, Guillotine: A blade or plate that moves up and down vertically in a breeching, its name inspired by appearance and movement.

Destructor: An incinerator satisfying the requirements of design and operation as specified in incinerator classifications III, IV, VI, and VII. These incinerator classifications are described in Sec. 17-3.

Drying Hearth: An area in the primary burning chamber of the incinerator on which wet materials are dried prior to being burned.

Expansion or Settling Chamber: A part of the incinerator in which the products of combustion lose some velocity and fly-ash particles may drop out of the gas stream on its way to the stack.

Gas Washer or Scrubber: An auxiliary unit, sometimes introduced into the flue gas system, employing water sprays to wash out particles and odors before the flue gas is released to the atmosphere.

Grates: Carrying the burnable material, the grates are designed to allow the passage of the combustion air through the bed of burning material. Grates will be met in several forms, such as fixed horizontal, movable horizontal, or inclined, either fixed or movable.

Mixing Chamber: This is a particular space provided between the primary and secondary furnaces in which the combustion products from the primary grates are turbulated by the introduction of obstructions to the gas flow, changes in its direction, and possibly checker-brick. See Fig. 17-1(a).

Spark Arrester: The danger from incinerator sparks in the setting of fires is widely recognized. In most installations, the providing of a spark arrester through which the flue gas must pass is essential. It is most frequently observed as a screen or cage atop the stack.

17-2. WASTE CLASSIFICATIONS

(a) ***Discussion.*** The troubleshooter will recognize that the design of individual incinerators is such as to qualify the unit particularly for a certain type of waste. Trouble calls may be arising from the simple fact of the incinerator being unsuited to the material it is being asked to burn.

Since successful incineration depends so heavily on the matching of the machine to the material, the Incinerator Institute has set up industry-recognized standards for classifying these two partners as an aid in mating them. The waste classifications range from 1 through 6 as follows:

(b) ***Type 1 Waste.*** This is rubbish composed of dry combustible waste, having a moisture content of not more than 25 percent by weight. Rubbish items include paper, rags, scraps of wood, cartons,

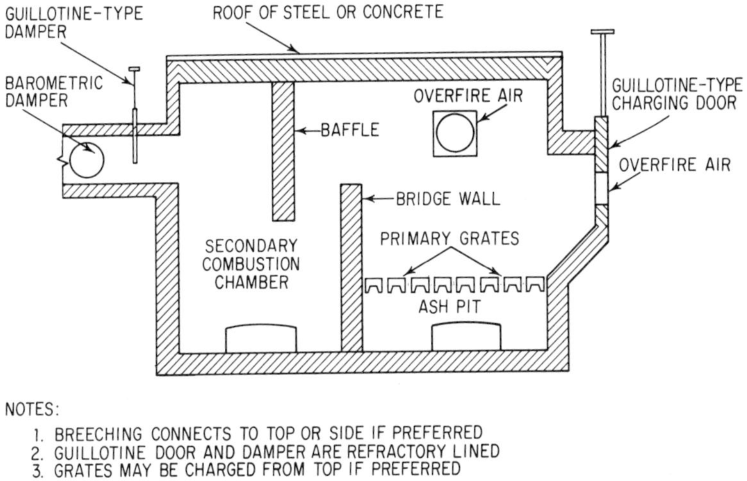

FIG. 17–1a. Parts of a heavy-duty destructor (no scale). (*Sargent, Inc.*)

floor sweepings, and similar materials. Type 1 waste should not have a Btu value in excess of 6500 Btu per lb or an ash of 10 percent or less.

(c) ***Type 2 Waste.*** Consists of about 50 percent garbage and 50 percent rubbish by weight. It is the common waste from residential activities. Calorific value usually does not exceed 4500 Btu per lb. The troubleshooter can expect to contend with this in his domestic and apartment trouble calls.

(d) ***Type 3 Waste.*** Moisture is predominant in this class, ranging from 60 percent of the total or more. Calorific value will average 2500 Btu per lb. Total ash and noncombustible remains should run about 8 percent.

(e) ***Type 4 Waste.*** If it is pathological material, animal or human tissue, it belongs in this classification. Incinerators for Type-4 materials are encountered at medical and surgical activities, however tissue is to be destroyed. Hospitals tend to favor the destructor form of incinerator with auxiliary fuel burners to achieve total destruction. These burners may be gas or oil. Automatic safety shut-offs are required for handling Type 4 waste, as on a flame failure of the auxiliary firing system. Since the burning capacity of these destructors may reach 500,000 Btu per hour, the unit must be located in a well ventilated area to provide ample combustion air. Also, clearances from combustible surroundings must be provided as designated by the manufacturer.

(f) ***Type 5 Waste.*** This waste may be gaseous, liquid, or semiliquid, generally being the by-product of an industrial or chemical process. The waste is sometimes noxious and corrosive and offers a real problem to the troubleshooter in areas where antipollution laws are in effect. Tars, paints, and solvents are typical Type-5 problems.

(g) ***Type 6 Waste.*** Like Type 5, Type 6 waste is a by-product of industrial activities, but this waste is solid or semisolid, and it may be toxic or dangerous from some form of contamination other than those which are not destroyed by burning. Nuclear reaction waste cannot be disposed of by incineration. Rubber and plastics fall into the Type-6 grouping.

17-3. INCINERATOR CLASSIFICATIONS

(a) ***Class I Incinerators.*** These are small, portable, packaged, off-the-shelf units with not over 5 cu ft storage or up to 25 pounds per hour burning rate for Type 1 or Type 2 waste. See Fig. 17-1(b).

(b) ***Class IA Incinerators.*** Similar to Class I units, but having a primary chamber of from 5 to 15 cu ft of space. Burning rate on Type 1 or Type 2 waste ranges from 25 pounds per hour for the 5-cu ft unit up to 100 pounds for the largest or 15-cu ft design.

(c) ***Class II Incinerators.*** Designed for Type 1 and Type 2 wastes, these are flue-fed units, the flue serving the double purpose of a charging chute and a chimney. This design is limited to 2 sq ft of burning area; consequently, it is common in residential installations.

(d) ***Class IIA Incinerators.*** These are the same as Class II units except for their larger size. A flue-fed incinerator with more than 2 sq ft of burning area belongs in the IIA group and has a separate flue for combustion.

(e) ***Class III Incinerators.*** Designed for Types 1 and 2 waste, these are direct-fed units with a burning rate over 100 pounds per hour.

(f) ***Class IV Incinerators.*** These are direct-fed incinerators designed for Type 3 waste with a burning capacity limited to 75 pounds per hour.

(g) ***Class V Incinerators.*** These are established in municipal service. The municipal installation is likely to offer the troubleshooter a complication of ailments. Some of these units are involved in waste heat reclamation, the burning of dry sewage, and charging problems. Municipal incinerators are individually designed to satisfy the particular requirements of the locality and are field-erected. A study of the plans and specifications by the troubleshooter, if he is not familiar with the unit, will be essential in reaching any conclusion and developing a corrective program.

(h) ***Class VI Incinerators.*** These are designed for pathological, Type 4 waste. Crematories are included in this class of incinerator.

FIG. 17–1b. Small, gas-fired, top-charge incinerator. (*Sargent, Inc.*)

(i) ***Class VII Incinerators.*** These will be the specially designed units for handling Type 5 or Type 6 waste. The troubleshooter needs to remember that for these types of waste, the incinerator has been specially designed. As with municipal installations, a study of the plans and specifications becomes essential before effective help can be given.

(j) ***Construction Standards.*** For the materials of construction, safety requirements, and limitations, refer to the National Fire Protection Association standard No. 82, 82A.

(k) ***Trash Burner.*** Units having a charging compartment of not over 10 bushels (12 cu ft) and set outdoors are considered domestic equipment. The troubleshooter should know that these waste burners, to be of approved design and construction, must withstand internal temperatures of at least 1200 F during intermittent operation. Also they must resist such alternate heating and cooling and withstand all kinds of weather. If warpage, cracking, or corroding appears within 2 years under normal operation, their construction is less than that desired.

17-4. INCINERATOR TESTING

(a) ***Smoke and Fly-ash Emission.*** The ASME has set up test procedures that the troubleshooter may use to guide him in his work. Duration of the normal test is established as three consecutive hours with the unit operating at its rated capacity and burning the type of waste for which it has been designed. The makeup of the test wastes are given as follows:

(b) ***Tests with Type 1 Waste.*** All of the test material shall be corrugated cardboard 1/8 to 3/16 in. thick and cut into 12-in. squares. Alternatively, a mixture of soft pine and cardboard may be used, provided that water is added to the mix in the amount of 25 percent by weight.

(c) ***Tests with Type 2 Waste.*** Fire a mixture of soft pine,

Table 17-1. Waste Production for Various Activities

Activity	Pounds per day
Apartment buildings	5 lb fixed allowance, + 1 lb per bedroom
Department stores	4 lb per 100 sq ft of floor area
Drug stores	5 lb per 100 sq ft of floor area
Homes, nurses or interns	3 lb per person
Homes, aged	3 lb per person
Homes, rest	3 lb per person
Homes, single residence	5 lb + 1 lb per bedroom
Hospitals	8 lb per bed
Hospitals, veterinary	Check type of animals to be handled
Office buildings	1 lb per 100 sq ft of floor area
Restaurants	2 lb per meal served
Hotels with dining room, luxury	3 lb per room + 2 lb per meal
Hotels, medium class	1½ lb + 1½ lb per meal
Motels	2 lb per room
Schools, grade	10 lb per room + ¼ lb per pupil
Schools, high	8 lb per room + ¼ lb per pupil
Schools, college level	Each school must be surveyed on its own merits
Shopping centers	Depends on kind of stores in the group
Supermarkets	9 lb per 100 sq ft of floor area
Trailer camps	6–10 lb per trailer
Warehouses	2 lb per 100 sq ft of floor area

NOTE: The weights given are average weights. Any given activity could exceed these figures or fail to equal them. Every project should be checked by a premises visit wherever possible, and a typical day's waste evaluated against these averages.

Table 17-2. Useful Volumes and Waste Weights

Volume and materials	Equivalents or weights
Cubic foot, U.S. std	7.50 gal
Cubic foot, U.S. std	0.80 or ⅘ std bu
Gallon, U.S. std	0.134 cu ft
Bushel, U.S. std	1.25 cu ft
Barrel, U.S. std	4.00 cu ft
Garbage can, 16 in. diam. × 22 in. high	2.50 cu ft
Garbage can, 18 in. diam. × 24 in. high	3.60 cu ft
Weight of garbage, 70 percent water	45 lb / cu ft
Weight of rubbish	10 lb / cu ft
Weight of loose paper (varies)	6 lb / cu ft, avg
Weight of scrap wood (varies)	12 lb / cu ft, avg
Weight of shavings (varies)	8 lb / cu ft, max
Weight of water, std, 39 F	62.4 lb / cu ft, max

cardboard, and shredded paper plus water in the amount of 50 percent by weight.

(d) ***Tests with Type 3 Waste.*** Use a charge composed of 35 percent cardboard and 65 percent water. Alternatively, the 35-percent component may be made up of cardboard and soft pine with the 65-percent balance entirely of water.

17-5. INCINERATOR TROUBLES AND THEIR CORRECTION

(a) ***General Discussion.*** The troubleshooter will face problems of design and construction from time to time that are involved in the relief of operating symptoms. Since every manufacturer of incinerators, whether package or field-erected, offers a distinctive design and his own choice of construction materials, the range of these design problems is boundless. Individual examination is not practical for this reason, and except for a few basic troubles common to all, the incinerator troubleshooter will be obliged to relate the symptoms to the cause strictly by his own reasoning powers. However, one powerful aid is available in the National Fire Protection Association's "Standard for Incinerators, Rubbish Handling" No. 82, 82A, or any latest edition thereof. This is a pocket-size brochure and should be in the pocket of every troubleshooter arriving at the scene of an incinerator trouble call.

This standard sets forth approved practice for all components of the incinerator system; a comparison with conditions as found can indicate the cause of the trouble, or at least pinpoint the area in which it is arising.

Where the trouble appears to be in the chimney, stack, or flue, assistance can be obtained from National Fire Protection Association's

Table 17-3. Natural-draft Stacks for Incinerators *

	Type 1 waste			*Type 2 waste*			*Type 3 waste*			*Type 4 waste*		
Burning rate,	Air supply,	Stack size		Air supply	Stack size		Air supply,	Stack size		Air supply,	Stack size	
lb/hr	cfm, 70 F	diam, in.	ht, ft	cfm, 70 F	diam, in.	ht, ft	cfm, 70 F	diam, in.	ht, ft	cfm, 70 F	diam, in.	ht, ft
100	350	14	30	250	12	25	200	12	30	180	10	25
150	525	16	30	375	14	30	300	14	30	270	12	25
200	700	18	35	500	16	30	400	14	35	360	14	25
300	1,050	20	35	750	18	35	600	16	40	540	16	30
400	1,400	22	40	1,000	20	40	800	18	40	720	18	30
500	1,750	24	40	1,250	22	40	1,000	20	45	900	20	30
600	2,100	26	40	1,500	24	40	1,200	22	45			
700	2,450	28	45	1,750	26	45	1,400	24	50			
800	2,800	30	45	2,000	28	45	1,600	26	50			
900	3,150	32	45	2,250	30	45	1,800	28	50			
1,000	3,500	34	45	2,500	32	45	2,000	30	50			

* These are minimum sizes as recommended by Sargent Building Specialties, Inc., manufacturers of incinerators. Heights are measured from base of side-charged incinerator, near sea level, with breeching not over 10 ft of straight run, less if bends are used. Increase height by 15% if incinerator is top-charged. Decrease height by 25% if stack is directly on top of incinerator, thus eliminating breeching or flue connection. Increase height 5% for each 1,000-ft elevation above sea level.

"Standard for Chimneys, Flues and Vents" No. 211. The spark arrester standard, compiled by the same authority, is NFPA No. 213, "Standard for Construction and Installation of Spark Arresters for Chimneys and Stacks."

All these standards are tools of the trade for the incinerator troubleshooter.

(b) ***Complaint.*** Incinerator smokes persistently.

Causes

1. Incinerator is being overcharged.
2. Primary air is not reaching the materials being burned in either the volumes or pressures necessary to promote thorough combustion.
3. There is insufficient secondary air in the combustion chamber; consequently, the turbulence that secondary air is responsible for creating is not occurring, and unburned particles are escaping up the chimney.
4. The material being fed to the incinerator is not suitable for the design of the equipment.
5. If combustion is being aided by auxiliary fuel burners, the burners are out of adjustment and are producing a smoky condition. This may be particularly true of oil fuel.
6. If combustion depends on mechanical draft, the draft may be insufficient.
7. Velocity of flue gases going through the after-chambers is too high to permit any significant settling out of particulates.

Recommended Actions

1. Weigh the materials being fed to the incinerator and compare this charging rate with that for which the machine is designed. If possible, reduce the weight of the individual charges but increase the frequency of the charging procedure. This should reduce smoke.
2. Examine the condition of the grate for possible clogging of the air holes. Check the draft over the grate. This should be slightly negative, and a positive reading indicates the presence of a downdraft, too much secondary air, a dirty or heavily sooted stack, or a lack of primary air.

Correction here involves the establishment of a positive primary air supply, removal of downdrafts by means of a barometric damper, possibly the installation of an induced draft fan, and a complete balancing between the primary and secondary air.

3. Secondary air inlets should be evaluated from the viewpoint of size and chamber draft. Velocity of the air is most desirable, and if the openings are too large, natural draft may be unable to bring in the secondary at a suitable velocity. Try adjusting the secondary air inflow, either up or down, and note affect on the smoke.

4. Examine the materials going into the incinerator in relation to the type of waste and the class of incinerator being used. Refer to Secs. 17-1 through 17-3.

5. Check the operating characteristic of the auxiliary burners. Note whether flame is clear or smoky. If smoky, adjust the burner and its combustion air, and if this is unsuccessful, change the fuel oil to a lighter one.

6. If mechanical draft fails to establish correct conditions in the firing chamber, look for obstructions in the stack, breeching, or the supply of primary or secondary air. Also, a fixed mechanical draft cannot be expected to handle satisfactorily an overcharged incinerator. Review the charging practices with the burning rate of the incinerator. Verify fan speed in rpm, capacity in cfm, and pressure, water gage.

7. For a condition of excessive velocity, damper down the draft. Install checkerbrick or baffles if practical. A barometric damper will lessen the draft and the velocity of gas movement through the incinerator firing chambers. Keep velocity below 9 ft per second.

(c) ***Complaint.*** Incinerator discharges particles over the area.

Causes

Same as causes for complaint (b) preceding; see numbers 1, 2, 3, 4, and 7.

Recommended Actions

Refer to recommended actions for Complaint 17-5(b), preceding.

Also, consider the mounting of a flue-gas washer or scrubber through which the flue gas must pass on its way to the stack. This unit may be wet or dry, the former employing water sprays, the latter, centrifugal force in separating the particles from the lighter gas.

(d) ***Complaint.*** Incinerator is emitting noxious fumes.

Causes

1. This is the result of burning certain chemical materials.
2. Odors are arising from the destruction of animal tissues and laboratory materials.

Recommended Actions

Correcting these troubles almost invariably will involve the addition to the incinerator of auxiliary equipment designed to absorb or adsorb the objectionable constituents of the gas. Washing and scrubbing with water will eliminate the contaminants in some instances. Other techniques may use activated carbon or mixtures of sawdust and iron oxide, through which the flue gases are forced before discharge to the atmosphere.

Depending on the contaminants in the flue gas, water washing or wet scrubbing may generate a consequent problem: the rapid corrosion or eating away of the washer piping and other construction materials.

For example, where sulfur is the contaminant, this writer has experienced very good results by the use of lead or piping of stainless steel. Simply as a subject for consideration, many of the present plastics are highly resistant to contaminants that attack ferrous or nonferrous metals. Plastics, therefore, are entitled to an evaluation in these situations.

(e) ***Complaint.*** Burning bits of paper and sparks flutter over the neighborhood.

Cause

Spark arrester is ineffective, or there is no spark arrester, a common complaint for trash burners.

Recommended Actions

1. Check the size, arrangement, and openings in the spark arrester against the NFPA "Standard No. 213" for spark arresters. Open-top arresters are not approved. For quick checking, the openings in the screen should not be so large as to pass anything larger than a $\frac{5}{8}$-in. sphere nor so small as to reject anything less than a $\frac{5}{16}$-in. sphere.

Also, the gross surface area of the arrester above the top of the chimney must be equal to, or exceed, twice the area of the chimney outlet.

These data are for domestic spark arresters.

For industrial stacks and chimneys, the mesh openings may pass a $\frac{7}{8}$-in. sphere but no larger and may not reject anything smaller than a $\frac{3}{8}$-in. sphere.

The gross area of the arrester above the top of the chimney or stack must be at least four times that of the stack outlet, if the stack or chimney is rectangular. For round or oval chimneys, the relationship is 6 to 1.

2. In the absence of a spark arrester, install one.

Table 17-4. Troubleshooter's Field Guide for Incinerator Installations

Probable causes	How to check for cause	Recommended ways to correct
COMPLAINT: Incinerator smokes persistently.		
REFERENCE: Sec. 17-5(b).		
1. Incinerator is being overcharged.	*a.* Weigh a typical charge that smokes.	*a.* Reduce the charge. Fire more frequently to meet needs.
2. Incinerator is of wrong classification for the type of waste.	*b.* Compare waste type with incinerator classification.	*b.* Replace incinerator with correct classification.

Table 17-4. Troubleshooter's Field Guide for Incinerator Installations (*Continued*)

Probable causes	How to check for cause	Recommended ways to correct
3. Draft is faulty; there is not enough primary air.	*c.* With draft gage, note condition in burning chamber.	*c.* Clean grate air openings. See Sec. 17-5(b) for details.
4. Auxiliary burners need adjusting.	*d.* Note condition of flame as clean or dirty. Adjust air-fuel mixture.	*d.* Adjust burners as necessary. Change fuel oil to a better grade.
5. Induced-draft fan is not delivering its rated cfm.	*e.* Measure fan discharge in cfm with velometer. Note whether this is equal to the nameplate rating. Check pressure with manometer, also check rpm.	*e.* Clean the fan. Adjust speed. Look for obstructions.
6. Excessive velocity through incinerator.	*f.* Check air speed through the settling chamber or other chamber. It should not exceed 8 or 9 ft per second.	*f.* Adjust draft fan or barometric damper to reduce pull through incinerator. Put in a baffle to promote settling.
7. Fine particles do not settle.	*g.* Note whether velocity is low enough for a normal settling for usual particles.	*g.* Consider a water washer or scrubber, mounted between incinerator and stack.

COMPLAINT: Particulate matter is being blown over the neighborhood.
REFERENCE: Sec. 17-5(c). For causes and checking, see Complaint 17-5(b) above.

COMPLAINT: Incinerator is emitting noxious fumes.
REFERENCE: Sec. 17-5(d).

1. Incinerator is being used for destruction of certain chemicals.	*h.* Get list of materials being incinerated. Note which ones give off the offensive fumes.	*h.* Water-wash or use activated carbon filters. Consult with chemists.

Table 17-4. Troubleshooter's Field Guide for Incinerator Installations (*Continued*)

Probable causes	How to check for cause	Recommended ways to correct
2. Animal tissues are being burned.	*i.* List kind of materials and weights.	*i.* Consult with manufacturer of incinerator. Water-wash or use carbon filter.
COMPLAINT: Burning bits and sparks flutter over the neighborhood. REFERENCE: Sec. 17-5(e).		
1. No spark arrester is provided; this is common with trash burners, domestic type.	*j.* Observe drifting of burning bits.	*j.* Fit unit with a suitable spark arrester.
2. Spark arrester is old and ineffective.	*k.* Observe drifting of burning bits. Take measurements as directed in Sec. 17-5(e).	*k.* Replace with a new approved arrester, to meet standard.

17-6. FIREPLACES AND THEIR TROUBLES

(a) *General Discussion.* Successful troubleshooting of a fireplace requires some knowledge of construction and design principles. For example, the fireplace opening has a distinct relationship in dimensions to the size of the room. Also it is affected by the wall in which it is placed, whether the short wall or the long wall. Table 17-5 gives recommended openings with these walls in mind.

Placement of the damper likewise affects the burning; if it is too low, there is likely to be smoke. From another trouble call, the fireplace is found to be too large for the flue; the area of the flue should be at least one-twelfth of the fireplace opening. Table 17-6 offers a quick check on the relationships of openings and their flues, as met in the field. These sizes in the table are considered to be safe and correct, and if they are, the trouble is of another source.

While fireplaces usually are simply that, with no facilities for an extended area of service, two modifications should be noted. First is the so-called "heatilator" fireplace. In this design, a sheet metal heat exchanger, shaped to fit the fireplace, provides a circulation of warmed air in the room. Cool air near the floor enters through a grille in each side

Table 17-5. Fireplace Openings and Room Sizes

Room dimensions, ft	*Fireplace widths* °	
	For short wall, in.	For long wall, in.
10 × 14	24	24–32
12 × 16	28–36	32–36
12 × 20	32–36	36–40
12 × 24	32–36	36–48
14 × 28	32–40	40–48
16 × 30	36–40	48–60
20 × 36	40–48	48–72

° For fireplace opening dimensions in width, height, and depth, see Table 17-6.

of the fireplace, passes upward through the passages of the heat exchanger, and then emerges as a warm stream from the two sides at about 7 ft above the floor.

The second modification is in the provision of a water back, or grid of

Table 17-6. Conventional Fireplace Dimensions

Opening width, in.	Opening height, in.	Depth, in.	*Flue sizes*	
			Rectangular, in.	Round, diam, in.
24	24	16	8 × 12	8
26	24	16	8 × 12	8
28	24	16	8 × 12	10
30	29	16	12 × 12	10
32	29	16	12 × 12	10
36	29	16	12 × 12	12
40	29	16	12 × 16	12
42	32	16	16 × 16	12
48	32	18	16 × 16	15
54	37	20	16 × 16	15
60	37	22	16 × 20	15
60	40	22	16 × 20	18
72	40	22	20 × 20	18
84	40	24	20 × 24	20
96	40	24	20 × 24	22

pipes, arranged against the back wall of the fireplace. Water heated in this device is then circulated either by gravity or a pump to various hot water radiators. Both of these forms will be examined in connection with their troubles.

The most detailed study of fireplaces known to this writer is "Book of Successful Fireplaces and How to Build Them," published by Donley Brothers Company, Cleveland, Ohio. Also recommended is the "Chimneys and Fireplace Circular," index number F7.0, issued by the Small Homes Council, University of Illinois. A third worthwhile publication is the Farmers Bulletin No. 1889, U.S. Department of Agriculture, titled "Fireplaces and Chimneys."

The troubleshooter is advised to obtain a copy of each publication. Each offers valuable data.

(b) ***Complaint.*** The fireplace smokes.

Causes

1. The fireplace is too large for its flue.
2. The damper is set too low. See Fig. 17-2.
3. There is insufficient combustion air, a condition resulting from tightly sealed doors and windows.
4. Downdrafts are being caused by trees or adjacent roofs higher than the chimney outlet.
5. Smoke emerging from another chimney is being deflected down the fireplace chimney.
6. Flue is off-center. This is an error by the original fireplace builder.

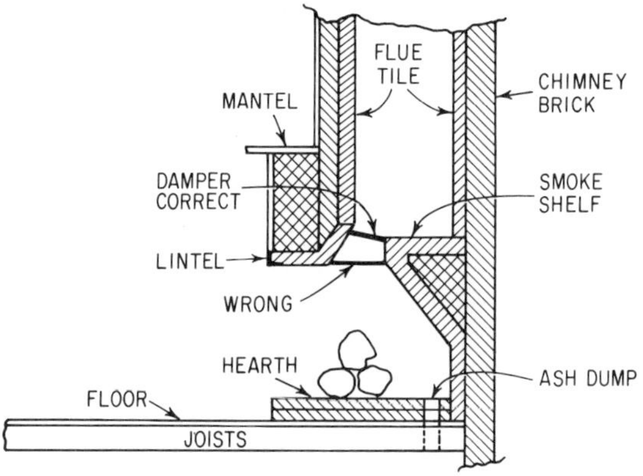

NOTES:

1. MINIMUM FLUE AREA TO BE AT LEAST 10 PERCENT OF FIREPLACE FRONT OPENING
2. SMALLEST ACCEPTABLE FLUE AREA TO BE 50 SQ IN.

FIG. 17–2. Damper, right and wrong (no scale).

7. Debris, such as tree leaves and branches, have fallen into the flue. If the flue is straight, they may be lying on the smoke shelf or be stuck in the damper.

8. Fireplace is too shallow. This also is an error by the original builder.

Recommended Actions

1. When the fireplace is too large for the flue, the opening can be made smaller by (a) laying one or two courses of brick on the hearth; (b) laying up new walls in order to narrow the opening; (c) installing a shallow metal hood under the fireplace breast.

2. Drop the top of the opening by means of a course of brick carried on a lintel of angle iron. This will counteract the effects of the low damper.

3. When a tight house prevents the flow of air into the fireplace, thereby reversing the normal air movement up the flue, an outpouring of smoke is inevitable. In most instances, opening an adjacent window a single inch will be enough to direct the smoke up the chimney.

4. Downdrafts can be eliminated by the simple expedient of adding a cap to the top of the chimney. This cap may be in the form of a flat slab, mounted in a way that provides a total discharge area under the cap equal to at least four times the area of the flue. Proprietary caps also can be used in this service.

5. When the flue is off-center, combustion products tend to loaf in the fireplace space most distant from the flue, and some of them wander out into the room. Correction involves either rebuilding the fireplace flue or so reducing the fireplace opening that it is brought into balance with the flue.

6. Clean out the flue and the smoke shelf.

7. Little can be done to improve a too shallow fireplace other than to reduce the size of the fire, perhaps by installing a small fire basket.

(c) ***Complaint.*** Smoke and flue gases are leaking from the joints in the flue tile.

Causes

1. Moisture and frost have set up breaks in the mortar.

2. The original job was not well-done.

Recommended Actions

1. Rebuild the chimney.

2. Refer to Fig. 17-3. Stuff a canvas bag with rags until it is of a size to fit snugly into the flue. Include one or more fire bricks in the bag to give it positive weight; then lower it down the chimney, stopping just below each joint. Pour grout down the flue, the grout being stopped by the bag and allowed to seep into the open joints. In this way, the entire flue may be swabbed with grout from one end to the other.

Material that slips past the bag may accumulate and harden on the smoke shelf or even spill into the fireplace itself. Because of these possibilities, bag grouting must be carried out with care, and any spillage on the shelf or hearth should be removed at once before it hardens.

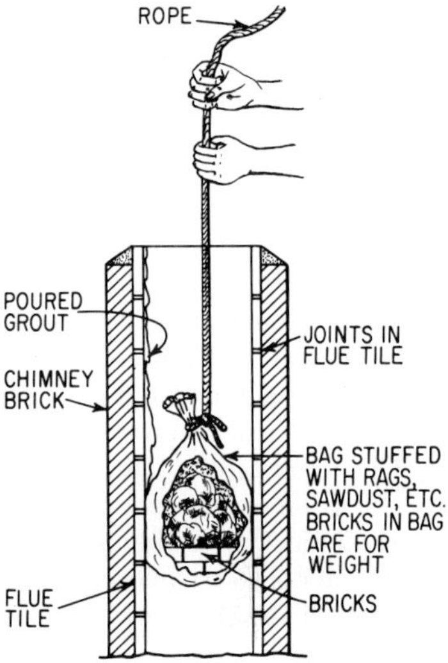

FIG. 17–3. Grouting leaky tile joints.

(d) ***Complaint.*** Fireplace does not burn well. Fire dies.

Causes

1. There is a lack of combustion air reaching the fuel. House is too tightly sealed.
2. Fuel logs are not arranged in an adequate number. At least three logs must be set on the hearth to form a continuing fire.
3. The flue is more or less obstructed.
4. The damper is closing off much of the throat.

Recommended Actions

1. Open a window. If the fire then revives, this pinpoints the trouble as a lack of air.
2. Make sure three or more logs are in position for burning.
3. Verify the condition of the flue and damper. If an obstruction is found on the smoke shelf, remove it through the fireplace. If the obstructing material is higher in the flue, lower a brick from the top of the chimney and swing it back and forth on a cord to loosen the obstruction.
4. Clean the flue by means of a vacuum nozzle.

(e) ***Complaint.*** Circulating fireplace creaks and crackles, especially after lighting off.

Cause

Metal surfaces expand with increases of their temperature and contract on cooling. This is a common sound phenomenon of gas-fired floor furnaces and space heaters, because of the swiftly concentrated heating of the exchange surfaces, and the newly fired fireplace is no different. Expansion is uneven in the heat-circulating fireplace; consequently, the metal must adjust itself in different directions, and the movement is accompanied by the crackling and creaking. Eventually, after a steady burning rate is established, the sounds disappear.

Recommended Action

Explain the reason for the noises. Spread the fire as evenly as practical in the fireplace. Do not promise to eliminate the sounds, as that is not practical. See Fig. 17-4 showing construction.

(f) ***Complaint.*** Warm air from the circulating fireplace has an odor.

Cause

House and room odors are naturally picked up by the air moving across the floor and into the heat exchanger grilles. As time passes, some of the dust adheres to the internal air passages through the exchanger, thus intensifying the odors.

Recommended Action

Clean the heat exchanger by inserting a vacuum nozzle through each of the grille openings. Occasionally, open an outside door or window in the room where the circulator is located, closing all openings to the remainder of the structure, and thus promote a sweep of fresh outside air through the exchanger. An air filter in the fireplace circuit is not prac-

FIG. 17–4. The "Heatsaver" air-circulating fireplace insert. (*Donley Brothers Company.*)

tical, since the movement of the air is strictly by gravity, and there is not enough motive force to force the air through a filtering material. Frequent vacuum cleaning of rugs, walls, and draperies is recommended, since all of these in time can become reservoirs of odor.

(g) ***Complaint.*** Upstairs radiators heat well from a water back in the fireplace, but the warm water does not enter the radiators on the ground floor.

Cause

In gravity circulation of hot water, the warm water rises to the upper spaces, the cold water moves into the lower parts of the system. Since the first floor radiators are on a level with the fireplace and the second floor units above it, the results as described are simply in accordance with natural laws.

Table 17-7. Troubleshooter's Field Guide for Fireplace Installations

Probable causes	How to check for cause	Recommended ways to correct
COMPLAINT: The fireplace smokes.		
REFERENCE: Sec. 17-6(b).		
1. Fireplace is too large for the flue.	*a.* Compare flue and opening dimensions with Table 17-6.	*a.* Lay a course of brick on hearth. Narrow the opening by additional side walls.
2. Damper is set too low.	*b.* Note location of damper in Fig. 17-2.	*b.* Add a metal hood to the lintel, thus reducing opening height.
3. Room is air tight. There is not enough combustion air.	*c.* Note draft up flue. If neutral or negative, open a window. See if draft develops.	*c.* Provide an opening for the entry of combustion air.
4. There is a downdraft.	*d.* Note proximity of trees and roofs to chimney outlet.	*d.* Mount a chimney cap to bar downdrafts. Consider extending chimney.
5. Flue is not centered.	*e.* Note whether flow of smoke to flue from fire is sluggish at one side of fireplace, brisk at the opposite side.	*e.* Rebuild the flue on center.

Table 17-7. Troubleshooter's Field Guide for Fireplace Installations (*Continued*)

Probable causes	How to check for cause	Recommended ways to correct
6. Flue is choked with debris.	*f.* Inspect. Lower brick on a cord to locate the obstruction.	*f.* Remove the obstruction.
7. Fireplace is too shallow to keep fire within the lintel boundary.	*g.* Note position of logs relative to back wall of fireplace.	*g.* Consider use of a small fire basket on the hearth.
COMPLAINT: Smoke and flue gases are seeping from the chimney into the room.		
REFERENCE: Sec. 17-6(c).		
1. Moisture and frost have produced cracks in the joints of the flue tile.	*h.* Note location of smoke.	*h.* See recommended action in Sec. 17-6(c).
COMPLAINT: Fireplace does not burn well. Fire dies or just smoulders.		
REFERENCE: Sec. 17-6(d).		
1. Lack of combustion air. Room is too tight.	*i.* Note absence of draft. See if opening a window restores draft.	*i.* Provide an adequate opening for the admission of air.
2. Fire logs are incorrectly laid.	*j.* Note number of logs and how they are arranged. Use not less than 3 in parallel to form a pyramid.	*j.* Relay the fire correctly.
3. Flue is obstructed.	*k.* Inspect.	*k.* Remove obstruction with vacuum nozzle or break loose with a brick lowered from top.
4. Damper is encroaching on throat excessively.	*l.* Inspect free opening past damper. Measure clearance.	*l.* Replace damper with one of right size.
COMPLAINT: Circulating fireplace creaks and crackles.		
REFERENCE: Sec. 17-6(e).		
1. Uneven expansion	*m.* Listen for evidence.	*m.* Explain reason. See

Table 17-7. Troubleshooter's Field Guide for Fireplace Installations (*Continued*)

Probable causes	How to check for cause	Recommended ways to correct
and contraction strains cause noise.		recommended actions in Sec. 17-6(e).
COMPLAINT: Air from circulating fireplace has odor. REFERENCE: Sec. 17-6(f).		
1. Odors collect in dust, and lint is eventually deposited in heat exchanger.	*n*. Note odor in atmosphere.	*n*. Clean the heat exchanger. Vacuum walls, floors, and drapes. Admit fresh air to exchanger.
COMPLAINT: First floor radiators do not heat with water from water back in fireplace. REFERENCE: Sec. 17-6(g).		
1. Gravity circulation sends hot water upstairs.	*o*. Note cold first-floor radiation; radiators hot upstairs.	*o*. Install motor-driven circulator and balance cocks.

Recommended Action

Install a motor-driven circulator pump in the water circuit, together with balancing cocks as needed to distribute the hot water to all radiators, regardless of elevation.

DIVISION 18

Piping and Miscellaneous

18-1. CORRECTING TROUBLES WITH INDOOR PIPING

(a) ***Discussion.*** For indoor piping functioning as a component of a heating system, refer to Division 4 for steam systems and Division 3 for hydronic installations. Compressed-air arrangements and services are examined in Division 13.

Indoor industrial piping includes steam lines, usually starting out at 125 psig or higher and being reduced as desired at the various points of delivery. Gas, water, and oil lines also may be encountered in industrial activities, these being of various sizes, pressures, and capacities.

Refrigerant lines may be involved occasionally.

Customarily these lines are color-coded; therefore the troubleshooter is readily able to distinguish the one in which he is interested.

Trouble calls, regardless of the fluid being carried, take one or more of these forms: (1) There is a leak. (2) The pipeline is noisy. (3) Pressure drop in the line is excessive, or there is an unexplained reduction in the delivery. These are examined in the complaints that follow.

(b) ***Complaint.*** A joint in the line leaks and refuses to remain tight.

Causes

1. The line is being hampered in its normal direction of expansion and is attempting to compensate by buckling slightly at this particular joint.

2. If it is a flanged joint, the gasket is defective or has been incorrectly placed. In a screwed joint, there is a defective thread or insufficient engagement of the threads. Also, the pipe dope may be reacting unfavorably with the fluid in the pipe.

Recommended Actions

1. Follow the pipeline in both directions from the point of leakage to the first significant change in direction. Note whether there is any obstruction to pipe movement, such as a brick wall, a steel or concrete beam, or an item of machinery. If all is free, then verify this freedom as far as the next change in direction. We have seen 12-in. lines spring their joints from a crooked passage through a solid brick wall. Compare the observed flexibility in changes of direction with the total expansion of the line between its anchors. Correct by the installation of an additional expansion bend or expansion joint, either slip or corrugated, to suit.

2. Reaction of the contained fluids with the joint dope usually shows itself by a discolored weeping at the joint. For example, a clear dope oozes as a deep brown.

(c) ***Complaint.*** The pipeline is noisy. Level of sound is shrill and so high as to be annoying. It produces an early fatigue in workers obliged to be in the area.

Cause

Fluid velocity is too high. Depending on the nature of the fluid, a maximum acceptable velocity should range between 4 and 10 ft per second.

Recommended Actions

1. Calculate the velocity in the line. If excessive, consider increasing the size of the pipeline. This is the best treatment.

2. If space permits, an enclosure of the pipe in a flume of lead sheets, or sound-absorbing material, will improve life in the area.

3. Creation of some competing background sounds, such as piped-in music, may be expected to help.

(d) ***Complaint.*** Delivery from the pipeline is substantially below its design capacity and is also showing an excessive pressure drop.

Causes

1. Cross-sectional area of the pipe is too small to accommodate the desired flow over the length of the line.

2. The viscosity of the liquid is greater than anticipated at the time the line was installed, and the head is insufficient to move the fluid freely.

3. Over a period of time, deposits have accumulated in the pipe, thus reducing the area of flow.

4. The pump is too small.

Recommended Actions

1. Check the size, capacity, and head of the pump. Replace with an adequate pump, if this one is found to be too small.

2. Try steam-tracing the line or use electric cable to heat the fluid sufficiently to produce a better viscosity.

3. Run an additional pipeline to supplement the existing line that is too small. Alternatively, replace the entire existing line with a larger one.

4. Open a joint somewhere along the line and inspect the inside surface for formations of tar, scale, or other choking material. Clean the line by chemical flushing as directed by a chemist who is experienced in this procedure.

(e) ***Complaint.*** Several high-pressure steam valves show porous bodies with steam seeping through to the atmosphere.

Cause

These valve bodies are defective.

Recommended Action

Replace such valves. They are potentially dangerous.

(f) ***Caution: Working on High-pressure Lines.*** Do not attempt to repack a high-pressure valve under pressure without making sure first that the valve design permits this.

Never strike a high-pressure pipeline, fitting, or valve with a wrench or hammer or any other implement. Cast-iron parts are particularly subject to failure from shock. We are well acquainted with the case of a plant workman who tapped a cast-iron high-pressure steam valve with a hammer in order to jar loose a stuck disk. He was buried the following day.

The failure of a high-pressure steam valve from water hammer is also known to have cost human life. When water hammer is present, the first care of the troubleshooter is to relieve it. This is discussed in Sec. 18-4.

18-2. CORRECTING TROUBLES WITH OUTDOOR PIPING

(a) ***Discussion.*** Outdoor piping is found variously supported above ground by towers and by piers of concrete or steel. It will be seen also in open trenches, as illustrated in Fig. 18-1. Fluids carried include steam, hot gases, cold gases, ammonia, other chemicals, water, and oils.

As with the indoor piping, the outdoor installations develop trouble in the form of leaks, they make too much noise, or they fail to meet the requirements of delivery or head. They also have a particular trouble of their own—the flow disturbances of freezing weather.

To troubleshoot leaks, noise, and skimpy delivery, refer to the suggestions given for the indoor piping. What to do about temperatures below 32 F will be shaped somewhat by the nature of the fluid in the pipe.

FIG. 18–1. Oil lines in an open trench on a tank farm.

For example, water freezes at 32 F; a propane gas line becomes empty at − 44 F because the liquid propane stops vaporizing at that temperature and remains liquid. While this may be an Arctic atmosphere, propane's sister fuel, butane, becomes similarly nonvaporizing at anything lower than +31 F, a winter temperature experienced by every state in the nation.

Outdoor pipelines, therefore, depending on their fluid, may require auxiliary treatment in the form of steam tracers or electric heating cables run with, and in contact against, the pipe, to counteract the effects of the temperature. What this means for the troubleshooter is that a condition of nondelivery suggests the immediate checking of the performance of the line tracer as a cause of this nondelivery; the fluid has congealed, and the gas has stopped gassing, and they will stay that way until heating is reestablished.

(**b**) ***Complaint.*** The fluid in a pipeline will not flow until the line or the fluid is heated.

Cause

As already noted, this is caused by subfreezing weather and an unheated fluid.

Recommended Actions

1. If the pipeline is not already heat-traced, arrange to have this done, using steam or electric cables as appropriate to the facilities available.

2. For a short outdoor run of pipe, try electrically heating the liquid tank as illustrated in Fig. 18-2(a) and (b). This type of heater warms up the body of the fluid and will be satisfactory if the fluid travel is brief enough to escape congealing before reaching its destination. For coils inside the tank, check their performance.

Caution: Do not attempt to spot-heat a line by means of a blow torch. Such heating usually ends up in generating a gas that sometimes develops sufficient pressure to burst the pipe and possibly kill the operator.

18-3. CORRECTING TROUBLES WITH UNDERGROUND PIPING

(a) ***Discussion.*** The troubleshooter will find his underground piping sometimes in a pipe tunnel, more often in a concrete or tile conduit, and where temperature control is not too critical, buried bare in the ground. Water and gas piping commonly fall in the last classification; steam and condensate pipe, in the first or second.

As with indoor and outdoor piping installations, the underground lines are subject to leak and inadequate delivery troubles. Noise is of no moment since what there is of it is in a place that annoys nobody.

Leakage is most important, and for conduit and bare burial arrangements, most expensive, since considerable excavation may be needed to reach the leak and provide working space for the repair. Pipe tunnel leaks are easiest to identify, but for the troubleshooter entering the tunnel, considerable danger is present. Danger is in the form of carbon monoxide or other poisonous or asphyxiating gases, and a short time in such an atmosphere can send one to the hospital, or worse. There is no odor to carbon monoxide, and if hot pipes are in the tunnel, the presence of a poisonous gas is practically unrecognizable in the heat.

The troubleshooter, therefore, is advised never to enter a pipe tunnel until the atmosphere has been checked by an explosimeter and a noxious gas meter. The first instrument can indicate the possibility of a quick death from a single spark; the other instrument, a slow death through the lungs.

Thorough ventilation of the tunnel also must precede an entry. If the atmosphere is breathable, the heat alone in some tunnels is enough to overcome the average worker.

(b) ***Symptoms of a Leak from Pipe in a Conduit or in the Ground***

1. For steam piping, excessive vapor in a manhole indicates a leak in

FIG. 18–2. (a) Sectional electric heating coils around a field tank. (b) Metal electric blanket may be wrapped around a small tank or drum. (*Dean Products, Inc.*)

the manhole or somewhere along the pipeline inside of its enclosing conduit, this steam following the pipe into the manhole.

2. The surface of the ground above a leaky steam pipe will be warm; snow melts in that particular area; and under some atmospheric conditions, fog-like vapors can be seen rising out of the soil.

3. For gas or gasoline piping, there will be a recognizable odor in the area of the leak. If an underground gasoline storage tank is in the vicinity, the leak possibly is in the tank, the gasoline seeping through the ground and finally emerging some distance from its origin.

4. Water pipe leaks may betray themselves by wet ground, but this is not to be counted on. Underground leaks in sandy soil often drain off into the area water table or into an adjacent ditch.

5. Oil from underground piping in time is likely to work its way to the surface of the ground. This is because it is lighter than any water in the soil and tends to move above this moisture.

Caution: Wherever a gas or gasoline leak is suspected and the odor is positive, check the flammability of the atmosphere with the explosimeter before undertaking any corrective work. Particularly be cautious in digging into the ground, since a spark from the shovel on meeting a stone may cause an explosion. Rubber shoes and sparkproof shovels are urged.

(c) *Complaint.* The pipeline is leaking, judging from symptoms.

Causes

1. Refer to Sec. 18-1(b)–(f).
2. Heavy traffic is passing over the pipeline.

Recommended Actions

1. Correct the leaks as recommended in Sec. 18-1(b)–(f).
2. If the location of the leak is in doubt, search it out from surface applications of the audimeter or other suitable tools and instruments as described in Divisions 1 and 2. Dig down and correct it.
3. Reinforce the cover over the line, as may be necessary to withstand the shocks and compression of heavy trucks passing over it.

(d) *Complaint.* Manhole is periodically flooded.

Causes

1. Surface wash is spilling down into the manhole.
2. The sump in the bottom of the manhole is too small, and the sump pump lacks capacity to meet the condition.

Recommended Actions

1. If possible, regrade the manhole cover area so that the street wash is guided to one side rather than into the manhole.
2. Increase the size of the manhole sump and the sump pump.

(e) *Complaint.* Iron pipe buried bare develops leaks from corrosion.

Causes

1. The soil is acid.

2. Stray electric currents are traveling the pipeline and leaving craters and pits where they leave the pipe.

Recommended Actions

1. Salt the pipe trench with slaked lime and sprinkle over the top of the pipe. Backfill with clean dry soil. Under no circumstance backfill with cinders.

2. Coat the pipe with a bitumastic paint. This writer has had very good success with the proprietary material, Apexior.

3. The planting of a sacrificial anode in the ground has been successful in many instances, the pipe being protected by the anode which serves as the point of departure for the current. A magnesium bar having one end in electrical contact with the pipe may be placed every 500 ft or so under this plan.

4. Unless the troubleshooter is experienced in corrosion control, a discussion of the situation with people in the business is recommended. This way, time and possibly false steps are saved in correcting the trouble.

Table 18-1. Troubleshooter's Field Guide for Piping Systems

Probable cause	How to check for cause	Recommended ways to correct
COMPLAINT: A joint in an indoor pipeline leaks persistently and cannot be stopped.		
REFERENCE: Sec. 18-1(b).		
1. Movements in expansion are being blocked.	*a.* Look for points of contact and near-contact of the line with beams or other structural erections.	*a.* If obstruction cannot be moved, relocate line or install additional expansion joints or bends.
2. Joint gasket is defective or badly placed.	*b.* Examine the leaky joint for evidence of a split or deformed gasket.	*b.* Replace the gasket. Make up new joint.
3. Joint is screwed, and seepage is chronic.	*c.* Open the joint. Look for insufficient thread engagement, a defective thread, or ineffective pipe dope.	*c.* Remake the joint, improving engagement, chase the threads to correct defects, and apply a suitable dope.

Table 18-1. Troubleshooter's Field Guide for Piping Systems (*Cont.*)

Probable cause	How to check for cause	Recommended ways to correct
COMPLAINT: Noise level of fluid in the pipe is excessive. REFERENCE: Sec. 18-1(c).		
1. Velocity of fluid in pipe is too high.	*d.* Calculate the velocity from the flow, in gpm, and the pipe internal area.	*d.* Reduce volume of flow. Replace the pipeline with a larger pipe. Enclose line in sound-absorbing material.
COMPLAINT: Delivery from pipeline is below the rated capacity, and discharge head is too low. REFERENCE: Sec. 18-1(d).		
1. Pipe is too small.	*e.* Compare flow and pipe size from a Hazen & Williams flow table.	*e.* Increase size of pipe.
2. Liquid viscosity is higher than expected.	*f.* Condition may be temporary because of cold weather.	*f.* Install a heat-tracing system around the pipe. Use electric cable, steam pipe, etc.
3. Pipe has been in service a long time and interior is fouled by deposits.	*g.* Open pipe and inspect internal condition.	*g.* Clean out with circulating acid, water, or other material suitable for loosening the deposits.
COMPLAINT: High-pressure steam valve has a porous body allowing steam to weep through. REFERENCE: Sec. 18-1(e).		
1. Valve body is defective.	*h.* Verify source of steam as the body of the valve.	*h.* Replace the valve with one that is sound.
COMPLAINT: Fluid in an outdoor pipeline will not flow unless it is heated. REFERENCE: Sec. 18-2(b).		
1. Cold weather.	*i.* Compare ambient temperature with analysis of the fluid, especially its pour point.	*i.* Install a heat-tracing system on the pipe with electric cable or a steam line. Heat the point of fluid origin,

Table 18-1. Troubleshooter's Field Guide for Piping Systems (*Cont.*)

Probable cause	How to check for cause	Recommended ways to correct
		such as a storage tank.

COMPLAINT: An underground pipe is leaking, based on the appearance of symptoms at grade level.
REFERENCE: Sec. 18-3(c).

1. Causes are detailed in Sec. 18-1(a)–(b).		
2. Heavy truck traffic over the line has started leaks by vibrating and deflecting the line.	*j.* Note intensity of ground disturbance on passage of heavy traffic.	*j.* Remove the conduit or earthen cover. Build a reinforcing structure over the conduit or pipeline.

COMPLAINT: Manhole for underground piping or cable system floods periodically.
REFERENCE: Sec. 18-3(d).

1. Heavy rains cause water to wash over the top of the manhole.	*k.* Check grade at manhole. Compare with adjacent surface. Pour water upgrade and note direction of flow.	*k.* Regrade around the manhole.
2. Sump and sump pump are too small.	*l.* Note capacity of sump in gallons and of pump in gpm. Check time of pump to empty sump.	*l.* Install larger sump and sump pump.

COMPLAINT: Iron pipe buried bare in the ground suffers from recurrent corrosion failures.
REFERENCE: Sec. 18-3(d).

1. Pipe is being attacked by an acid soil.	*m.* Check the nature of the soil with litmus paper. Paper turns pink if acid character is significant.	*m.* Salt the pipe trench with slaked lime. Backfill with neutral or alkaline soil. USE NO CINDERS.

Table 18-1. Troubleshooter's Field Guide for Piping Systems (*Cont.*)

Probable cause	How to check for cause	Recommended ways to correct
2. Stray electric currents are creating craters and pits.	*n*. Test pipeline for presence of an electric current or field.	*n*. Paint pipe with bitumastic such as Apexior. Install magnesium sacrificial anodes. Consult experts in the corrosion field.

18-4. HANDLING WATER HAMMER IN PIPELINES

(a) ***Discussion.*** Water hammer may be defined as the blow struck against the interior of a pipeline by a fast-moving column of water that is abruptly stopped in its forward movement by the closing of a valve in that line. After the initial impact against the closure, the water bounces backward, this reaction continuing to the nearest larger pipe, tank, shock absorber, or other suitable space for the release of this energy of movement.

The amount of energy observed in a water hammer is a product of the length of the pipeline before the closure (since this determines the weight of water that rams the closure) and the velocity of flow at the time of the closure. Experiments by various researchers have developed a rule of thumb for evaluating the pressure exerted by the hammer as 54 psi per ft per second of the velocity of flow. Thus the abrupt closing of a valve on water flowing through a pipeline at say 10 ft per second can be expected to generate an instantaneous pressure of 540 lb per sq in. Since the piping and valves rarely are heavier than for a working pressure of 125 psig, the destructive danger of the hammer is obvious.

Detroit Edison Company, which operates an underground steam distribution system not far from being the world's largest, after having a number of valve and fitting failures from condensate hammer in the steam lines, carried on a lengthy series of experimental tests, including the use of glass piping and a moving picture camera for charting the course of the hammer wave. A pressure transducer set up on the end of a 16-in. main indicated pressure waves as high as 7,000 lb per sq in. but lasting only 0.0005 second. It is no surprise that failures occurred under such thumps as these.

Figure 18-3 shows a government installation of 8-in. steam piping, operating at 125 psig saturated. After months of loud hammer about

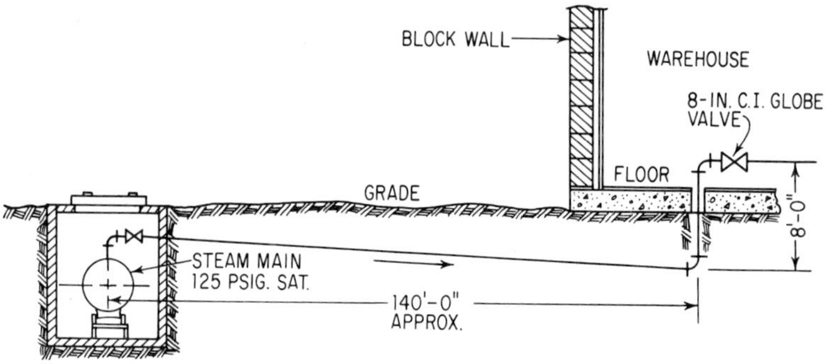

FIG. 18–3. Steam service piping, existing (no scale).

which nothing was done, the 8-in. cast-iron valve exploded, killing one man and critically injuring another. The line should have been constructed as shown in Fig. 18-4.

A conclusion reached by Detroit Edison was that any line which hammers is no place for a cast-iron valve or cast-iron fittings, since this material is not immune to failure from shock. For the troubleshooter, this means eliminate the hammer or replace all cast-iron components with steel. The first alternative of course is the preferred one.

Although originally and usually associated with water, the phenomenon of the hammer can occur with any incompressible fluid.

(b) ***Effect of Piping Arrangements.*** The development of a hammer cannot be predicted infallibly from any given arrangement of the piping system. Figure 18-5 pictures a trap and bypass arrangement with the trap discharging into a condensate return line. This is expected to hammer because the sudden drop in pressure over the trap

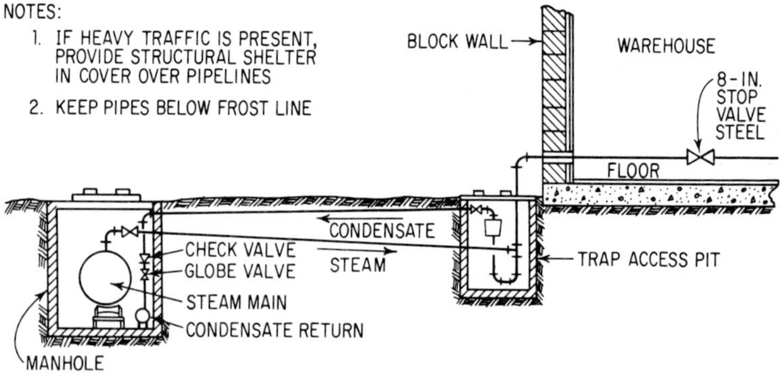

FIG. 18–4. Steam service piping, recommended (no scale).

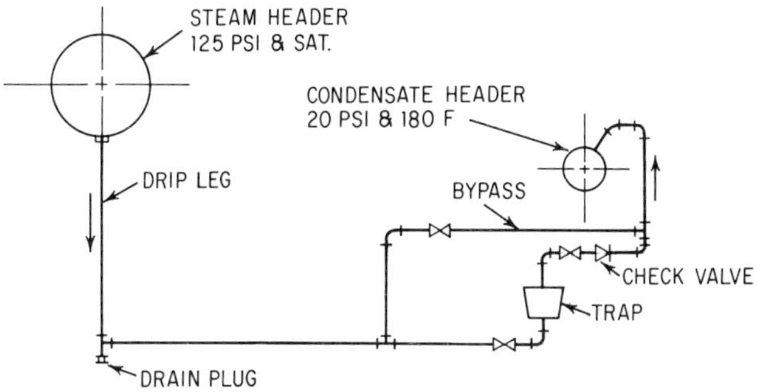

NOTE:
TRAP DISCHARGE TO LOWER PRESSURE AND TEMPERATURE WILL FLASH BUT HAMMER DOES NOT ALWAYS OCCUR IF DISCHARGE VELOCITY IS LOW. USE OVERSIZE PIPE AND TRAP SEAT. THIS MAY STOP HAMMER

FIG. 18–5. Arrangement likely to hammer (no scale).

will permit a flashing of the trap discharge and set up shock waves in the condensate line.

The likelihood of a hammer with this arrangement can be demonstrated mathematically, but the fact is that, in a number of reported similar installations, no hammer occurred.

This uncertainty of design consequences is the reason troubleshooters encounter hammers; particular operating conditions have a powerful influence and thus become the troubleshooter's problem. How these situations may be corrected will be shown in the subsequent sections.

The basic guidelines for hammer corrections may be stated as follows:

1. *Reduce the speed of flow.* This can be accomplished by installing a different capacity of faucet, one smaller than at present.

2. *Wherever check valves* are in the line, especially of the free-swing check-type, replace them with units that close more slowly. Check valves especially are considered to be potential sources of noise.

3. *Steam lines* should be designed and pitched to pass steam and condensate in parallel flow. Attempting to drain condensate in a direction contrary to the steam flow is a prolific source of noise. Also condensate should be removed promptly from a steam header. The less water in the line the less reason for noise.

4. *Mount an air chamber* on the pipeline close to the faucets. This is perhaps the oldest way of compensating for hammer, but it has the weakness of needing recharging with air from time to time.

5. *Water-hammer arresters* certified by the Plumbing and Drainage Institute (PDI) should be mounted on the offending lines, likewise close to the point of flow stoppage. This is a standard the trouble-

shooter should familiarize himself with and adopt as practical. Certified arresters have the advantage of containing a permanent charge of gas or air. Recharging, barring physical accident, is never necessary.

(c) ***Complaint.*** A steam trap discharge line, Fig. 18-3, is producing a water hammer in the condensate line into which the trap is discharging. Condensate temperature is less than 200 F.

Cause

The high-pressure trap discharge, at 125 psig and approximately 340 F, flashes into steam on entering the low-pressure areas of the condensate return line. The expansion of the discharge creates a condition of very high velocity. Impacts and noise may be expected to follow at the first interference with the condensate flow, as in a 90-deg elbow, at check valves, or at any complete closure.

Recommended Actions

1. Lead the trap discharge to a flash tank. This will then require a sump pump in the tank from which the condensate can be pumped back into the condensate return line. Do the same for a receiver.

2. An interesting possibility is to use the discharge as the motivating force with an injector, as in Fig. 18-6. This scheme mixes the discharge with the cooler condensate from the condensate line before disposing of it into the line.

(d) ***Complaint.*** Severe pounding is observed on various water supply lines to washing machines, toilets, and lavatories.

Causes

1. Valves are shutting off the supply abruptly.
2. Supply pipes are too long.

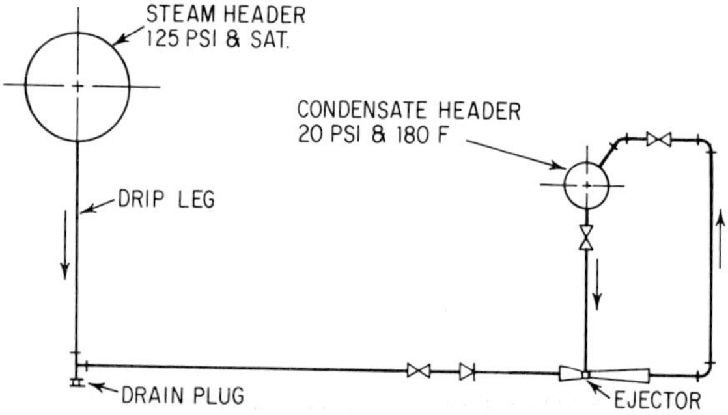

NOTES:

1. MIXTURE FROM EJECTOR IS COOL ENOUGH TO AVOID FLASHING
2. SIZE EJECTOR TO SUIT AMOUNT OF STEAM HEADER CONDENSATE

FIG. 18–6. Arrangement to avoid hammer (no scale).

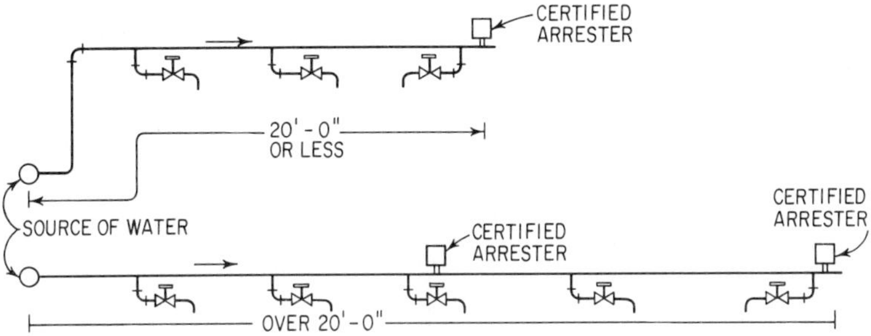

FIG. 18–7. Location guide for arresters (no scale). *Note:* Refer to PDI Standard PDI-WH201 for sizing and placing arresters.

3. There is high velocity in the supply. Ten feet per second should not be exceeded, with 6 to 8 ft per second better.

Recommended Action

Install certified hammer arresters as illustrated on the drawing in Fig. 18-7. Note that in all cases of multiple shutoff valves, the arrester is located as close to the last valve or faucet as may be practical. Sizes of the arresters for each installation can be taken from the "Standard PDI-WH201" published by the Plumbing and Drainage Institute, Oak Park, Illinois. This standard should be in the reference file of every troubleshooter.

18-5. SEPTIC TANK TROUBLES AND CURES

(a) Strictly viewed, septic tank troubles are in the field of the sanitary engineer rather than the mechanical. However, the subject is being examined herein because the mechanical troubleshooter sooner or later finds himself asked to explain why an indicated septic tank is overflowing and odorous and what can be done to correct the situation.

First consideration for the troubleshooter is to understand what goes on inside the tank. Sewage from commodes, urinals, sinks, and other producers of waste enters the tank as a mixture of liquids and solids. In the stagnant waters of the tank, the solids gradually settle out, collecting on the bottom as a sludge that periodically must be removed. These solids and sludge are subjected to decomposition by a form of bacteria called "anaerobic," which thrive in the absence of free oxygen and thus reduce the volume and weight of the materials in the tank. The word "septic" refers to this putrifaction process.

With the scum floating on top of the tank contents and contained between the baffles, the ultimate effluent entering the disposal field is relatively clean and should percolate and become lost in the ground. That this satisfactory ending for the liquid does not always result creates one

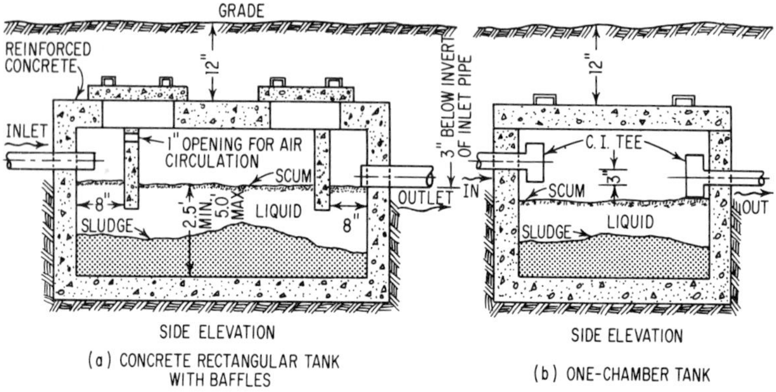

NOTE:

WHEN MEASURING DEPTH OF SLUDGE, WRAP BOTTOM OF MEASURING STICK WITH WHITE TURKISH TOWELING. TOWELING TURNS BLACK WHERE IT IS SUBMERGED IN SLUDGE TO SHOW DEPTH.

FIG. 18–8. Septic tanks (no scale).

of the troubleshooter's septic tank problems, and it will be discussed in a later section.

Figure 18-8 is a sketch of a septic tank in elevation, with its various functions identified.

When the disposal field, usually of tile with unsealed joints, involves several parallel lines of tile, the placement of a distribution box, as in Fig. 18-9, is required in order to distribute the effluent evenly through the entire field. Otherwise a given line may "hog" the load or shirk its rightful share.

The troubleshooter may be faced with the need to evaluate the capacity of the tank as a possible source of the observed trouble. For domestic installations, customary practice is to size the tank on the basis of the number of bedrooms. Table 18-2 gives these data.

In the opinion of this writer, a bedroom basis for sizing might be suit-

Table 18-2. Bedrooms and Septic Tanks (Includes garbage grinders, washers, etc.)

Number of bedrooms	Minimum tank capacity, gal	Equivalent per bedroom, gal
2 or less	750	375
3	900	300
4	1,000	250
Over 4	1,000 + 250 for each bedroom over 4	

able to a family not sleeping more than two persons per bedroom. For additional sleepers, each of whom will need and use the sanitary facilities, an addition of 25 percent to the gallons per bedroom of Table 18-2 is recommended. This addition is particularly desirable where showers provide the usual manner of bathing. The appearance of surface ground water over a septic tank during a morning of showering is not unusual, although the capacity of the tank is entirely adequate at other times.

Sizing on a person basis, as developed by the Small Homes Council, University of Illinois, is recommended for study by the troubleshooter. Quite possibly, a tank suitable for one family will be too small for a larger family moving later into the house.

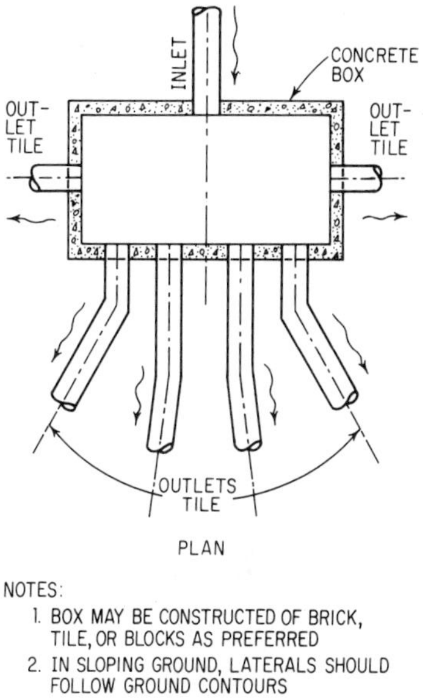

FIG. 18–9. Distribution box (no scale).

Septic tanks are available of rectangular preformed concrete and of steel, cylindrical in shape. Alternatively, tanks may be built of bricks or concrete blocks.

(b) *Good Sewage and Bad.* Domestic grease traps, as found in the conventional small home, are not necessary in many instances. Grease rises to the top in the tank and, in time, if not merged with the scum and sludge, may get into the tile field and clog the moisture passages in the soil. Simply as a rule of thumb, if the sludge is removed every 2 to 3 years, the likelihood is that any dangerous accumulation of grease goes with it.

Small amounts of caustics or lye usually are not inhibitive to the digestion process in the tank. Neither are the usual commercial soaps, detergents, bleaches, and drain cleaners commonly found in domestic activities.

Disinfectants tend to affect adversely the bacterial action and are best kept out of the tank. Clogging can result from the disposal of paper towels, sanitary napkins, newspaper, bags, and rags by way of the septic tank, the clogging occurring in the tile of the percolation field.

A relatively clean source of water, such as roof drains, or large volumes of water from other sources, although running only for brief and irregular periods, must be excluded from the septic tank. Such large-volume influxes can sweep sludge and solids into the outlet pipe and produce clogging in the field.

(c) ***Septic Tank Location.*** In suburban areas where potable water is available from a piped municipal source, setting the septic tank is mainly a problem of having enough ground area to accommodate the percolation system and of making sure that the percolation system will not produce surface water and malevolent odors on a neighbor's side of the fence.

Since subsurface travel, like surface drainage, is a downhill movement, the property of a neighbor at a lower level than one's own has been known to interfere with neighborly friendships. The troubleshooter can help in unruffling this situation.

In rural areas, where water from a well or nearby spring is a common source of domestic supply, care must be taken to keep a septic tank sufficiently far from the water source to prevent pollution from underground seepage. The basic rule is to place the septic tank at least 50 ft distant from the well or spring; actually the proximity to the well or spring should be evaluated in reference to the percolation field. Pollution may be stored in the tank, but it certainly is distributed from the tile field. This means to provide the 50-ft clearance between the water source and the nearest tile disposal line, if possible.

Caution points out that infectious materials and disease-producing pathogens are not removed from the effluent by actions within the tank and are not killed until they filter out into the ground. If that filtering is not complete before seepage reaches a well or spring, sickness and disease can follow. Typhoid fever is one disease known to be spread by the use of contaminated water or milk.

(d) ***Seepage Pits.*** The troubleshooter who encounters trouble with a seepage pit should consider replacing the pit with a conventional disposal field. The pit is simply a dug pit, or sewage well, that is lined with open-jointed bricks, stones, concrete blocks, or other materials that will allow the effluent from the septic tank to leach, or percolate, into the surrounding ground.

Seepage pits can be quite dangerous to subsurface water levels and should never descend nearer than 4 ft to the water table in the area. They are sometimes used to dispose of the septic tank effluent in situations where the area needed for a tile disposal field is not available.

(e) ***Percolation Testing.*** To perform a percolation test, the troubleshooter will preferably need an auger sized to dig a hole 4 in. in diameter and to a depth equal to the bottom level of the tile disposal field. In the absence of an auger, the holes may be dug with a spade, but the auger is quicker, easier, and neater. Also needed will be a measuring stick about 1½ in. wide and a batter board to be laid across the top of the hole, against which the water level in the hole can be referenced. Figure 18-10 illustrates the arrangement. (If a spade must be used to dig the hole, a larger hole will be the probable result.)

After cleaning out the test hole of all loose material, pour about 2 in. of coarse sand into the hole to form a sandy bottom. This is to keep sediment from getting into the soil during the test and perhaps causing a false reading.

Now pour in water to a depth of at least 12 in. over the sandy bottom and maintain this depth by adding water if necessary for a period of at least 4 hours. If water can stand in the hole overnight, so much the better as a method of saturating the adjacent soil and simulating conditions that the septic tank must contend with during times of rainy weather.

Start measuring the percolation rate on the day following the preparation of the hole and after it was first charged with water. At this time the water in the hole should be 6 in. deep, this depth being established by either adding or removing water from the hole. Next, set the measuring stick in the hole and against the batter board and note the level of the water on the stick, as well as the time. Wait 30 minutes; then again note the level of the water with the measuring stick. This drop is the basis for computing the rate of percolation.

Table 18-3 presents the meaning of the percolation rate in terms of square feet of disposal-field area needed. These data are for domestic or residential installations. Larger activities, such as institutions and commercial establishments, are examined in later sections.

(f) ***Trench Spacing.*** Because percolation lines, when set too

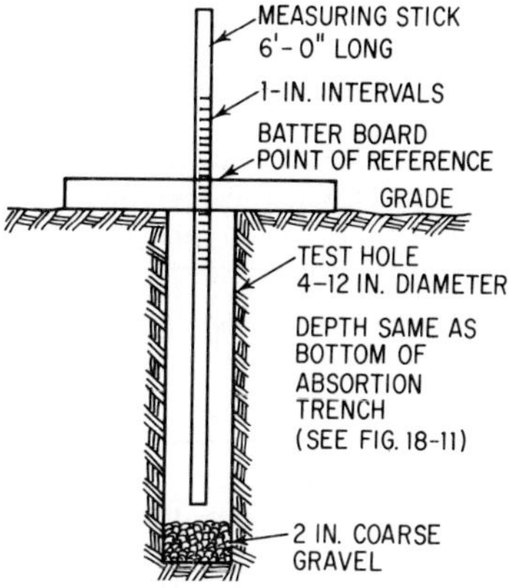

FIG. 18–10. Percolation test hole (no scale).

Table 18-3. Percolation Test Data

Time for drop of 1 in. of water level in test hole, minutes	Disposal area required per bedroom, sq ft
1 or less	70
2	85
3	100
4	115
5	125
10	165
15	190
20	215
30	250
45	300
60	330

close together, affect unfavorably the performance of the field, the troubleshooter finding surface wetness above the lines should verify the line spacing, referring to Table 18-4.

(g) *Soil Requirements*

1. Percolation time should not exceed 1 hour per in. of drop in water level. See Table 18-3.

2. Ground water table should not approach closer than 4 ft to the surface of the ground.

3. Rocks and impervious clays should not be encountered closer than 4 ft to the bottom of the trenches. If a seepage pit is part of the system,

Table 18-4. Minimum Spacing of Disposal Field Trenches *

Width of trench, in.	Distance, center line to center line, ft
12–18	6
19–24	6½
25–30	7
31–36	7½

* Depth of trench must not be less than 18 in. In selecting the depth, remember the frost line and the possibilities of freezing if the tiles are too close to the surface. Figure 18-11 shows a typical trench arrangement.

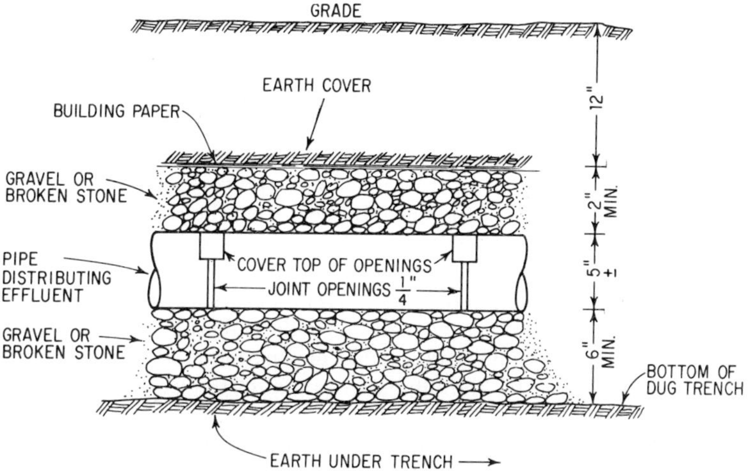

FIG. 18–11. Typical absorption trench (no scale).

this 4-ft clearance to rock or clay is measured from the bottom of the pit.

These are minimum requirements for a satisfactory performance of the septic tank system. The troubleshooter is advised to evaluate the conditions as found against these standards.

(h) ***Complaint.*** The septic tank system is flooding, with puddles of malodorous water appearing from time to time on the surface of the ground.

Causes

1. The septic tank is being overloaded by too many persons using the system.
2. The septic tank is being overloaded but the overloading is from roof runoff or other source of water in volume.
3. The disposal field is suffering from clogging. In winter, and for a shallow field, ice may be present.
4. The removal of sludge from the tank is overdue.
5. The soil characteristics include clay or rock and are unfavorable for percolation of the effluent.
6. Roots have wormed into the tile lines and have blocked the normal percolation procedure.

Recommended Actions

1. In sequence, tabulate the following data: (a) capacity of tank; (b) number of persons and bedrooms; (c) date of last sludge removal; (d) source of all liquids entering the tank, including rainwater and floor drains.

2. Walk the tile lines and look for possible tree roots in proximity of the lines.

3. Check for clogging by spot-opening of the tile trenches.

4. Measure the water depth in the tank between the top of the sludge and the level of the scum. When this distance equals or exceeds one-half of the total depth of the tank, a cleaning should be ordered.

5. If all the above checks are normal, run a percolation test.

6. Evaluate the site of the tank and field from the viewpoint of possible drainage from adjacent properties. This can be rainwater or seepage from the neighbor's septic tank system.

Surface water coming in from adjacent properties may be diverted by the erection of a dike along the high side of the suffering area. Alternatively, a drainage ditch instead of a dike may perform the same function. Either way, a place for the ultimate disposal of the trespassing water must be located.

Also, there may be legal possibilities of requiring the offending owner to erect the dike or dig the ditch, *if* the condition has been developed *after* the plaintiff had established himself and proved the trouble was of his neighbor's creation.

7. If percolation is being checked by a stratum of clay, ascertain the thickness of this stratum. In some instances, the thickness is limited to a few feet, the undersoil being porous. Percolation in these conditions can be expedited by driving holes through the clay and into the porous undersoil, thus providing an additional percolation field.

8. Clogging roots of trees or shrubbery may be dug out and the tile replaced in that localized area.

9. Clogging of the tile field from sludge that entered the field from an overfilled tank and has closed the pores of the ground cannot be cleaned and must be replaced. This usually means providing a new and virgin disposal area, always expensive and frequently difficult if the ground area in which the tank is set is limited.

10. The use of chemical cleaners introduced into a septic tank system, may or may not produce adequate results. Reports from the field are conflicting. This is natural since chemical reactions are strengthened or weakened as a consequence of the chemical characteristics of the materials in the septic system on which the cleaner must operate. Thus a given cleaner working well in one system may be more or less ineffective when applied elsewhere.

(i) ***Complaint.*** In the area of the septic tank there appears to be a more-or-less permanent unpleasant odor.

Causes

1. The odor is coming from tank effluent that has oozed up to the surface of the ground.

2. Normally, the septic tank vents through the fixture vents in the

building. Since these extend at least 6 in. above the roof, odors from here are most unlikely. However, siphonage may be occurring somewhere in the system on a fixture trap, thus allowing odors from the tank to back up into the building.

Recommended Actions

1. Examine the ground over the disposal tiles for evidence of effluent oozing. If observed, refer to the recommended actions given for correcting the complaint described in Sec. 18-5(h).

2. Check all waste lines in the building for a defective or missing trap seal. Replace trap if defective and alter waste piping as necessary to prevent siphonage of the sealing water in the trap.

(j) ***Septic Tanks for Institutions.*** Under this title may be classed schools, motels, trailer parks, restaurants, hotels, and camps. The problems here, in addition to those discussed in Sec. 18-5(a)–(e), are those of spillage of sewage into the tank from multiple sewer lines. In trying to track down a clogged disposal field, the troubleshooter has a large area to explore. And if the clogging is from sludge that has been carried over, the question of where to locate the replacement line might require a compromise answer.

Table 18-5 provides estimating data on the amount of sewage per day per person that may be expected in the large activities as noted. Table 18-6 will help in comparing the disposal area actually in use with the optimum gallons per square foot of area, based on the percolation rates. By the use of these two tables, the troubleshooter should be able to develop a good, overall picture of the system characteristics with which he must deal.

(k) ***Complaint.*** Individual sewage lines from homes in a mobile-home park discharging into the septic tank system frequently become clogged and result in a backing up of sewage into the home fixtures.

Causes

1. The occupants are using the sewer system for carrying away types of waste for which the system is not suited. Specific examples are paper towels, sanitary napkins, newspapers, and rags.

2. This condition is also possible if the clogging is from sludge in the disposal field. When this occurs, liquids cannot follow the normal path of flow through the tank and into the percolation areas, since the soil will not absorb the effluent.

Recommended Actions

1. If the stoppage is in a single line, probabilities are that foreign materials are in the piping between that particular mobile home and the septic tank. Clear the line with a "snake" or by opening affected fixture traps for removal of the obstructing materials. Post a notice prohibiting disposal of such materials into the sewer system.

2. For sludge blockage of the disposal tile and surrounding soil, the

Table 18-5. Institutional Sewage *

Institutional activity	Gallons per day per person
Apartment houses	60
Rooming houses	40
Hotels, 2 persons per room, private baths	60
Restaurants, toilet and kitchen wastes	10
Trailer parks, central bathhouse	35
Mobile-home parks	50
Construction camps	50
Day schools, cafeteria, gymnasium, showers	25
Boarding schools	100
Hospitals	250 (sometimes more)
Office workers	15
Factories	35
Swimming pools and bathhouses	10
Country clubs	100 (if resident)
Country clubs	25 (day patron)
Motels, per bed (no kitchen)	40
Motels, per bed (with kitchen)	50
Self-service laundries, per washer	50 (per customer)
Service stations	10 (per vehicle)

* These are estimated figures. For a more accurate evaluation of an establishment, the troubleshooter could refer to the establishment's water meter, and then deduct from the total gallonage brought in such volumes as are not being discharged into the septic tank.

symptom would be a general backing up of sewage in all homes connected to the septic tank. Take samples of the soil around the tile and examine for a condition of soil-sludge amalgamation. Percolation is impossible in such amalgams and the cure is to install a new tile field in virgin earth. Make a percolation test of the new location and size the length of the lines accordingly.

18-6. COMPLAINTS FROM OUTDOOR SWIMMING POOLS

(a) ***Discussion.*** Simply as background knowledge, the troubleshooter should recognize the three forms of swimming-pool construction. These are the all-concrete pool, the all-metal pool, and the vinyl-lined designs. Each form is likely to develop its own particular trouble as time passes, but the mechanical troubleshooters' operations will be in correcting the performance of the various mechanical components of the pool installation. These are water recirculating pumps; filters, sand, or, alternatively, diatomaceous earth; various tanks for pump suc-

Table 18-6. Sewage Allowance for a Disposal Field *

Time for drop of 1 in. in water level in test hole, minutes	Maximum per square foot of disposal area per day, gal
1 or less	5.0
2	3.5
3	2.9
4	2.5
5	2.2
10	1.6
15	1.3
20	1.0
30	0.9
45	0.8
60	0.6

* Adequacy of sanitary handling of sewage is more important and more critical in institutional projects, because of the number of persons more or less continuously exposed to disease, particularly typhoid fever and dysentery. Under no circumstances should septic-tank effluent be permitted to appear on the surface of the ground.

tion and refilling and for slurries; chlorinators; hair interceptors and water-heating equipment, either direct-fired with gas or oil or as a heat exchanger.

Fundamental information that the troubleshooter should take with him in responding to a trouble call includes the following:

1. The rate of water heating commonly ranges from ½ to 1 deg F per hour. Obviously, the latter, higher rate calls for a higher Btu per hour output of the heating equipment.

2. The rate of filtering in a filter of diatomaceous earth is normally from 2 to 4 gal per minute per sq ft of the filtering surface. For a sand filter, this is too fast, and the flow should be kept below 2 gal per minute per sq ft of filtering surface.

3. For public pools and those operated by a hotel, motel, or community, both design and operation may be stipulated by a local or state board of health. The troubleshooter should have on hand a copy of current rules and regulations as they apply to the pool presently in trouble.

4. Rates of flow into the pool from each inlet are limited by good

Table 18-7. Troubleshooter's Field Guide for Septic Tank Systems

Probable causes	How to check for cause	Recommended ways to correct
COMPLAINT: Effluent from the septic tank is appearing as a shallow pond on the surface of the ground.		
REFERENCE: Sec. 18-5(h) and (j).		
1. Rainwater from roof and water from floor drains are being led to septic tank.	*a.* Check every source of input to the septic tank. Tank is for sewage only.	*a.* Disconnect nonsewage inputs to the tank. Dispose elsewhere.
2. Load is too great for tank.	*b.* Compare size and capacity of tank with the calculated sewage input. Use Tables 18-2, 18-3, 18-4, and 18-5.	*b.* Construct a supplementary tank on discharge of present tank. Size from tables.
3. Tank needs cleaning.	*c.* Measure depth of liquid between top of sludge and the scum level. If more than one-half depth of tank, cleaning is needed.	*c.* Authorize cleaning.
4. Disposal field is clogged by roots.	*d.* Measure distance from trees and shrubbery to nearest tile line. If within 4 ft, open up trench.	*d.* Cut out the roots.
5. Disposal field is clogged by sludge.	*e.* Spot-test for presence of sludge in soil.	*e.* If extensive, relocate the tile lines.
6. Disposal area is rocky or underlined with clay.	*f.* Run a percolation test as described in Sec. 18-5(e).	*f.* Look for a new and suitable field.
7. Surface water is draining over the septic system from higher ground.	*g.* Check the grade or run a level over the adjacent area to identify the direction of drainage.	*g.* Consider dike for high side. Detour the drainage away from disposal field.

Table 18-7. Troubleshooter's Field Guide for Septic Tank Systems (*Continued*)

Probable causes	How to check for cause	Recommended ways to correct
COMPLAINT: An unpleasant odor is present in the area much of the time. REFERENCE: Sec. 18-5(i).		
1. Tank effluent is reaching surface by oozing or is puddling in a low spot.	*h.* Walk the field. Look for wet spots or sink holes.	*h.* See recommended actions for Sec. 18-5(h).
2. A trap seal has defaulted somewhere on the building waste system; this allows odors to escape into the structure.	*i.* Check the status of each fixture seal.	*i.* Replace defective trap.
COMPLAINT: In a mobile-home park, sewage lines from the individual homes become stopped up, and sewage backs up into the fixtures. REFERENCE: Sec. 18-5(k).		
1. Home occupants are using sewage line for general waste-disposal purposes.	*j.* Clear line with a plumber's snake. Withdraw a sample of obstructing material, if possible, for identification.	*j.* Post notices prohibiting sewer disposal of paper towels, sanitary napkins, etc.
2. Tile disposal field is clogged.	*k.* Make spot checks of soil over and under tile for evidence of sludge amalgamation.	*k.* Impervious condition of soil requires arranging a new field, elsewhere.

practice (and perhaps by local regulation) to 10 gpm for each 1-in. inlet, 20 gpm for a $1\frac{1}{4}$-in. inlet, 30 gpm for a $1\frac{1}{2}$-in. inlet, and 50 gpm for a 2-in. inlet. Outlet capacities of course must handle volumes equal to the rate of inflow.

5. For checking the head losses of pipe friction, the recirculating pump suction friction loss should not exceed 6 ft per 100 ft of equivalent piping length. Equivalent length includes the valves and fittings on the line, generally taken as an additional 50 percent of the linear piping. Pump discharge head loss is considered satisfactory if it is less than 12

ft per 100 ft of equivalent pipe length. Another way of looking at this phase of a problem is to consider a flow velocity of less than 10 ft per second as being acceptable. A rate-of-flow indicator on the discharge of the recirculating pump is likely to be mandatory on all public pools and will help the troubleshooter in determining both pump and filter performance.

6. The chlorination system should be capable of feeding 1 lb of available chlorine in a 24-hour period, for every 3,000 gal of water in the pool.

Caution: Large installations in which chlorine is handled as a gas will have the apparatus housed in a practically gas-tight enclosure. A ventilating fan in this enclosure is required to produce at least one air change per minute. Switches for this fan's motor and for lights in the enclosure will be outside the door. A gas mask, also set outside the enclosure, should be used by the troubleshooter if there is any possibility of escaped gas being present.

Figure 18-12 presents a flow diagram that should be helpful to a troubleshooter in identifying the relative positions of the mechanical equipment in an installation. Figure 18-13 shows the sketch of the internal arrangement of the filtering media of a diatomite filter. Figure 18-14 shows a sand filter.

(b) *Complaint.* Water temperature fails to reach the temperature stated in the installation contract.

Causes

1. The temperature differential between the outside air ambient temperature and the water temperature professed in the contract may be considerably greater than the contract differential. In this case, the

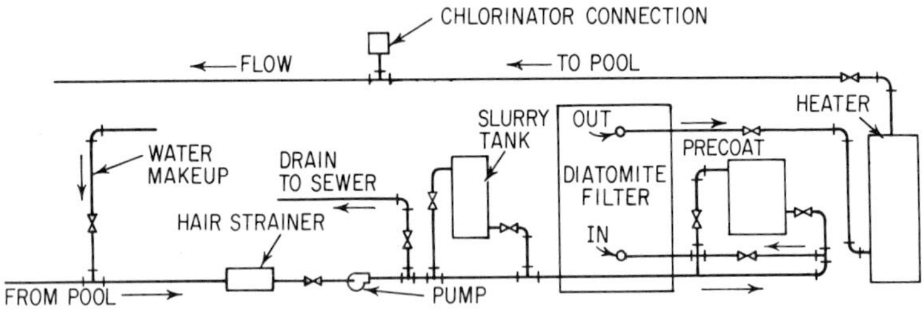

NOTES:

1. PIPING CONNECTIONS ARE TO SUIT CONDITIONS. TRACE ON EACH JOB
2. BACKWASH PIPING NOT SHOWN. TRACE IN FIELD
3. SUCTION OR BALANCE TANKS ARE INCLUDED IN SOME SYSTEMS
4. SAND FILTERS FLOW DOWN ON FILTERING CYCLE, UP WHEN BACKWASHING
SEE FIGURE 18-14

FIG. 18–12. Flow diagram for a swimming pool (no scale).

capacity of the heating equipment may be insufficient to meet the larger demand.

2. The combustion efficiency of the direct-fired heater may be less than it should and can be. This means the heating unit is not making available the heat it is designed to generate.

3. The calorific value of the fuel, whether oil or gas, is less than the design requires to meet its rated heat output.

4. Volume of water being recirculated through the heater is less than required.

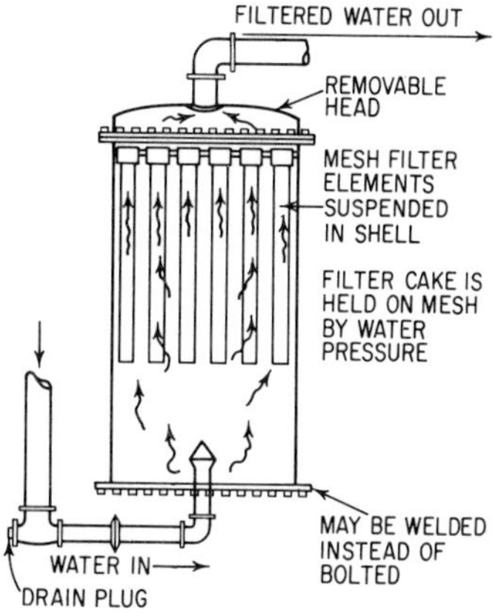

FIG. 18–13. Diatomite filter (no scale).

Recommended Actions

1. Note the temperature differential between the water leaving the heater and the recirculated water entering the heater. Compare this with the design differential. Compare the ambient temperature over the pool area with that specified in the contract.

2. Compare the observed Btu output of the heater per hour with the output specified on the nameplate of the heater.

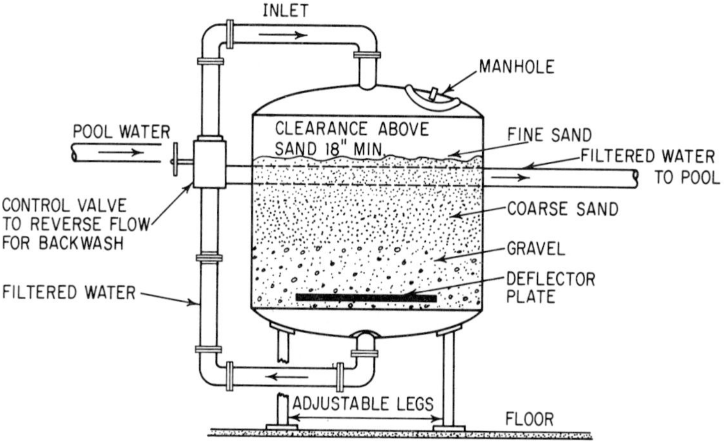

NOTE:

BACKWASH PIPING NOT SHOWN. IT IS CONNECTED TO THE CONTROL VALVE WHERE FLOW REVERSALS ARE ACCOMPLISHED EITHER MANUALLY OR AUTOMATICALLY

FIG. 18–14. Sand filter (no scale).

3. Calculate the Btu per hour to elevate the pool-water temperature to the design level. Correct this for heat loss by evaporation from the surface of the pool; then compare the corrected figure with the heater capacity. Conduction losses from the pool to the ground, both bottom and sides, will be relatively unimportant, since temperature stability of the ground is soon realized. However, if the pool is very large and exact calculations are desired, heat loss to the earth should be considered. Normal heating-up time from 40 to 80 F is usually taken as 48 hours.

4. Note the flue-gas temperature from the heater, also the percentage of CO_2. These figures will indicate the efficiency of the combustion. The appearance of smoke shows defective combustion also.

5. Obtain a fuel analysis from the fuel supplier, including its calorific value. Compare this with the analysis on which the heater capacity is based. If current fuel is inferior in any way, replace with the fuel specified in the equipment performance contract. Measure the rate of fuel consumption, in gallons per hour for an oil burner and cubic feet per hour for a gas burner. What efficiency must be maintained by the burners to release the Btu per hour required to meet the water-heating needs of the pool?

6. Check the pump head and discharge in gpm of the recirculating pump. If less than the rating, put a tachometer on the driving motor and note whether speed is less than normal and how much. For correction of pump troubles, refer to Sec. 15-2(b) for reciprocating pumps, Sec. 15-3(b) for rotary pumps, and Sec. 15-4(b) for centrifugal pumps.

(c) ***Complaint.*** Filtration of pool water is unsatisfactory. Water in pool remains unclear.

Causes

1. For a system with a sand filter, filtration rate probably is too high. If water is passing through the filter faster than 2 gpm, channeling through the filter bed is probable. Also the coagulant is being carried through on the high-velocity water and is creating turbidity in the pool.

2. For a diatomite filter (Fig. 18-13), a possible reduction in water pressure entering the filter or a stoppage of the pump permits the filter cake to fall off the mesh of the filter elements. Thereafter, until the filter cakes have been restored, the water going through will be less filtered than desirable, and turbidity in the pool is likely to be seen.

Recommended Actions

1. Just to be sure the sand filter is right for the job, check the size of the filter sand, gravel, quartz, or whatever is being used. Good practice uses filtering particles of 0.4 mm, approximately, which are free of clay. At least 18 in. of clear space above the filter bed is desired in order to let the bed float upward in backwashing periods but not so far as to be washed out with the backwash water.

2. Turbidity after a sand filter also results from reactions of alum in the pool, producing a floc there instead of on the sand filter where it belongs. If there is a surge tank on the system, the alum should be injected there. Recommend the installation of a surge tank if none is present.

3. For the diatomite filter that operates on an on-and-off schedule, the installation of a small, circulating pump is suggested to operate in a closed circuit, or bypass, with the filter. By maintaining pressure in the filter, it holds the filter cakes against the filtering elements. Then when the main circulator brings in water from the pool, it will find filter cakes waiting and ready to do their job.

(d) ***Complaint.*** To solve some space problems, a diatomite filter was installed in place of a sand filter. Turbidity appears from time to time soon after the recirculating pumps are started.

Causes

1. Stopping of the recirculating pumps removes the pressure and allows the filter cake to fall off the filtering elements.

2. Diatomite filters create a pressure loss up to 35 psig, compared with the 15 psig or less of sand filters. If the pump originally installed for the sand filter has not been replaced by one with a higher discharge head, probabilities are that it cannot hold the filter cake against the elements even when running.

Recommended Actions

1. For an intermittently operating filter, install a bypass pump for keeping pressure on the filter, as described in Sec. 18-6(c), provided the pressure as shown on the gage is inadequate. This may be run continuously.

2. For a filter being operated 24 hours a day, the recirculating pump should be changed to provide sufficient pumping head to hold the filter cakes against the elements.

18-7. COMPLAINTS FROM INDOOR SWIMMING POOLS

(a) ***Discussion.*** Indoor pools experience temperature and turbidity troubles no different from the outdoor unit. For correcting these complaints, the troubleshooter is referred to the preceding sections where directions are given in some detail.

In addition, the indoor pool presents the problem of condensation control. Wherever a wall, window, or ceiling of the pool area develops a surface temperature below the dew point of the room air, water appears on that surface, stains and marks it in some cases, and promotes decay and deterioration almost always.

The owner will have a choice of several possibilities for ending this condition, no one of which is free of either investment or operating expense. The best way of solving the problem is to avoid it by proper combinations of insulation, double glazing, vapor barriers, ventilation, and recirculation of the room air, all initiated when the pool installation is being designed and the housing is still on the drawing board. When the troubleshooter is called in, the time for such preparations is past; he can only point out the available remedies for the situation as he finds it. In accordance with the established method of presenting complaints, this one is examined as follows:

(b) ***Complaint.*** Excessive condensation on the walls and ceiling of a community indoor swimming pool has caused the fall of several ceiling panels, the growth of rust on window sashes, and stains on walls and floors.

Cause

Condensation is resulting from the combination of high humidity in the pool atmosphere and a sub–dew point temperature of the surfaces where the condensation is appearing.

Recommended Actions

1. The simplest treatment is to bring in relatively dry outside air, warm it to the temperature of the inside air, and mix it with the inside air. The result is a lowering of the humidity in the pool area and consequently a drop in the dew point. If this drop is to a level below the temperature of the cold walls and windows, the condensation will be stopped.

Equipment involved includes a fan or blower for bringing in the dry, cold air from outdoors and an air heater of some kind, such as electric resistance coils or steam or hot water coils. Also needed will be an exhauster fan to keep the mixture moving through the pool area and then out, thus making space for more of the dry supply.

The amount of outside air brought in depends on the relative conditions of this air and the inside air, as well as on what inside dew point must be maintained. Accurate calculations are complicated because of the many variables, but an approximation is readily made by use of a psychrometric chart, Figure 12-1. To illustrate, suppose these conditions:

Inside air, 82 F dry bulb and 70 percent relative humidity, having a moisture content from the chart of 114 grains per lb.

Outside air, 20 F dry bulb and having 10 grains of water vapor per lb of dry air.

Desired inside dew point of 45 F, achieved by reducing the vapor content of the air from 114 to 44 grains per lb.

Thus by removing $114 - 44 = 70$ grains per lb of the air over the pool, we can establish dry surfaces in the area, so long as these surfaces are not colder than 45 F. To dispose of these 70 unwanted grains, we can expose them to absorption by the cold, dry outside air, every pound of the latter being capable of accepting 44 − 10 = 34 grains per lb. Getting the condition we want, therefore, of a 45-F dew point requires bringing in $^{70}/_{34}$ = 2.06 lb of outside air to be mixed with each pound of the inside air.

The total weight of outside air is then computed from the volume in cubic feet of the pool enclosure, divided by the cubic feet per pound, and multiplied by 2.06 lb. This total weight, multiplied by 12.1 cu ft per lb of 20-F air, divided in turn by 60 minutes in one hour, produces the approximate cfm required of the fans.

2. This outside air must be heated to the 82-F level before being either released to the pool area or allowed to make contact with the interior surfaces. Recommended arrangements see that the warmed outside air is directed across the glass and other surfaces of exterior walls for a fast drying up of the condensation nuisance. For a hung ceiling, it might be possible to use the space over the panels as a plenum with diffuser discharges downward and into the pool area.

3. The use of chemical drying systems or refrigeration to eliminate excess humidity is possible, but it is not likely to be as economical as the ventilation arrangements described.

4. In some instances, the addition of storm glazing to windows will raise the dew-point temperatures of the glass and thus produce a saving in the ventilating and heating system. If these glass surfaces are of considerable area, the troubleshooter should give some attention to the possible economies by such glazing.

5. If the addition of insulation to the walls is contemplated, care must be exercised in the selection of the insulation to avoid hygroscopic materials. We have seen frame siding torn loose from its studding by the swelling of an insulation poured into the wall behind the siding. The moisture that promoted the swelling came from ambient atmospheric humidity and from water vapor migrating upward from the ground beneath the house.

6. The reasonableness of the pool conditions, as found by the troubleshooter, can be estimated by comparing them with generally accepted practice. Water temperatures tend to range from 75 to 78 F and the air over the pool from 80 to 83 F. The differential of water and air temperatures is usually considered satisfactory at 5 deg.

(c) ***Complaint.*** Pool accumulates debris during periods when it is out of use.

Recommended Action

Cover the pool with a polypropylene cover or, alternatively, one of aluminum. The latter is available in accordion construction and so made as to be spread or retracted simply by pushing a button. Usually this retractable cover is operated by a 12-volt storage battery; however, it can be arranged for house current.

18-8. GENERAL SAFETY PRECAUTIONS FOR TROUBLESHOOTING

(a) ***Discussion.*** Throughout the handbook, the attention of the troubleshooter to certain hazards indigenous to some jobs has been promoted. Ignoring recommended precautions, such as ventilating underground tunnels or handling of gas cylinders carefully, can be fatal. The need to observe these precautions cannot be emphasized too strongly.

However, other dangers of the kind that materialize as the result of once-in-a-while conditions likewise can be deadly, and the troubleshooter, in subsequent sections, is warned against them. Safety in relationship to these conditions has neither beginning nor end; one minute we have it, the next it is gone. These circumstances press on the troubleshooter the need for continuing alertness. Reasons for this need will become clear as the particular dangers are examined and the ways to avoid them suggested.

(b) ***Combustion Dangers, Gas.*** A gas leak always is dangerous as the source of a flash fire or explosion. The presence of a leak should be tracked down immediately wherever suspected, using soap and water or other suitable indicators.

Liquefied petroleum gas is heavier than air, and although generally required by various codes to be odorized, may be slow in betraying the existence of a leak because of its nature to hug the floor. We just may not smell it as soon as desirable. In any event, wherever a leak is suspected, open fire, sparks, and even some lighting fixtures, should be extinguished immediately. There is one exception: If opening the switch of an electric light tends to spark, do not open it. Let that possible spark remain uncalled.

In checking with the human nose for a gas leak, make the full circle of the suspected point of leakage. This is to be sure and detect gas movement being directed by a draft.

Prove a boiler furnace gas-free and effectively vented before attempting to relight an extinguished burner, if the pilot also has been extinguished.

(c) ***Combustion Dangers, Oil.*** After any period of interrupted combustion during which fuel oil may have leaked into the furnace, examine the furnace floor for the presence of an oil puddle. Subsequent lighting of a burner, also igniting the oil pool, may develop such high temperatures as to enfire soot in the chimney or breeching. Also the intense heat can produce warpage of various metal surfaces exposed to it.

Table 18-8. Troubleshooter's Field Guide for Swimming Pools

Probable cause	How to check for cause	Recommended ways to correct
COMPLAINT: Water temperature in the pool will not heat to a comfortable swimming temperature. REFERENCE: Sec. 18-6(b).		
1. Air- and water-temperature differential may be wider than was contemplated under the pool contract.	*a.* Take air temperature and water temperature. Compare with contract range.	*a.* Increase the output of Btu in the pool heater. Compare needed output with nameplate rating.
2. If a direct-fired heater, the combustion is defective.	*b.* Measure input of fuel in gpm or cfm and compare with rating. Check flue-gas temperature. Determine what is normal CO_2 rating and CO_2 as taken by troubleshooter's meter.	*b.* Clean burners and gas passages in the heater. Adjust air to fuel flow.
3. Unfired heat exchanger is dirty or too small.	*c.* Inspect all heat-exchange surfaces and clean. Evaluate the heat-exchange rating with what is needed.	*c.* If exchanger is too small, refer owner to the supplier.
4. Pump circulation through heater is deficient.	*d.* Record pump discharge head and the flow in gpm. Compare with nameplate data. Note rpm, using a shaft tachometer.	*d.* Open pump and repair as found necessary. Restore rated heat, speed, and capacity.
COMPLAINT: Pool water stays turbid despite regular filtering schedule. REFERENCE: Sec. 18-6(c).		
1. For a sand filter, rate of flow may be too high.	*e.* Check flow rate. Limit to 2 gpm per sq ft of sand surface.	*e.* Adjust flow rate to value set by filter manufacturer.
2. For a diatomite filter, water pressure	*f.* Take pressure of water by gage. If less	*f.* Check speed in rpm of circulating pump.

Table 18-8. Troubleshooter's Field Guide for Swimming Pools (*Continued*)

Probable causes	How to check for cause	Recommended ways to correct
is not sufficient to hold the filter cake against the screen element.	than 35 psig, this is likely the trouble. Note pressure difference between suction and discharge of pump.	If pressure difference over pump is excessive, open pump, balance, and repair. See Division 15 for appropriate type of pump trouble and correction.
3. On a sand filter, alum coagulant may be reacting in pool rather than in a surge tank. Floc is promoting turbidity.	*g*. Note where coagulant is being added to water. Should be ahead of filter.	*g*. Adjust feeding and place of feed of the coagulant.
4 This is a diatomite filter operating on an off-and-on schedule. Result is a drop in pressure that allows filter cake to fall off the elements.	*h*. Inspect bottom of filter for an accumulation of filter cakes. Observe whether elements are properly coated.	*h*. Consider an auxiliary pump for continuous operation or for operation when main circulator is not running.
COMPLAINT: For an indoor pool, condensation is forming on walls, ceiling, and windows. Metal is rusting, and several ceiling panels have fallen.		
REFERENCE: Sec. 18-7(a) and (b).		
1. Combination of high humidity in pool enclosure and sub–dew point temperatures of surfaces in this area.	*i*. Condition is readily observed. Take temperature and humidity readings of pool atmosphere and outside air. Also take surface temperatures by contact thermometer.	*i*. Follow directions given in the test in Sec. 18-7(b).

Whenever oil pools are discovered, do not attempt to relight the furnace until the pool has been mopped reasonably dry with paper or rags.

(d) ***Water Level, Out of Sight***

1. Stop firing immediately if burning oil or gas. If coal fuel,

cover fire with ashes. Meanwhile, if boiler is fitted with forced draft, induced draft, or both forms of fans, allow fans to run and vent gases and heat from furance.

2. Do not attempt to restore water level by adding additional makeup water until the boiler cools. An injection of relatively cold water into the hot boiler may produce an immediate failure of the metal as the result of acute thermal shock.

(e) ***Pressure Out of Control.*** If the steam pressure in a steam boiler is seen to be rising steadily above the relief setting of the relief valve, whether or not the relief valve is spilling steam, kill the fires immediately as described in the preceding section, 18-8(d)1.

A venting relief valve with the pressure still climbing indicates either a valve of too small capacity to override the steam-generating ability of the firing equipment as it should or a proper valve that has become degraded by lime accumulations or from some operating defect. Either condition calls for a reassessment of the boiler-operating program and its control.

(f) ***Unfamiliar Valves or Controls.*** To be certain he is not facing the need for chancy actions with unfamiliar equipment, the troubleshooter should study carefully the boiler and firing system instructions. Guessing may be easier than studying, but it also may cost the troubleshooter his job and the boiler owner expensive repairs. Only when he knows exactly what every component does, should he feel qualified to light off a power boiler.

(g) ***Water Hammer in a Steam Line.*** The danger of water hammer in a steam line (Section 18-4) becomes present each time a steam boiler is cut into a main header already carrying steam from other units. The recommended way to avoid the danger is to open all the header drain valves while the steam valve on the boiler scheduled for cutting in remains closed. This technique removes water from the path of the freshly introduced steam. The troubleshooter should remember this precaution.

(h) ***Danger in Hot Piping***

1. A power plant employee swung off the top of a steaming boiler and, to steady himself while his feet searched for the top of the ladder, firmly grasped a gray-colored, bare, 1-in. pipe. He thought it was a lead, cold water line. Actually it was carrying steam at 125 psig. His hands became, in that instant, a hospital case.

Never touch any bare pipe until its fluid is identified, and its harmlessness known for sure. Steel pipe glows in the darkness at about 875 F. Surfaces at lower temperatures are unlikely to offer any visual warning.

2. A troubleshooter should never tap the body of a valve to free a

sticking disk or gate. This action is particularly hazardous if the valve body is of cast iron, a material that is vulnerable to shock. The penalty for disregarding this warning is, at best, a ruined valve and, at worst, scalding from a burst casting.

(i) ***Will Your Eyes Be in Danger?*** If the job involves chipping, scaling, riveting, grinding, calking, or compressed-air blasting, be prepared with goggles, a face mask, and sometimes, an industrial gas mask. Gloves and helmets also will be desirable at times. A good idea is to have these protective devices "in stock" on the tool and instrument truck as a regular custom. We cannot always know what we are running into until arrival at the job.

(j) ***Fire Danger in Air Compressors.*** Since the compression cylinders require lubrication, there is potential danger of fire and explosion should the combination of oil mist and temperature be right for ignition. In addition, if the lubricating oil has been fed in excessive quantities, and particularly if its viscosity has been high, carbonization may grow, harden, and eventually become incandescent. These glowing particles can start the fire.

In returning a compressor to service after a period of inspection or repair, the troubleshooter should be careful to ensure the use of a lubricating oil approved by the compressor manufacturer. He will also see that no carbon deposits are left within the cylinder or in the piping to the air-receiver tank.

Table 18-9 shows temperature of air after compression to various

Table 18-9. Air-compression Temperatures for Single-cylinder Compression *

Pressure, psig	Before cooling, deg F
50	340
100	485
110	505
120	530
125	540
130	550

* Starting from a free ambient of 60 F and atmospheric pressure at sea level. Aftercooling for single-stage machines will reduce these temperatures substantially, depending on the design of the cooler and the temperature of the cooling medium. For a two-stage compressor, this aftercooler becomes the intercooler, the aftercooler then being set to accept the compressed air from the second stage.

pressures. Because of these high temperatures, intercooling and aftercooling are valuable in reducing the dangers from internal ignition.

(k) *The Atmospheric Dangers*

1. On arriving at a job, the presence of a strange odor is a possible warning to the troubleshooter of danger to him in the area. Certain odors such as creosote, animal renderings, fish oils, hydrogen sulfide, mercaptans, and waste decomposition can produce more or less violent nausea in some individuals. Chemical odors may produce irritation of the throat, lungs, and eyes. In some instances, going ahead with the work may be impossible until the wind changes or the process emitting the odor is stopped.

Until strange odors are identified for causing potential reactions, the troubleshooter should either postpone the work or proceed very cautiously. The striking of a match in the presence of an unidentified odor has been known to this writer to cause a flash that killed five men and put a dozen badly burned individuals into the hospital. Use the explosimeter whenever even slightly doubtful.

2. Never enter an underground tunnel or an empty tank until the air in these spaces is tested and proved harmless. In these same listed spaces, check the atmosphere for explosive and inflammable atmospheres as well as for carbon monoxide. Leave all naked lights out until sure of the safety of the conditions.

3. Welding in a confined space calls for complete ventilation during the entire operation. Flexible ducts may be dropped into confined areas, delivering a continuous supply of fresh air. Insufficient ventilation of a welding procedure may generate respiratory irritations in some individuals.

(l) ***Handling Compressed-gas Cylinders.*** In handling acetylene tanks, store them always in a vertical rather than horizontal position. The reason is that acetylene is rather unstable, and to handle it safely, it is carried in a solution of acetone. If the tank is used to furnish gas from a horizontal position, the acetone is lost with the acetylene, and a dangerous condition possibly will develop.

Exercise care in handling all compressed-gas cylinders. Some are more sensitive to shocks than others, but in carrying on a job, there is no time to inquire of the cylinder's degree of sensitivity.

Caution: Do not exhaust an acetylene cylinder to any pressure below 50 psig. Send it back to its source at this pressure and begin on a new one.

Remember that oxygen reacts violently with grease and oil. Keep it away from your clothing and your gloves.

(m) ***Look Out for Relief Valves and Blowdowns.*** This danger is easy to incur. Just stand, in the course of your actions, in the line of exhaust from a relief valve or a blowdown valve. The first may scald you automatically; the latter is opened by someone without checking the point of discharge.

(n) ***Set up a Watchman.*** Station a watchman beside the manhole of any boiler or tank into which you, as the troubleshooter, must go. Without him to monitor your condition, you may never come out alive. The heat or gases may overcome you, or someone who cannot tolerate open manholes will close up yours without checking the possibility of somebody being in there. There have been cases of a boiler being closed up, filled with water, and lighted off, without reference to the unexplained absence of an employee. He later showed up as animal fat in the gage glass.

(o) ***In an Unfamiliar Plant.*** Going into an unfamiliar plant, especially if the lighting is poor, can be dangerous to the troubleshooter whose mind is on the solution of the problem rather than where he puts his feet. For example: an engineer in a poorly lighted power plant basement, attempting to walk across what he took for a wet area on the cement floor, stepped into a water filled pit left from the removal of a large condenser circulating pump. Fortunately he could swim.

The pit was not guarded in any way because, the regular personnel explained, they all knew the pit was there.

(p) ***Electricity Is for Electricians.*** Unless the mechanical troubleshooter is also qualified in electricity, he should leave electrical equipment strictly alone. Opening and closing the switches for small motors should be the limit of his activities electrically. Even these switches should not be manipulated until the effects elsewhere in the building have been checked and the interruptions or restorations of service explained to, and the results approved by, those affected.

(q) ***References of Interest***

1. California (and probably other states) has published its "General Safety Orders." The field is well covered for machinery of various types, gases, and hazardous chemicals. It is available from the Division of Industrial Safety, 455 Golden Gate Avenue, San Francisco, California, 94102. This is an excellent general guide, wherever you are.

2. "Handbook of Industrial Safety Standards," latest revision. Published by the Accident Prevention Department, Association of Casualty and Surety Companies, 60 John Street, New York, New York. Recommended for all troubleshooters.

3. "Safety Manual for Marine Oil-fired Water Tube Boilers." Published by National Safety Council, Chicago, Illinois. Helpful for land-based boilers as well.

4. "Safety in Industry," "Shipboard Ventilation for Hazardous Atmospheres." Issued by U.S. Department of Labor, Bureau of Labor

Standards, Washington, D.C. One of the best treatises on the ventilation of closed vessels, such as tanks and holds.

5. "Public Safety Programs for Atomic Energy," published by the U.S. Atomic Energy Commission, Washington, D.C. This booklet could be useful to people working in this special branch of troubleshooting.

6. "Air Pollution Control: The Field Operations Manual," published by the U.S. Department of Health, Education, and Welfare. If the problem has anything to do with air, such as the handling of offensive or dangerous odors and vapors, this publication of great size can tell the troubleshooter a great deal about how to clear up the condition.

7. Manufacturers' catalogs are always useful. Keep the up-to-date ones in your field of work. Especially desirable are those brochures issued by specialists in hazardous conditions.

General References

In addition to the manufacturers' catalogs and data sheets accumulated from handling particular jobs, the troubleshooter needs a permanent library of reference materials in which he can find answers to questions of basic design and construction.

The following publications have been found useful at various times in troubleshooting work by the author. Titles out of print may be available at the local library.

"ASHRAE Guide and Data Book," American Society of Heating, Refrigeration, and Air Conditioning Engineers.

"Audels Pumps, Hydraulics and Air Compressors," Frank D. Graham, Theodore Audel & Company.

"Audels Questions and Answers Engineers and Firemen's Examination," Theodore Audel & Company.

"Automatic Control of Heating and Air Conditioning," J. E. Haines, McGraw-Hill Book Company.

"Basic Refrigeration and Air Conditioning," Emerick, Prentice-Hall.

"Cast Iron Soil Pipe and Fittings Handbook," Cast Iron Soil Pipe Institute.

"Causes and Preventions of Failure of Septic Tank Percolation Systems," Federal Housing Administration (FHA 533).

"Combustion," Combustion Engineering Company.

"Compressed Air Handbook," Compressed Air Institute.

"Corrosion, Causes and Prevention," Frank N. Speller. McGraw-Hill Book Company.

"Dairy Engineering," Arthur W. Farrall, John Wiley & Sons.

"Dehydration of Food," T. N. Morris, D. Van Nostrand Company, Inc.

"Diesel Engines, Operation and Maintenance," American Technical Society.

"Diesel and Gas Engine Power Plants," Glenn C. Boyer, McGraw-Hill Book Company.

"Diesel Operation and Maintenance," Orville Adams, Prentice-Hall.

"Diesel Engineering Handbook," Diesel Publications, Inc.
"District Heating Handbook," National District Heating Association.
"Domestic Oil Burners," Burkhardt, McGraw-Hill Book Company.
"Electric Testing and Troubleshooting," Phillip T. Green, Industrial Press.
"Engineering Contracts and Specifications," Abbett, John Wiley & Sons.
"Evaporative Cooling," John R. Watt, Industrial Press.
"Fans, Theodore Baumeister, Jr., McGraw-Hill Book Company.
"First Aid to the Ailing House," Roger C. Whitman, Pocket Books, Inc.
"Gas Turbines," R. Thomas Sawyer, Prentice-Hall.
"Heat Engines," Allen and Bursley, McGraw-Hill Book Company.
"Heating Handbook, A Manual of Standards, Codes and Methods," Emerick, McGraw-Hill Book Company.
"Individual Sewage Disposal Systems," U.S. Department of Health, Education, and Welfare.
"Industrial Piping," Charles T. Littleton, McGraw-Hill Book Company.
"Kent's Mechanical Engineers Handbook," Kent, John Wiley & Sons.
"Lightning Book," Peter E. Viemeister, Doubleday & Company.
"Manual of Septic Tank Practice," U.S. Department of Health, Education, and Welfare.
"Marks Standard Handbook for Mechanical Engineers," McGraw-Hill Book Company.
"Mechanical Specifications Handbook for Buildings and Plants," Emerick, McGraw-Hill Book Company.
"Modern Electric and Gas Refrigeration," Althouse and Turnquist, Goodheart-Willcox, Inc.
"National Electrical Code Handbook," McGraw-Hill Book Company.
"National Fire Protection Association Handbook," National Fire Protection Association.
"National Plumbing Code Handbook," Vincent T. Manas, McGraw-Hill Book Company.
"Odors, Psychology and Control," McCord and Witheridge, McGraw-Hill Book Company.
"Petroleum Refinery Engineering," W. L. Nelson, McGraw-Hill Book Company.
"Pipe Fitters Handbook," Forrest R. Lindsey, Industrial Press.
"Plant Operator's Manual," S. Elonka, McGraw-Hill Book Company.
"Power Plant Engineering and Design," Morse, D. Van Nostrand Co., Inc.
"Practice of Sanitation," Hopkins and Schulze, Williams & Wilkens Co.
"Principles of Combustion in the Steam Boiler Furnace," Arthur D. Pratt, Babcock & Wilcox Company.
"Pump Operation and Maintenance," Tyler G. Hicks, McGraw-Hill Book Company.
"Radiant Heating," Adlam, Industrial Press.
"Recommended Minimum Requirements for Plumbing," U.S. Bureau of Standards.
"Standard Refrigeration and Air-Conditioning Questions and Answers," Elonka and Minich, McGraw-Hill Book Company.

"Standards," As appropriate for the job at hand, published by the National Fire Protection Association; U.S. Department of Commerce (Commercial Standards); American Insurance Association, Formerly the National Board of Fire Underwriters.

"Steam Plant Operation," Woodruff and Lammers, McGraw-Hill Book Company.

"Ventilation Manual," Jordan and Morse, Edwin A. Scott Publishing Co.

"Vitrified Clay Pipe Handbook," Southern Clay Pipe Institute.

Index

Index of Troubleshooters' Field Guide

Field guides:
- cast iron boilers, 59
- centrifugal pumps, 405
- coal firing, 79
- comfort air conditioning systems:
  - commercial absorption, 253
  - commercial mechanical, 248
  - domestic, 245
- compressed air equipment and systems, 336
- diesel engines, 355
- direct-current motors, 427
- direct-fired domestic hot water, 155
- electric domestic water heaters, 167
- electric space heaters, 144
- fireplaces, 451
- forced circulation heating systems, 109
- fuel and ash handling systems, 204
- gas burners, 69
- gas turbines, 373
- gasoline engines, 376

Field guides:
- gravity hydronic heating systems, 100
- humidity control, 320
- incinerator installations, 443
- induction motors, 425
- piping systems, 462
- reciprocating pumps, 388
- rotary pumps, 395
- septic tank systems, 480
- steam heating systems:
  - from district steam mains, 133
  - from high-temperature, high-pressure water, 126
  - premises, steam, 121
- steam engines, 346
- steam turbines, 363
- swimming pools, 489
- synchronous motors, 430
- unfired domestic water heaters, 165
- ventilation systems, 300
- warm air heating systems, 183
- well pumps and water systems, 419

Subject Index

Absolute humidity, 1
Absolute pressure, 1
Absolute temperature, 1
Absolute zero, 1
Absorption system of refrigeration, 232
Activated alumina, 1
Activated carbon, 1
Aerosols:
  defined, 1
Air binding, 1
Air fuel ratio, 2
Algae, 2
Alkalinity, 2
Altitude gage, 25
Ambient temperature, 2
Ammeter, 29
Ammonia:
  aqua, 233
  anhydrous, 233
Ammonia regulator, 233
Analysis: proximate, 2
  ultimate, 2
Angle of repose, 2
Ash-free analysis:
  coal fuel, 2
Ash handling, 2, 204
Aspirating burners, gas, 65

Bagasse: fuel, 2
Barometric pressure, 3
Beading of tube ends, 3
Belling of tube ends, 3
Blowback: defined, 3
Blowtorch effect, 3
Boilers:
  domestic: cast iron, 37, 59
      cracked sections, 37
      excessive rusting, 38, 106
Boilers:
  domestic: furnace puffs, 38
      protection against corrosion, 106
      smoking, 38
    steel, 41
      caustic embrittlement, 41
      rusting, 60
      soot accumulations, 41
  high pressure: fire tube, 46
      excessive smoking, 48, 49
      failure to meet rating, 48
      loss of pressure at peaks, 46
      wet steam, 47
    water tube, 49
      designs: typical, 50, 51
      tube troubles: corrosion of, 52
        distortion, 53
        leaks at drums, 52
        tumors: internal, 54
      vibration, 51
  horsepower defined, 3
Boilers licensed operators:
  cities where required, 54
  states where required, 53
Brine cooler:
  refrigeration, 233
Brine pump, 233
Burners:
  combination gas and oil, 73
  gas fuel, 63
  oil fuel, 55
  venturi design, 64

Caking coals, 3
Calorie:
  defined, 4
Calorific value, 4

Capacitor:
  tester for, 34
Capacitor replacement kits:
  field use, 34
Carbon dioxide, 257
Carry-over:
  defined, 4
Centigrade temperature scale, 4
Change of state (opposite of steady state), 4
Chimney effect, 4
Circulator:
  noisy, 99, 111, 152
Coal gas:
  defined, 4
Coal firing:
  hand, 78, 79
  pulverized, 78
  stokers, 76
Coefficient of performance:
  explained, 4
Cold storage plants, 267
Colloidal fuel:
  defined, 5
Comfort air conditioning, 211
  commercial: absorption, 230, 253
    mechanical, 216, 248
  domestic, 212, 245
Compound gages, 25
Compressors: refrigeration: centrifugal, 219
    rotary, 220
Condensate:
  disposal, 131
Condensation at registers, 240
Condenser:
  noise control, 239
Conductivity:
  defined, 5
Controls:
  thermostatic, 226
Convection:
  defined, 5

Deaeration:
  feed water, 5
Deep well pumps, 313
Degasification, 5
Degree day defined, 5
Degrees:
  superheat, 5
Desuperators, 5
Detector: leaks: refrigeration, 33
Dew point:
  defined, 5
Dial indicator, 29
Domestic hot water:
  hard quality, 149
  iron rust in, 148
  liming conditions, 151
Domestic hot water heaters:
  direct fired, 155
  electric, 167
  unfired, 161, 165
Downfeed heating system:
  described, 5
Draft gages, 25
Draft hoods:
  testing of, 81
Drafts, 182
Dry steam:
  defined, 5
Dulong's formula for heating values of fuels, 6
Dutch oven, 6

Eductors, 6
Efficiency:
  steam boilers, 6
Ejecters, 6
Electric heating:
  via heat from light, 139
  infrared, 141
Electric heating systems, resistance type, 137
Electric water heaters, 166
Engines:
  Diesel, 355
  gasoline, 376
  steam, 346

Enthalpy:
defined, 6
Evaporative cooling, 236
Evasé stack, 6
Expansion tanks, 86
Explosimeter:
description of, 26

Fahrenheit temperature scale, 6
Fan performance, use of curves, 7
Field guides (*see* Index of Troubleshooters' Field Guides)
Feed water treatment, 7
Field instruments for trouble shooters, 22
Field tools for trouble shooters, 31
Fire point:
defined, 7
Fireplaces:
dimensions of, 446
smoking troubles, 75
troubles, other, beside smoking ones, 451
Flame detector:
safety control, 7
Flame mirror:
uses for, 30
Flame retention oil burner:
described, 56
Flame speed, 56
Flarebacks:
furnace, 38, 61
Flash point:
oil fuel defined, 7
Flashbacks:
aspirating gas burners, 66
Flooded system of refrigeration, 7
Fly ash:
described, 7
Foaming:
boilers, cause of, 7
Forced draft firing, 65
Fouling:
scale and soot, 8
Freeze-up of expansion valve, 8
Freons:
characteristics, 257
Fuel and ash handling, 193, 204, 206
Fuels:
coals, 74, 79
gas: city: manufactured: water gas, 63, 201
blast furnace gas, 3
butane, (l.p.), 63
natural, 9, 198
propane, (l.p.), 63
oil, API gravity formula, 7
burners, types of, 8
commercial grades, table of, 8
Fumes:
afterburners for, 8
Furnaces: warm air heating systems:
flarebacks, 39
smoke and odors, 38, 41
troubles, miscellaneous, 42, 43
Fusible plug:
description of, 9

Gages:
various types, 25
Gas:
burners, 63, 69
various forms, characteristics of, 9, 261
Gas safety in handling, 199, 200
Gas turbines, 373
Grates:
coal, hand fired, 74
Gravity:
specific, defined, 9
Grindability:
defined, 9
(*See also* Hardgrove scale of grindability)

Hardgrove scale of grindability, 10
Hardness of water:
defined, 10

Head:
dynamic, 10
net positive suction, 10
Heat balance, 10
coefficient of transmission, 11
exchangers, 68, 231
latent, 10
from light, 139
sensible, 10
specific, 10
Heating systems:
hydronic, forced circulation, 94
gravity circulation, 83, 100
conversion to forced, 92, 93
steam, from district street mains, 128
generated premises, 115, 121
from high temp.-high press. water, 124
modified to hot water, 135
vacuum (vapor), 20
warm air, forced circulation, 176
gravity circulation, 175
gravity to forced conversion, 189
Hermetically sealed motors, 273, 274
Higher heat value:
fuel, defined, 11
Hogged fuel:
defined, 11
Hot water pressure problems, 170
Humidistats, 26
Hydronic systems piping, 107
heat pad arrangements for tanks, 196

Ice making, 255, 260, 263
Ice tank for air conditioning, 244, 245
Incinerators, classification of, 436, 438
installation, 443
stacks and draft, 440
troubles and corrections, 439
Infrared heating systems:
electric, 141
Instruments:
field list, 1
Insulation:
economics, 138
Ion:
definition, 11
Isothermal:
defined, 12

Joints:
expansion, bellows, 12
slip type, 12
Jumper tube:
field use of, 12

Kelvin temperature scale, 12
Kerosene:
fuel, 12

Lagging:
defined, 12
Leak detectors:
refrigerant, 12, 33
Leaks:
condenser, 266
Licensing:
boiler operators, 53, 54
Ligament:
defined, 12
Lignite:
fuel, 12
Liquid system of ash removal, 12
Load factor, 12
Locker plants:
cold storage, 269, 272
Low side:
refrigeration circuit, 13
Lubrication of motors:
time schedule, 425

Manifold, refrigeration:
for testing, 26
Mechanical equivalent of heat, 13
Megger (megohmmeter):
use of, 30

Meta-anthracite coal:
described, 13
Micron:
defined, 13
Miniature boilers:
dimensions of, 13
Mollier diagram:
explained, 13
Motors:
hermetically sealed, servicing of, 273

Natural circulation of fluids, 14
Nitrogen:
inert gas, use and storage, 14

Ohmmeters, 30
Oil burners:
safety precautions, all types, 40
troubles with, 38–40, 58, 59, 61, 68, 72
Orifice:
steam, limiting, 130, 134
Outlets:
warm air, smoke issuing from, 181
Overfire air:
defined, 14
Oxygen:
atmospheric component, 14

Packaged air conditioners, 14, 219
Packaged steam generators, 14
Particle size and filter efficiency, 14
Peak load:
defined, 15
pH scale of water condition described, 15
Pilot light problems:
gas burners, 68, 72, 154, 155
Piping:
miscellaneous services, 455
outdoor locations, 457
water hammer in, 465
Preheated air, 15
Pressure-atomizing oil burners, 8
Propane gas:
properties of, 63
Properties of refrigeration brines, 264
Psychrometric chart:
explained, 15
Pumps, troubles with, centrifugals, 383
deep well, 413
plunger, 415
reciprocating, 383, 388
rotary, 391
shallow well, 412
vertical, 414

Quality of steam:
described, 15

Radiant floors:
troubles, 104, 105, 112
Radiation:
equivalent direct, 16
Radiators:
various, troubles, 89, 91, 96, 98
Refrigerant characteristics:
average charge, ice making, 264
properties of calcium brine, 264
properties of sodium brine, 264
safety in handling, 258
temperature-pressure relations, 256
Refrigerant cylinders:
charging, 33
control devices, 223
safe handling of cylinders, 259, 260
Refrigeration:
steam jet, 235
Relative humidity:
defined, 16
Residual fuel, 17
Returns from heating system:
hydronic, 17
Ringlemann smoke scale, 17

Rusting of boilers, 113

Safety:
handling gases, 199
Safety precautions for troubleshooters:
general, 486
Septic tank systems, 469, 470, 474
percolation test, 474
Sewage handling:
institutional, 478, 479
Sludge in oil piping:
cleaning of, 197
Solar water heating, 168–170
Soot:
control of, 68, 72
Space heaters:
direct fired, 45, 137
electric, 141
floor furnaces, 45
Specific heat:
defined, 18
various foods, 269
Spectroscopic analysis:
defined, 18
Stack temperatures:
excessive, 71
Steam heating systems (*see* Heating systems, steam)
Steam jet refrigeration, 235
Stokers:
coal, 76, 77
Sulphur dioxide:
characteristics, 258, 259
Suppliers:
instruments, listing of, 36
Swimming pools:
indoor, 485
outdoor, 478

Tachometers:
use of, 30
Temperatures:
air, compressed, table of, 492
industrial, optimum, table, 313
various scales defined, 19
Thermometers:
various types, described, 22, 23
Tools:
troubleshooters', 1
Turbines, gas, 365, 369
steam, 357, 360

Ultrasonic leak detector:
described, 29
Underground piping:
troubles with, 459
Underwriters Loop:
described (Hartford Loop), 20
Unfired water heaters, 161, 165

Vacuum heating systems (vapor), 20
Valves:
radiator, vent sputters, 132
Vapor barrier:
described, 20
Ventilation systems:
make-up air required, 277, 289
Vertical pumps, 414
Viscosity:
defined, 21
Volume:
copper tubing, table, 274

Wastes:
classification of, 434
production of, 438
Water:
corrosive qualities, 238, 239
hammer in pipe lines, 21, 465
hardness of, 238
refrigeration systems, 237
Water gas:
carburetted, described, 21
Weights and volume equivalents, 439
Weights and volumes of air:
table, 335
Welding:
enclosed spaces, ventilation of, 285
Wells:
water, troubles with, 418, 419